Dynamic Behavior of Advanced Materials and Structures

Dynamic Behavior of Advanced Materials and Structures

Guest Editors

Weidong Song
Lijun Xiao
Xianfeng Yang

MDPI

Basel • Beijing • Wuhan • Barcelona • Belgrade • Novi Sad • Cluj • Manchester

Guest Editors
Weidong Song
School of Mechatronical Engineering
Beijing Institute of Technology
Beijing
China

Lijun Xiao
School of Mechatronical Engineering
Beijing Institute of Technology
Beijing
China

Xianfeng Yang
School of Aeronautic Science and Engineering
Beihang University
Beijing
China

Editorial Office
MDPI AG
Grosspeteranlage 5
4052 Basel, Switzerland

This is a reprint of the Special Issue, published open access by the journal *Materials* (ISSN 1996-1944), freely accessible at: www.mdpi.com/journal/materials/special_issues/L5TY9M4583.

For citation purposes, cite each article independently as indicated on the article page online and as indicated below:

Lastname, A.A.; Lastname, B.B. Article Title. *Journal Name* **Year**, *Volume Number*, Page Range.

ISBN 978-3-7258-4584-2 (Hbk)
ISBN 978-3-7258-4583-5 (PDF)
https://doi.org/10.3390/books978-3-7258-4583-5

Contents

materials

MDPI

Editorial

Dynamic Behavior of Advanced Materials and Structures

Weidong Song * and Lijun Xiao

State Key Laboratory of Explosion Science and Safety Protection, Beijing 100081, China; xljbit@bit.edu.cn
* Correspondence: swdgh@bit.edu.cn

Following the rise in applications of materials and structures in complex environments, such as high-speed impacts and explosions, research on the dynamic response of materials and structures is becoming increasingly important [1–3]. Studying their mechanical behavior of advanced materials and structures, such as mechanical metamaterials [4–8], composites [9–11], and lightweight sandwich structures [12–14], under dynamic loads and gaining a thorough understanding of their behavioral characteristics under extreme conditions are vital for both material and structural optimization, as well as ensuring their safety and reliability in service. This Special Issue (SI), "Dynamic Behavior of Advanced Materials and Structures", presents recent theoretical, experimental, and simulation research findings regarding the dynamic behavior of advanced material and structures. This Editorial summarizes the ten publications (nine research articles and one review article) included in this SI.

Unlike traditional materials that expand laterally under compression, auxetics display a distinct negative Poisson's ratio characteristic, which enhances their energy absorption capacity under dynamic loading [15–17]. Accordingly, auxetics have attracted extensive research attention in recent years [18–20]. Galea Mifsud et al. [21] summarized the research progress on analyzing the mechanical properties and deformation mechanisms of auxetic systems through finite-element simulations.

Carbon-fiber-reinforced polymers (CFRPs) have a variety of advantages properties in comparison to metals, including light weight, high strength, high stiffness, corrosion resistance, and fatigue resistance [22,23], and have been widely used in fields of aerospace and traffic control. However, the dynamic performance and failure mechanism of CFPRs remain unclear due to their complex mesoscopic structure. Qiang et al. [24] presented the low-velocity impact response and residual compressive properties of carbon fiber–epoxy resin (CF/EP) composites, revealing the impact damage on the compression failure mechanisms of the materials. Xi et al. [25] reported that the energy absorption capacities and failure modes of the open-section, thin-walled composite structures under axial crushing loading could be regulated by changing the geometry of the cross-section. Stachyra et al. [26] introduced a novel analytical description for coupled mode shapes of a cantilever composite beam. Lan et al. [27] demonstrated that the preloading value and the impact velocity affected the delamination behavior of CFRP laminated plates significantly. The resistance to out-of-plane displacement of the laminated plates could be enhanced by biaxial tensile preload.

With the advancement of technology, lightweight materials have become widespread in the development of protective structures. Sandwich structures filled with lightweight materials, such as foams and honeycombs, have been widely applied in energy-absorbing devices for collision protection in fields including aerospace, rail transportation, and the automotive industry, due to their light weight, high specific strength, and high specific stiffness [28,29]. How to design high-efficiency sandwich structures for energy absorption is a widely studied topic. Mahgoub et al. [30] experimentally investigated the dynamic response of a sandwich beam filled with a stepwise gradient polymethacrylimide (PMI) foam core under low-velocity impacts, demonstrating that a negative gradient core is beneficial for energy absorption.

In military and public security safety fields, materials and structures are often subjected to explosive shock waves. Bian et al. [31] investigated the dynamic response of non-metallic annular protective structures made by the continuous winding of PE fibers under internal blasts, indicating that the blast resistance of the structure can be improved by adding polyurethane foam in the inner layer. The dynamic response of a rectangular steel plate subjected to shock waves was examined by Li et al. [32] through the wavelet transform (WT) and the improved ensemble empirical mode decomposition (EEMD) methods.

Steel automotive wheel rims typically experience dynamic wear and tear during service, which significantly affect their service life. Borecki et al. [33] proposed new methods for identifying the technical condition of steel car wheel rims and predicting the approaching end of wheel rim service life with limited parameters.

Electromagnetic springs, which act as an active vibration isolation system, have been widely researched due to their fast response, non-contact, and adjustable stiffness [34,35]. In the study by Zheng et al. [36], an improved Bouc–Wen Model was proposed to describe the dynamic characteristics of a toothed electromagnetic spring, which enhanced the accuracy of predicting the hysteresis behavior in electromagnetic spring active isolators.

Nowadays, with the rapid development of advanced manufacturing technologies such as additive manufacturing, an increasing number of advanced materials and structures have been developed, including multi-component alloys [37–39], interpenetrating phase composites [40–42], and nanomaterials [43,44]. Future research needs to pay more attention to the dynamic behavior of these novel materials. Simultaneously, the dynamic behavior of materials often depends on their microstructures, necessitating more extensive application of multi-scale analysis methods for investigation [45–47]. Additionally, integrating artificial intelligence technologies with mechanical design and the prediction of materials/structures [48–50] also requires further investigation.

Conflicts of Interest: The authors declare no conflict of interest.

References

1. Meyers, M.A. *Dynamic Behavior of Materials*; John Wiley & Sons: Hoboken, NJ, USA, 1994.
2. Silberschmidt, V. *Dynamic Deformation, Damage and Fracture in Composite Materials and Structures*; Woodhead Publishing: Cambridge, UK, 2016.
3. Shukla, A.; Ravichandran, G.; Rajapakse, Y. *Dynamic Failure of Materials and Structures*; Springer: Berlin/Heidelberg, Germany, 2010.
4. Yu, X.; Zhou, J.; Liang, H.; Jiang, Z.; Wu, L. Mechanical Metamaterials Associated with Stiffness, Rigidity and Compressibility: A Brief Review. *Prog. Mater. Sci.* **2018**, *94*, 114–173. [CrossRef]
5. Xiao, L.; Shi, G.; Song, W. Machine Learning Predictions on the Compressive Stress–Strain Response of Lattice-Based Metamaterials. *Int. J. Solids Struct.* **2024**, *300*, 112893. [CrossRef]
6. Zheng, X.; Lee, H.; Weisgraber, T.H.; Shusteff, M.; DeOtte, J.; Duoss, E.B.; Kuntz, J.D.; Biener, M.M.; Ge, Q.; Jackson, J.A.; et al. Ultralight, Ultrastiff Mechanical Metamaterials. *Science* **2014**, *344*, 1373–1377. [CrossRef]
7. Song, W.; Li, R.; Feng, G.; Xiao, L. Large Dynamic Mechanical Behaviors and Deformation Mechanism of Hybrid Triply Periodic Minimal Surface Structures. *Int. J. Impact Eng.* **2025**, *203*, 105359. [CrossRef]
8. Feng, G.; Li, S.; Xiao, L.; Song, W. Mechanical Properties and Deformation Behavior of Functionally Graded TPMS Structures under Static and Dynamic Loading. *Int. J. Impact Eng.* **2023**, *176*, 104554. [CrossRef]
9. Zhu, T.; Ren, Z.; Xu, J.; Shen, L.; Xiao, C.; Zhang, C.; Zhou, X.; Jian, X. Damage evolution model and failure mechanism of continuous carbon fiber-reinforced thermoplastic resin matrix composite materials. *Compos. Sci. Technol.* **2023**, *244*, 110300. [CrossRef]

10. Wang, A.; Xu, G.; Liu, X. Effect of polyurea coating on low-velocity impact properties of unidirectional carbon fiber-reinforced polymer composites plates. *Structures* **2024**, *61*, 106090. [CrossRef]
11. Chen, P.; Li, Y.; Yin, B.; Li, S.; Jia, W.; Lao, D.; Wang, H.; Liu, J. New design of bismuth borate ceramic/epoxy composites with excellent fracture toughness and radiation shielding capabilities. *Mater. Today Commun.* **2023**, *35*, 106102. [CrossRef]
12. Zhu, Y.; Sun, Y. Dynamic response of foam core sandwich panel with composite face sheets during low-velocity impact and penetration. *Int. J. Impact Eng.* **2020**, *139*, 103508. [CrossRef]
13. Acanfora, V.; Zarrelli, M.; Riccio, A. Experimental and numerical assessment of the impact behaviour of a composite sandwich panel with a polymeric honeycomb core. *Int. J. Impact Eng.* **2023**, *171*, 104392. [CrossRef]
14. Li, J.; Zhang, W.; Wang, Z.; Wang, Q.; Wu, T.; Qin, Q. Dynamic response and failure of CFRP Kagome lattice core sandwich panels subjected to low-velocity impact. *Int. J. Impact Eng.* **2023**, *181*, 104737. [CrossRef]
15. Jin, X.; Wang, Z.; Ning, J.; Xiao, G.; Liu, E.; Shu, X. Dynamic Response of Sandwich Structures with Graded Auxetic Honeycomb Cores under Blast Loading. *Compos. Part B Eng.* **2016**, *106*, 206–217. [CrossRef]
16. Zhang, X.; Hao, H.; Tian, R.; Xue, Q.; Guan, H.; Yang, X. Quasi-Static Compression and Dynamic Crushing Behaviors of Novel Hybrid Re-Entrant Auxetic Metamaterials with Enhanced Energy-Absorption. *Compos. Struct.* **2022**, *288*, 115399. [CrossRef]
17. Zhang, Y.; Ren, X.; Han, D.; Cheng, X.; Jiang, W.; Zhang, X.; Zhang, X.; Xie, Y. Static and Dynamic Properties of a Perforated Metallic Auxetic Metamaterial with Tunable Stiffness and Energy Absorption. *Int. J. Impact Eng.* **2022**, *164*, 104193. [CrossRef]
18. Guo, M.; Yang, H.; Ma, L. 3D Lightweight Double Arrow-Head Plate-Lattice Auxetic Structures with Enhanced Stiffness and Energy Absorption Performance. *Compos. Struct.* **2022**, *290*, 115484. [CrossRef]
19. Duncan, O.; Shepherd, T.; Moroney, C.; Foster, L.; Venkatraman, P.D.; Winwood, K.; Allen, T.; Alderson, A. Review of Auxetic Materials for Sports Applications: Expanding Options in Comfort and Protection. *Appl. Sci.* **2018**, *8*, 941. [CrossRef]
20. Kolken, H.M.A.; Zadpoor, A.A. Auxetic Mechanical Metamaterials. *RSC Adv.* **2017**, *7*, 5111–5129. [CrossRef]
21. Galea Mifsud, R.; Muscat, G.A.; Grima-Cornish, J.N.; Dudek, K.K.; Cardona, M.A.; Attard, D.; Farrugia, P.-S.; Gatt, R.; Evans, K.E.; Grima, J.N. Auxetics and FEA: Modern Materials Driven by Modern Simulation Methods. *Materials* **2024**, *17*, 1506. [CrossRef]
22. Sridharan, S.; Pankow, M. Performance Evaluation of Two Progressive Damage Models for Composite Laminates under Various Speed Impact Loading. *Int. J. Impact Eng.* **2020**, *143*, 103615. [CrossRef]
23. Liu, S.; Luan, Y.; Li, Y.; Su, Q.; Guo, Z.; Song, W. A 3D printed continuous carbon fiber reinforced composite with function of self-detecting and self-healing of internal damages. *Compos. Sci. Technol.* **2023**, *243*, 110264. [CrossRef]
24. Qiang, X.; Wang, T.; Xue, H.; Ding, J.; Deng, C. Study on Low-Velocity Impact and Residual Compressive Mechanical Properties of Carbon Fiber–Epoxy Resin Composites. *Materials* **2024**, *17*, 3766. [CrossRef] [PubMed]
25. Xi, X.; Xue, P.; Liu, X.; Bai, C.; Zhang, X.; Li, X.; Zhang, C.; Yang, X. Energy Absorption and Failure Modes of Different Composite Open-Section Crush Elements under Axial Crushing Loading. *Materials* **2024**, *17*, 3197. [CrossRef] [PubMed]
26. Stachyra, G.; Kloda, L.; Szmit, Z. Coupled Modal Analysis and Aerodynamics of Rotating Composite Beam. *Materials* **2023**, *16*, 7356. [CrossRef]
27. Lan, K.; Wang, H.; Wang, C. Delamination Behavior of CFRP Laminated Plates under the Combination of Tensile Preloading and Impact Loading. *Materials* **2023**, *16*, 6595. [CrossRef] [PubMed]
28. Li, J.; Gao, G.; Yu, Y.; Zhuo, T. Experimental and numerical study on the lightweight design of load-bearing energy absorption structure for subway train. *Thin-Walled Struct.* **2024**, *197*, 111542. [CrossRef]
29. Zhao, X.; Wei, L.; Wen, D.; Zhu, G.; Yu, Q.; Ma, Z. Bending response and energy absorption of sandwich beams with novel auxetic honeycomb core. *Eng. Struct.* **2021**, *247*, 113204. [CrossRef]
30. Mahgoub, M.; Liu, C.; Tan, Z. Study on Dynamic Mechanical Properties of Sandwich Beam with Stepwise Gradient Polymethacrylimide (PMI) Foam Core under Low-Velocity Impact. *Materials* **2024**, *17*, 2099. [CrossRef]
31. Bian, X.; Yang, L.; Wang, T.; Huang, G. Numerical Investigation on Anti-Explosion Performance of Non-Metallic Annular Protective Structures. *Materials* **2023**, *16*, 7549. [CrossRef]
32. Li, Z.; Xu, W.; Wang, C.; Liu, X.; Sun, Y. Investigation on Vibration Characteristics of Thin-Walled Steel Structures under Shock Waves. *Materials* **2023**, *16*, 4748. [CrossRef]
33. Borecki, M.; Rychlik, A.; Zan, L.; Korwin-Pawlowski, M.L. Steel Automotive Wheel Rims—Data Fusion for the Precise Identification of the Technical Condition and Indication of the Approaching End of Service Life. *Materials* **2024**, *17*, 475. [CrossRef]
34. Yu, S.; Wang, A.; Zhao, L. Characteristics test study on electromagnetic actuator. *Mach. Des. Manuf.* **2011**, *1*, 136–138.
35. Zhang, B. Research on Hybrid Isolator Design Technique. Master's Thesis, Harbin Engineering University, Harbin, China, 2011.
36. Zheng, X.; Zhang, C.; Lou, Y.; Xue, G.; Bai, H. Dynamic Characteristic Analysis of a Toothed Electromagnetic Spring Based on the Improved Bouc—Wen Model. *Materials* **2023**, *16*, 4889. [CrossRef]
37. Liu, S.; Hu, M.; Xiao, L.; Feng, G.; Song, K.; Song, W.; Qiao, J. Effects of Strain Rate and Low Temperature on Dynamic Behaviors of Additively Manufactured CoCrFeMnNi High-Entropy Alloys. *Mater. Sci. Eng. A* **2024**, *913*, 147100. [CrossRef]
38. Medvedev, A.E.; Maconachie, T.; Leary, M.; Qian, M.; Brandt, M. Materials & Design Perspectives on Additive Manufacturing for Dynamic Impact Applications. *Mater. Des.* **2022**, *221*, 110963.

39. George, E.P.; Curtin, W.A.; Tasan, C.C. High Entropy Alloys: A Focused Review of Mechanical Properties and Deformation Mechanisms. *Acta Mater.* **2020**, *188*, 435–474. [CrossRef]
40. Song, W.; Mu, K.; Feng, G.; Huang, Z.; Liu, Y.; Huang, X.; Xiao, L. Mechanical Properties of 3D Printed Interpenetrating Phase Composites with TPMS Architectures. *Thin-Walled Struct.* **2023**, *193*, 111210. [CrossRef]
41. Xiao, L.; Mu, K.; Liu, S.; Song, W. Experimental Study on the Fracture Behavior of 3D Printed Interpenetrating Phase Composites with Triply Periodic Minimal Surface Architectures. *Thin-Walled Struct.* **2025**, *208*, 112847. [CrossRef]
42. Zhang, M.; Zhao, N.; Yu, Q.; Liu, Z.; Qu, R.; Zhang, J.; Li, S.; Ren, D.; Berto, F.; Zhang, Z.; et al. On the Damage Tolerance of 3-D Printed Mg-Ti Interpenetrating-Phase Composites with Bioinspired Architectures. *Nat. Commun.* **2022**, *13*, 1–13. [CrossRef] [PubMed]
43. Crook, C.; Bauer, J.; Guell Izard, A.; Santos de Oliveira, C.; Martins de Souza e Silva, J.; Berger, J.B.; Valdevit, L. Plate-Nanolattices at the Theoretical Limit of Stiffness and Strength. *Nat. Commun.* **2020**, *11*, 1–11. [CrossRef]
44. Wang, Y.; Zhang, X.; Li, Z.; Gao, H.; Li, X. Achieving the Theoretical Limit of Strength in Shell-Based Carbon Nanolattices. *Proc. Natl. Acad. Sci. USA* **2022**, *119*, 1–11. [CrossRef]
45. Liu, S.; Sun, T.; Kou, Z.; Han, X.; Gao, Q.; Zhang, J.; Liu, X.; Zhang, L.; Orava, J.; Song, K.; et al. Promising Pathways for Balancing Strength and Ductility in Chemically Complex Alloys with Medium-to-high Stacking Fault Energies. *Int. J. Plast.* **2025**, *190*, 104358. [CrossRef]
46. Lei, M.; Aditya, R.; Liu, L.; Wu, M.; Wang, J.; Zhou, K.; Yao, Y. A Multi-Scale Constitutive Model for AlSi10Mg Alloy Fabricated via Laser Powder Bed Fusion. *Int. J. Solids Struct.* **2025**, *306*, 113111. [CrossRef]
47. Tang, Y.; Li, D. Dynamic Response of High-Entropy Alloys to Ballistic Impact. *Sci. Adv.* **2022**, *8*, 1–9. [CrossRef]
48. Yu, G.; Xiao, L.; Song, W. Deep Learning-Based Heterogeneous Strategy for Customizing Responses of Lattice Structures. *Int. J. Mech. Sci.* **2022**, *229*, 107531. [CrossRef]
49. Li, X.; Wang, P.; Zhao, M.; Su, X.; Tan, Y.; Ding, J. Customizable Anisotropic Microlattices for Additive Manufacturing: Machine Learning Accelerated Design, Mechanical Properties and Structural-Property Relationships. *Addit. Manuf.* **2024**, *89*, 104248. [CrossRef]
50. Liu, Q.; Wu, D. Machine Learning and Feature Representation Approaches to Predict Stress-Strain Curves of Additively Manufactured Metamaterials with Varying Structure and Process Parameters. *Mater. Des.* **2024**, *241*, 112932. [CrossRef]

materials

MDPI

Article

Study on Low-Velocity Impact and Residual Compressive Mechanical Properties of Carbon Fiber–Epoxy Resin Composites

Xueyuan Qiang [1], Te Wang [1], Hua Xue [1], Jun Ding [2,*] and Chengji Deng [2]

1 National Center for Composites Quality Testing & Inspectionl (Hubei), Hubei Zery Testing Technology Co., Ltd., Xiangyang 441000, China
2 The State Key Laboratory of Refractories and Metallurgy, Wuhan University of Science and Technology, Wuhan 430081, China
* Correspondence: dingjun@wust.edu.cn

Abstract: Room temperature drop hammer impact and compression after impact (CAI) experiments were conducted on carbon fiber–epoxy resin (CF/EP) composites to investigate the variation in impact load and absorbed energy, as well as to determine the residual compressive strength of CF/EP composites following impact damage. Industrial CT scanning was employed to observe the damage morphology after both impact and compression, aiding in the study of impact-damage and compression-failure mechanisms. The results indicate that, under the impact load, the surface of a CF/EP composite exhibits evident cratering as the impact energy increases, while cracks form along the length direction on the back surface. The residual compressive strength exhibits an inverse relationship with the impact energy. Impact damage occurring at an energy lower than 45 J results in end crushing during the compression of CF/EP composites, whereas energy exceeding 45 J leads to the formation of long cracks spanning the entire width of the specimen, primarily distributed symmetrically along the center of the specimen.

Keywords: CF/EP composites; low-velocity impact; impact-damage mechanism; compression after impact; failure mechanism

1. Introduction

Resin matrix composites have broad development prospects due to their brilliant fatigue resistance, corrosion resistance, and shock absorption. In addition, resin matrix composites possess the advantages of a simple forming process, strong plasticity of the material structure, and performance, and are widely used in the automotive industry, aerospace industry, and coal-mining industry [1–4]. Carbon-fiber-reinforced composites, a pivotal category within resin matrix composites, find extensive utilization in the aerospace sector, shipbuilding, and other fields owing to their higher strength, lower density, and better fatigue resistance [5–10]. However, with the continuous improvement of material performance requirements, particularly when confronted with dynamic impact loads (for example, the effect of bird flocks, hail, and other foreign objects on aircraft during flight), the research on the impact resistance and damage mechanism of carbon-fiber-reinforced plastics (CFRP) has become crucial [11–17].

In practical applications, carbon-fiber composites may face a variety of impact loads from collisions, impacts, and accidents. Therefore, many scholars have implemented studies on their impact resistance and damage mechanism [18–21]. Liu et al. [22] prepared hybrid composite laminates composed of carbon-fiber unidirectional layers and braided layers by the resin-impregnation method and carried out drop-weight impact experiments on different laminated samples. The results show that the laminated samples affect the impact response of the composite laminates by changing the overall bending stiffness Kumar et al. [23] prepared SiC/Al composites and explored the influence of SiC content on its shock resistance and wear resistance. The results indicate that the impact strength and specific wear of the composites with SiC volume fraction is 9%. Some studies have already proven that glass fiber possesses strong impact resistance. Therefore, to further enhance the properties of CFRP in low-velocity impact tests, some researchers contemplated adding a glass-fiber layer to it [24–27]. Lei et al. [28] investigated the low-velocity impact and post-compression behavior following the impact of various fiber composites and flame-retardant epoxy resin. The findings prove that introducing glass fiber can change the impact-damage mode of laminates and effectively improve the impact resistance and compressive impact strength of laminates. When it comes to the study of the damage behavior of carbon-fiber-reinforced resin matrix composites, in addition to routine performance testing and cross-sectional morphology analysis, researchers can also utilize finite-element methods to simulate the behavior of carbon-fiber composites under impact loads and predict their damage mechanisms [29–31]. Researchers [32–38] have carried out research on the mechanical properties of materials with single/mixed reinforcements for low-velocity impact or/and post-impact damage and analyzed the material failure mechanisms.

In summary, researching the impact resistance and damage mechanism of CFRP is not only of great significance for improving the performance and reliability of materials but also provides theoretical guidance and technical support for its widespread applications in aviation, automotive, and other fields [39–41]. Most studies adopt ultrasonic and ray damage research means and fewer of them use the industrial CT method. The existing literature on T700 carbon-fiber-reinforced epoxy-resin matrix composites' low-speed impact and post-impact damage research appears to be insufficient, while the utilization of industrial CT to study the impact damage is little reported. The existing research mainly focuses on the influence of reinforcing materials and preparation methods on the properties of composites, but the damage mechanisms of composites under varying loads may turn out to be corresponding changing forms. Therefore, this study used T700 carbon fiber as the reinforcing material, QC350 epoxy resin was used as the matrix, and composites were prepared by the vacuum introduction molding method. Through low-speed impact and post-impact compression tests, based on the non-destructive characterization method of industrial CT, the damage failure mechanism of laminated plates was analyzed, and the effect of impact energy on residual compressive strength was investigated.

2. Materials and Methods

2.1. Materials

The reinforcement material is T700 carbon-fiber multidirectional cloth (Toray, Tokyo, Japan), the matrix material is epoxy resin QC350, and the ratio of QC350 is QC350A:QC350B = 100:30. The plywood layup is [(45/−45)/(0/90)]. The nominal thickness of the single layer is 0.29 mm, and the thickness of the specimen is 5.7 mm. The performance parameters of T700 carbon-fiber multidirectional cloth are detailed in Table 1, and the performance parameters of epoxy resin QC350 are detailed in Table 2.

Table 1. Property parameters of T700 carbon fibers.

Tensile Strength/MPa	Young's Modulus/MPa	Elongation/%	Fineness/(g/km)	Density/(g·cm^{-3})	Fiber Volume Fraction/%	Thickness/mm	Diameter/μm
4900	2300	2.1	800	1.80	54.5	0.42	7

Table 2. Property parameters of QC350 epoxy resin.

Mixed Viscosity/(CPS)	Glass-Transition Temperature/°C	Tensile Strength/MPa	Modulus/MPa
250~300	75~85	65~75	2800~3200

CF/EP composites were prepared by the vacuum introduction molding method, and the preparation schematic is shown in Figure 1. In accordance with the prescribed lay-up order, the prefabricated body was laid flat to the mold surface, and the auxiliary materials, such as the infusion net and the inlet–outlet hose, were laid in order to strengthen the flow performance of the epoxy resin. A vacuum-bag film with a length and width of 400 mm larger than the sample was laid flat, and the mold cavity was sealed with sealing tape. At room temperature, the vacuum system was used to achieve the pressure of 0.1 Pa between the vacuum bag and the mold cavity, and the pressure was maintained for 1 h. With the assistance of the atmospheric pressure effect, the epoxy resin slowly entered the cavity along the infusion network until the resin filled up the entire mold and impregnated the preformed body, and then it was cured at 80 °C.

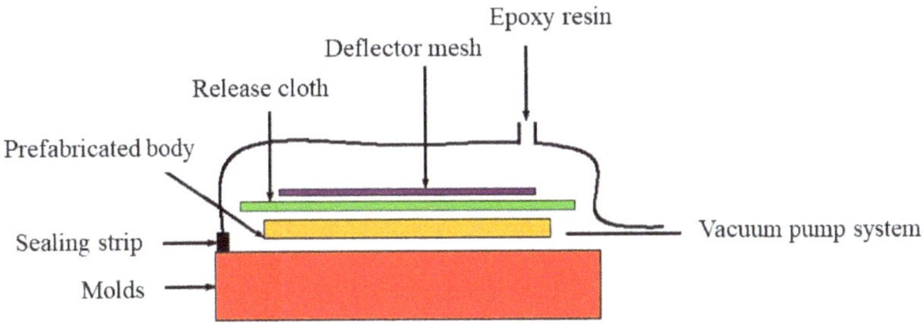

Figure 1. Schematic diagram of vacuum infusion molding.

2.2. *Testing and Characterization*

According to ASTM D7136/D7136M-2020 [42], the low-velocity impact test was carried out by using a φ16 mm hemispherical punch with a weight of 4.3 kg, and the weight of the whole drop-weight system was 5.482 kg (the equipment model and manufacturer are CEAST 9350, Instron, Pianezza, Italy). The sample should be placed on the impact fixed base, and the position of the sample should be adjusted appropriately. In that case, the end of the punch could accurately impact the center of the sample. Put down four plastic calipers in turn (as shown in Figure 2) to ensure that the calipers can firmly fix the sample, and the clamping force on the sample is basically the same so as to prevent the sample from slipping during the drop hammer impact process, affecting the accuracy of the test results. According to the test requirements under different amounts of energy, the test parameters were set in turn. The impact energies were sequentially administered at 25 J, 35 J, 45 J, 65 J, 85 J, 105 J, and 125 J, resulting in impact velocities of 3.02 m/s, 3.57 m/s, 4.05 m/s, 4.87 m/s, 5.57 m/s, 6.19 m/s, and 6.75 m/s, respectively. At the same time, the drop hammer impact test in turn should be carried out using the anti-secondary impact device. Through the data acquisition system, the key performance indexes, such as energy absorption and bearing load of the sample during the impact process, are obtained, and the impact resistance of the material is evaluated.

According to the ASTM D7136 standard, the velocity $v(t)$, displacement $\delta(t)$ and absorbed energy $E_a(t)$ of the punch varying with time can be obtained and are shown in Equation (1), Equation (2), and Equation (3), respectively.

$$v(t) = v_i + gt - \int_0^t \frac{F(t)}{m} dt \quad (1)$$

where $v(t)$ is the velocity of the punch at moment t in m/s. v_i is the initial velocity of the punch in m/s. g is the acceleration of gravity in m/s^2.$F(t)$ is the contact force at moment t in N. m is the mass of the punch in kg.

$$\delta_t = \delta_i + v_i t + \frac{gt^2}{2} - \int_0^t \left(\int_0^t \frac{F(t)}{m} dt \right) dt \quad (2)$$

where $\delta(t)$ is the punch displacement in mm. δ_i is the starting displacement of the punch at t = 0 in mm.

$$E_a(t) = \frac{m\left(v_i^2 - v(t)^2\right)}{2} + mg\delta(t) \quad (3)$$

where $E_a(t)$ is the absorbed energy at time t in J.

The compressive damage load of the plywood obtained from the compression test can be calculated from Equation (4) to obtain the compressive strength F^{CAI}.

$$F^{CAI} = P_{max}/A \quad (4)$$

where F^{CAI} is the compressive strength in MPa. P_{max} is the maximum load in N at the time of specimen destruction. A is the cross-sectional area of the specimen in mm^2.

Figure 2. Clamping diagram of low-velocity impact test specimen.

The residual compression test under different impact energies was carried out by an electronic universal testing machine (Instron 5989, Instron; the physical picture of the device is shown in Figure 3). The sample after the impact test was clamped to the compression fixture to ensure that the sample and the fixture were aligned, and the damage position of the sample emerged at the center of the fixture. At the same time, the fixture was equipped with an anti-bending device to prevent the introduction of bending loads other than axial compression during the compression process. The fixture specimen was positioned within the pressure plates of the testing apparatus and subjected to a compression load applied longitudinally along its axis at a rate of 1.25 mm/min until reaching its peak, where the load decreased by 30%. The apparent morphology of the test sample was observed by industrial CT.

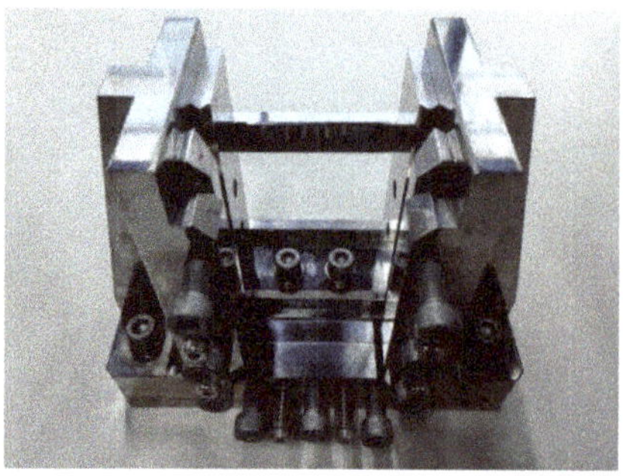

Figure 3. Schematic diagram of the remaining compression experimental fixture after impact.

3. Results and Discussion

3.1. Low-Speed Impact Experiment

Figures 4–6 show the non-destructive CT inspection patterns of CF/EP composites on the front and back of the specimens after impacts at different energies. It can be seen that round craters of different sizes and depths are produced on the impact surface, and long cracks along the axial direction appear on the back surface. It can be observed from Figure 4 that, when the impact energy is 45 J, while there are only shallow pits on the surface of the sample, there are only a few short cracks and inconspicuous damage. At an impact energy of 105 J, an increasing incidence of extensive internal cracks becomes discernible within the specimen (Figure 5). Figure 6 shows the surface and back of the specimen after the 125 J impact energy impact test, from which the back of the specimen is affected by tensile stress can be seen, resulting in fiber fracture, but the damage does not penetrate the entire specimen. It can be acknowledged that, with the increase of impact energy, the longer the crack, the greater the impact damage will be [43–45].

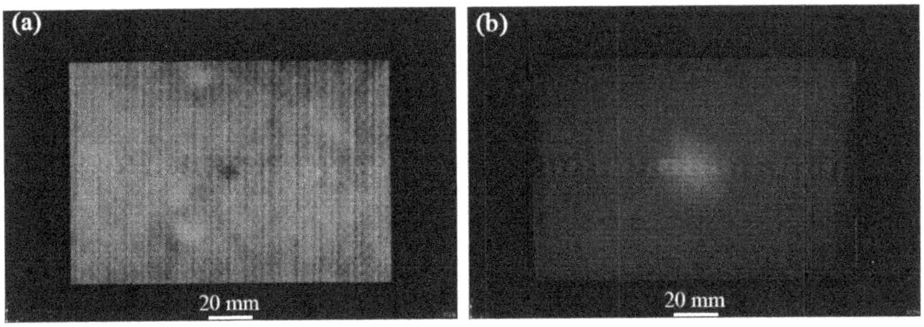

Figure 4. CT damage atlas of sample after 45 J impact energy (**a**) specimen face, (**b**)specimen back.

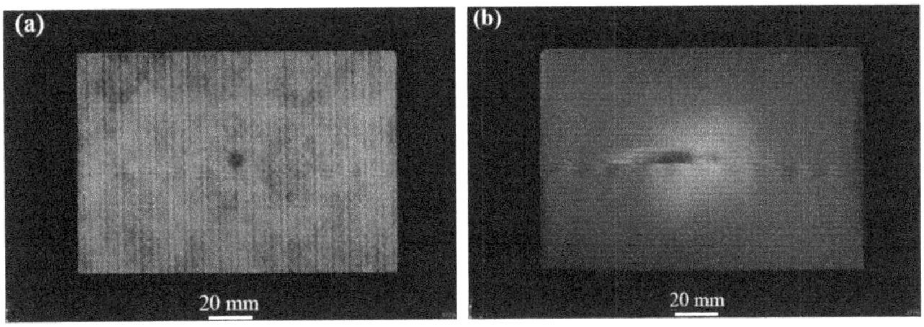

Figure 5. CT damage atlas of sample after 105 J impact energy (**a**) specimen face, (**b**)specimen back.

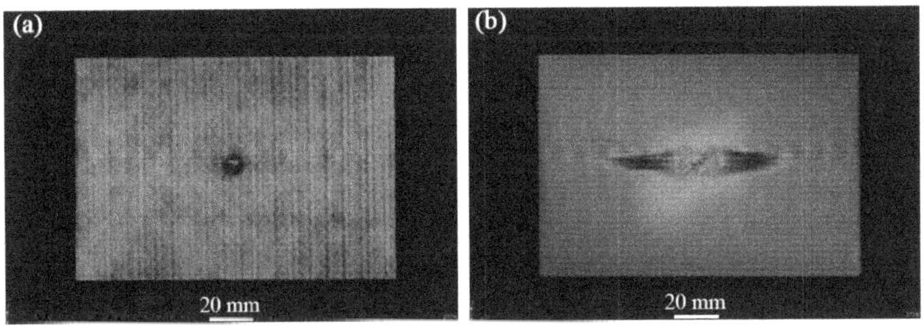

Figure 6. CT damage atlas of sample after 125 J impact energy (**a**) specimen face, (**b**)specimen back.

Figure 7 demonstrates the load-time curve, energy-time curve, and load-displacement curve of CF/EP composites under different energy impacts. It can be observed from Figure 7a that the curves display three stages: linear loading, platform, and nonlinear unloading. Under low-energy impact, the sample undergoes a short platform stage and then enters the unloading stage, indicating that there is a debonding of the sample matrix from the fibers and a small amount of fiber breakage. Because of the rapid impact process, in a very short period of time, when the phenomenon of debonding and fiber breakage is obviously intensified, there is an outstanding downward trend at the curve until the specimen is damaged. The composite's failure modifies from matrix failure to fiber fracture, and the fiber is capable of resisting the punch load. Therefore, when the energy exceeds 45 J, the load increases rapidly, which is the "critical value" of the composite-failure mode transformation. With the energy reaching 125 J, the load value of the platform stage fluctuates significantly, indicating that a wide area of fiber and matrix failure becomes evident on the back of the sample [46,47]. When the integrity of the laminated plate structure is destroyed, its bearing capacity experiences a corresponding decrease, and the impact force holds a downward trend in the second half of the platform stage.

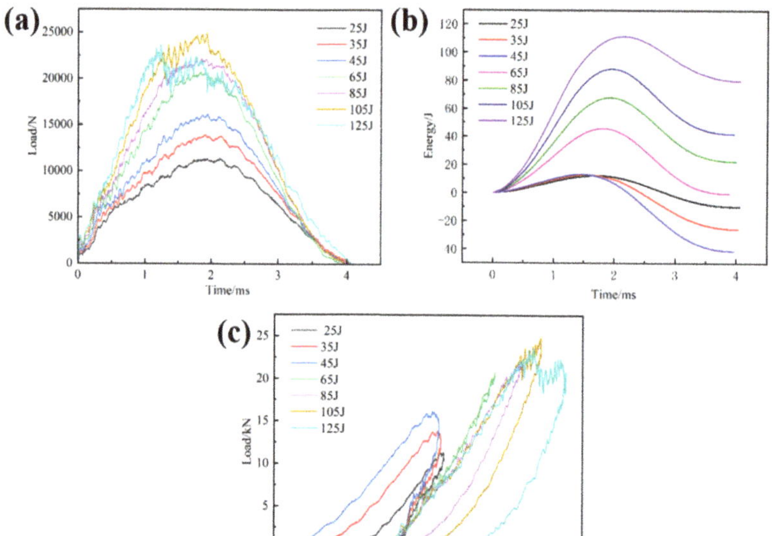

Figure 7. Load/energy-time curves and load-displacement curves of specimens under different impact energy levels; (**a**) load-time curves, (**b**) energy-time curves, (**c**) load-displacement curves.

It is evident from Figure 7b that the absorbed energy (dissipated energy) in the composite sample is less than the initial impact energy. This is owing to the damage, such as fiber breakage, in the sample consuming part of the energy. When the energy increases to 45 J, the peak absorption energy undergoes a leapfrog increase, and the primary mechanism of energy dissipation transforms from matrix cracking to fiber fracture progressively, which is consistent with the conclusion of Figure 7a. With the increasing impact energy, the composites will absorb more energy.

It can be indicated from Figure 7c that the whole load-displacement curve shows two stages, namely the linear loading stage and the nonlinear unloading stage. However, at low energy impacts, three phases are presented, which are the linear loading phase, the nonlinear loading phase, and the nonlinear unloading phase, which are more obvious. During the loading stage, the curve is linear and accompanied by a small jagged fluctuation.

It shows that, during the impact-loading process, the composite laminate contains a large number of staggered carbon fibers to continuously resist the pressure applied by the punch, resulting in a continuous reciprocating 'pressure-resistance' phenomenon, and the stiffness change is small or even ignored. During the loading process, the majority of the punch's kinetic energy is stored as elastic strain energy. As the punch applies sustained pressure to the laminate, the composite experiences matrix failure accompanied by progressive fiber breakage until the load reaches its peak capacity. When the impact force reaches the peak, the punch rebounds under the action of accumulated elastic strain energy, and the load-displacement curve also enters the nonlinear unloading stage. When the load is unloaded to zero, the displacement on the curve does not return to the initial position, which is caused by the impact-induced irreversible damage of the laminate and a small part of the energy dissipation loss [37,48,49].

3.2. Compression Experiment after Impact

The residual compressive strength and damage that form the CF/EP composites under different impact energies at room temperature are listed in Table 3. It can be seen from Table 3 that the damage mode of the sample is end crushing, while the impact energy is less than 45 J. Compared to Figure 4, it can be observed that, after the sample is impacted by low energy, the impact surface forms pits of different depths, and micro-damage, such as surface matrix cracking, occurs. The impact-damage area is small, and the residual bearing capacity of the sample is sufficient. When the impact energy exceeds 45 J, the damage mode of the specimen changes to LDM (lateral multi-layered expansion) damage, and the residual compressive bearing capacity of the specimen is significantly reduced or even fails, which is consistent with the findings in the previous section. The residual compressive strength at the maximum energy of 125 J is 165.62 MPa, which is 37% lower than that at 25 J. According to the content of Table 3, the curve of impact energy surplus compressive strength shown in Figure 8 is drawn. Figure 8 indicates that, with increasing impact energy, the residual compressive strength of the sample decreases gradually, and the residual bearing capacity of the sample decreases progressively.

Table 3. Residual compressive strength and damage form of CF/EP composites under different impact energies at room temperature.

Specimen Number	Impact Energy/J	Residual Compressive Strength/MPa	Failure Mode
1	25	264.37	End failure
2	35	247.70	End failure
3	45	243.52	Valid LDM
4	65	222.16	Valid LDM
5	85	220.40	Valid LDM
6	105	191.14	Valid LDM
7	125	165.62	Valid LDM

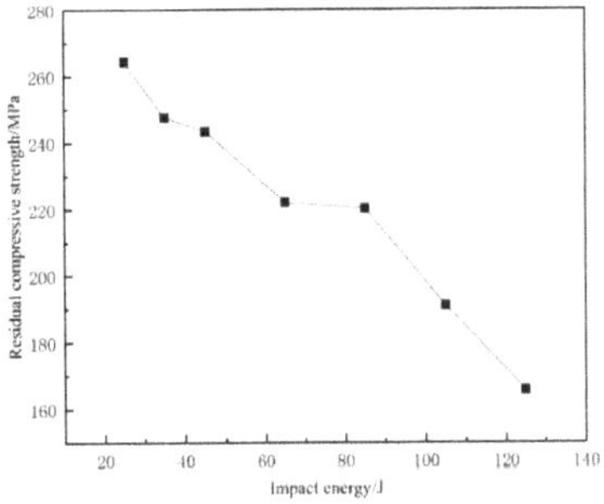

Figure 8. Residual compressive strength curves at different impact energies.

Figure 9 illustrates the damage morphology of the CF/EP composites compressed after 35 J energy impact. The specimen shows obvious end-crushing damage. The specimen is subjected to compressive loading, and the compression surface shows delamination and cracks of different lengths, and even narrow delamination pores. The front and back surfaces showed flat cracks extending along the entire width direction, and the rest of the plane showed unnoticeable visual compression damage. At the end of the specimen, along the thickness of the long side, there is a compression damage of about 90°, and the fibers show destructive fracture. Because the end of the specimen is subjected to a concentrated compression load, the adhesive strength between the end fibers and resin decreases, and debonding delamination damage easily occurs. With the further increase of the load, the specimen end is subject to concentrated contact stress. The load is insufficient to transfer through the specimen end to the impact of the small damaged area. The end of the damage is further aggravated. The fiber cannot effectively carry the compression load, resulting in the end of the fiber fracture, and the matrix cracks and delaminates. The specimen at the end of the compression collapses.

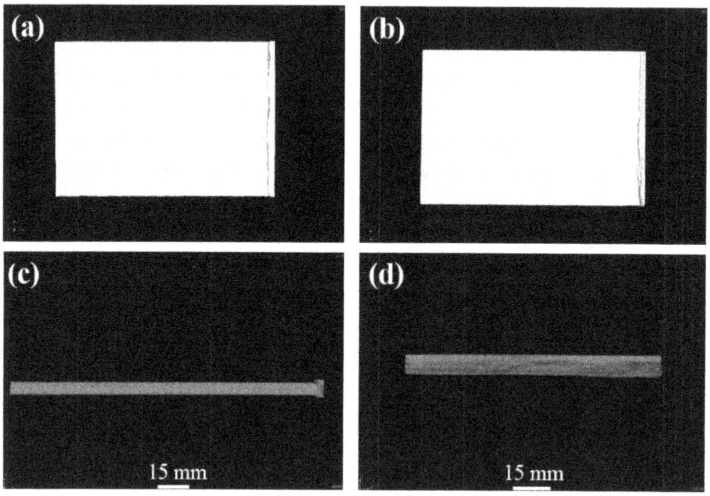

Figure 9. CT atlas of compressive test specimen after 35 J impact energy; (**a**) specimen face, (**b**) specimen back, (**c**) specimen length direction end, (**d**) specimen width direction end.

Figures 10–12 display the damage picture of compression specimens of CF/EP composites after 45J, 105 J, and 125 J energy impacts. Under the axial compression load, matrix cracks first appeared in the center part of the impact surface, accompanied by a little delamination and fiber pullout. On the front side, long cracks across the entire width were common, with surface fiber debonding and delamination, and the surface matrix appeared to be wrinkled and raised. Compression damage on the back of the specimen is more prominent, such as long cracks, fiber breakage, debonding delamination, and other injuries, with transverse and longitudinal fiber partially pulled out and fractured, and the specimen is severely buckled, with even the formation of "arch bridge" delamination as well. There are cracks along the transverse expansion, with the formation of long cracks across the width of the specimen, and a basically symmetrical distribution along the center of the specimen but, in the axial direction, only held a small distance, resulting in catastrophic damage failure. From the thickness direction, clear and obvious intralaminar cracks and delaminations appeared inside the specimen, and there was an inclined shear fracture of about 45° spanning almost the entire thickness direction. Even a long strip of porosity and compression collapse were also present. Deeper observation of the internal damage at the location of the impact point of the specimen proves that the damage is even more severe.

Observation from the long edge direction shows that the center of the long edge of the specimen shows significant delamination, fiber breakage, and other destructive damage, and the back of the specimen is raised. The findings from the wide edge direction reveal a heightened incidence of debonding and delamination characterized by varying degrees of delamination severity, forming delamination damage of various lengths and sizes, as well as oblique cracks and flat cracks, and the oblique cracks will penetrate through the multilayered fibers and resins to form shear-type damage. Because of under-compressive loading, the stress formed at the end of the specimen is continuously transferred to the vicinity of the impact defect. The stress is redistributed and concentrated, making cracks and delamination in the middle of the damaged region expand significantly and form a penetrating fracture, which is the specimen's serious failure.

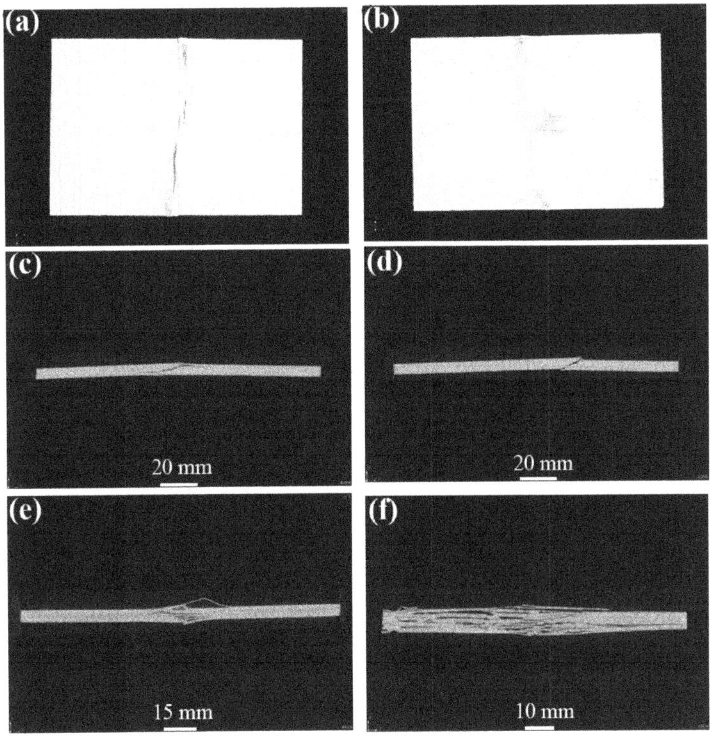

Figure 10. CT atlas of compressive test specimen after 45 J impact energy; (**a**) specimen face, (**b**) specimen back, (**c**,**d**) specimen length direction side, (**e**) specimen center of length direction, (**f**) specimen center of width direction.

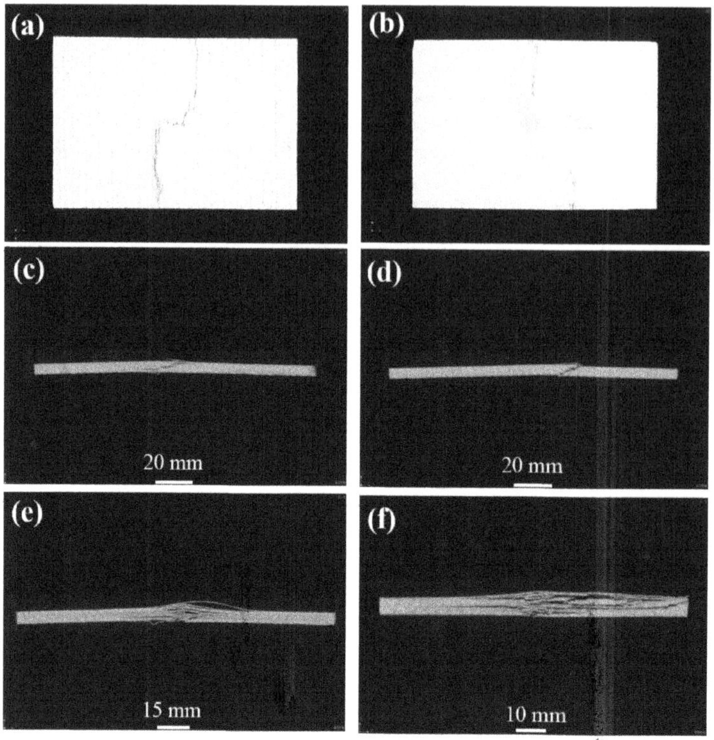

Figure 11. CT atlas of compressive test specimen after 105 J impact energy; (**a**) specimen face, (**b**) specimen back, (**c**,**d**) specimen length direction side, (**e**) specimen center of length direction, (**f**) specimen center of width direction.

As shown in Figure 13, the overall CAI load-displacement curves at different energies show a linear relationship. In the initial stage, with the onset loading of the load, the displacement of the indenter into the end of the specimen increases slowly and has a nonlinear relationship. With the further increase in the compression load, the load-displacement curve appears to be obviously linear, and finally, the load is presented as falling off a cliff. At this time, the specimen with an initial impact damage has reached the load-bearing limit, and the specimen appears to have clearly visible fiber pullout, delamination, matrix cracking, and other serious damage and has been destroyed. Through observation, it can be seen that sudden brittle fracture occurs in the specimens when the impact energy is below 45 J. The compression load reaches its limit and then suddenly decreases, which is due to the smaller impact energy, lighter damage, and the need to withstand more load on the fiber and matrix under the compression load [50,51]. Before the damage occurs at the interface, the load-carrying capacity of the specimen has reached its limit and thus brittle fracture occurs. When the specimen is above 45 J, because the specimen failed to appear along the width direction of the complete fracture, the specimen has produced localized damage, and the surface layer of fibers has been visually fractured or pulled out and cannot effectively carry any more compressive load. So, the bearing capacity is reduced, and the matrix has become the main bearer of the compressive load to strengthen the generation of irreversible plastic deformation [52].

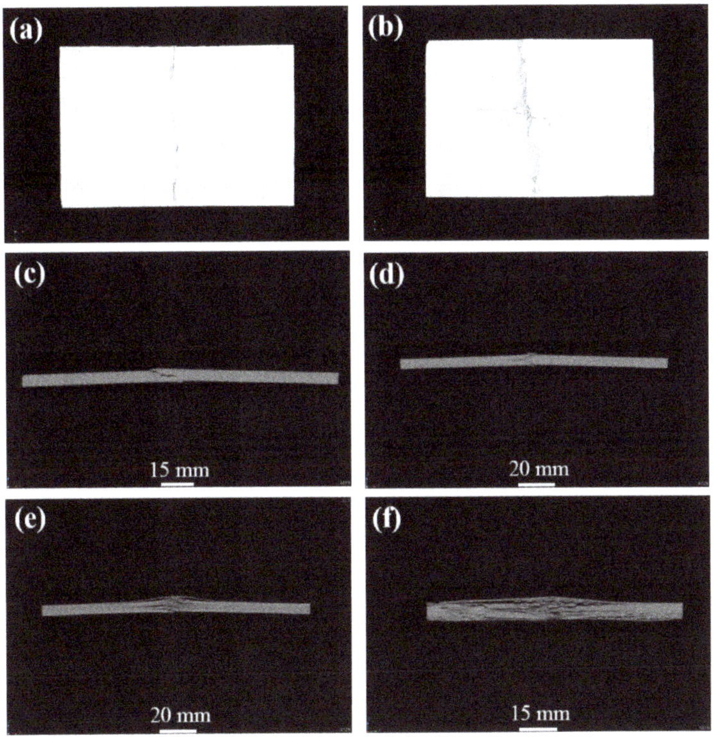

Figure 12. CT atlas of compressive test specimen after 125 J impact energy; (**a**) specimen face, (**b**) specimen back, (**c**,**d**) specimen length direction side, (**e**) specimen center of length direction, (**f**) specimen center of width direction.

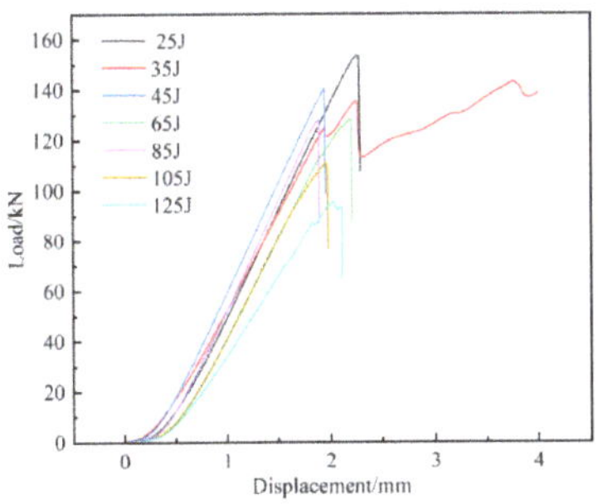

Figure 13. Residual compressive load-displacement curves of CF/EP composites.

4. Conclusions

(1) After the CF/EP composites were impacted by low speed, the specimens would have pits on the front side. The pits will be deeper with the increase of impact energy. In addition, in terms of the back side of the matrix-cracking phenomenon, the greater the energy is, the more significant the back damage will be;

(2) Impact damage has a great influence on the residual compressive strength of CF/EP composites. The residual compressive strength decreases with the increase of impact energy. The critical energy value is 45 J, and the end crushing occurs after compression of the sample below 45 J. The delamination propagation (LDM) damage occurs from the

middle to the lateral side of the sample after compression of the sample above 45 J, and the residual compressive strength decreases sharply when the impact energy is over 45 J. The residual compressive strength at the lowest energy of 25 J is 264.37 MPa, and the residual compressive strength at the highest energy of 125 J is 165.62 MPa, which is 37% lower than the residual compressive strength at 25 J;

(3) After the CAI test, the impact surface of CF/EP composites produced cracks along the length and width directions, and the cracks gradually expanded transversely to the edge of the sample from the end of the impact cracks as the impact energy increased.

Author Contributions: X.Q., T.W., J.D. and C.D.: Conceptualization, Methodology, Investigation; X.Q., T.W., H.X., J.D. and C.D.: Data curation, Writing—Original draft preparation; J.D. and C.D.: Supervision; J.D.: Writing—Reviewing and Editing. All authors have read and agreed to the published version of the manuscript.

Funding: This research received support from the Key Research and Development Program of Hubei Province (2023BAB106).

Institutional Review Board Statement: Not applicable.

Informed Consent Statement: Not applicable.

Data Availability Statement: Data are available upon request due to restrictions on privacy. The data presented in this study are available on request from the corresponding author.

Acknowledgments: The authors acknowledge the financial support from the Key Research and Development Program of Hubei Province (2023BAB106).

Conflicts of Interest: Authors Xueyuan Qiang, Te Wang and Hua Xue were employed by the company Hubei Zery Testing Technology Co., Ltd. The remaining authors declare that the research was conducted in the absence of any commercial or financial relationships that could be construed as a potential conflict of interest.

References

1. Liu, M.; Hou, D.Y.; Zheng, K.K.; Gao, C.H. Characterization of friction and wear of phenolic resin matrix composites reinforced by bamboo fibers of alkaline and $LaCl_3$ treatment. *Mater. Today Commun.* **2023**, *35*, 106361. [CrossRef]
2. Wu, W.J.; Zou, Y.; Li, C.H.; Li, Y.W.; Wang, Z.Y.; Chang, N.; Shi, Y.S. Effect of impregnated phenolic resin on the properties of Si–SiC ceramic matrix composites fabricated by SLS-RMI. *Ceram. Int.* **2023**, *49*, 1624–1635. [CrossRef]
3. Zou, Z.Y.; Qin, Y.; Fu, H.D.; Zhu, D.; Li, Z.Z.; Huang, Z.X. ZrO_2f-coated CF hybrid fibrous reinforcements and properties of their reinforced ceramicizable phenolic resin matrix composites. *J. Eur. Ceram. Soc.* **2021**, *41*, 1810–1816. [CrossRef]
4. Niu, J.W.; Wang, Z.F.; Liu, H.; Ma, Y.; Pang, H.X.; Wang, X.T. Response surface optimization of pitch phase change densification using composite phenolic resin co-carbonization to prepare high performance carbon refractories. *J. Anal. Appl. Pyrol.* **2023**, *175*, 106144. [CrossRef]
5. Zhu, T.Q.; Ren, Z.Y.; Xu, J.; Shen, L.L.; Xiao, C.L.; Zhang, C.; Zhou, X.; Jian, X.G. Damage evolution model and failure mechanism of continuous carbon fiber-reinforced thermoplastic resin matrix composite materials. *Compos. Sci. Technol.* **2023**, *244*, 110300. [CrossRef]
6. Wang, A.N.; Xu, G.W.; Liu, X.G. Effect of polyurea coating on low-velocity impact properties of unidirectional carbon fiber-reinforced polymer composites plates. *Structures* **2024**, *61*, 106090. [CrossRef]
7. Chen, P.; Li, Y.B.; Yin, B.; Li, S.J.; Jia, W.B.; Lao, D.; Wang, H.L.; Liu, J.F. New design of bismuth borate ceramic/epoxy composites with excellent fracture toughness and radiation shielding capabilities. *Mater. Today Commun.* **2023**, *35*, 106102. [CrossRef]
8. Yan, M.L.; Liu, Y.C.; Jiang, W.G.; Qin, W.Z.; Yan, Y.; Wan, L.Y.; Jiao, W.C.; Wang, R.G. Mechanism of matrix influencing the cryogenic mechanical property of carbon fibre reinforced epoxy resin composite. *Compos. Commun.* **2022**, *33*, 101220. [CrossRef]
9. Cugnoni, J.; Amacher, R.; Kohler, S.; Brunner, J.; Kramer, E.; Dransfeld, C.; Smith, W.; Scobbie, K.; Sorensen, L.; Botsis, J. Towards aerospace grade thin-ply composites: Effect of ply thickness, fibre, matrix and interlayer toughening on strength and damage tolerance. *Compos. Sci. Technol.* **2018**, *168*, 467–477. [CrossRef]
10. Liu, H.P.; Lei, W.; Tong, Z.M.; Guan, K.K.; Zhang, H.J. Enhanced diffusion kinetics of Li ions in double-shell hollow carbon fibers. *ACS Appl. Mater.* **2021**, *13*, 24604–24614. [CrossRef]
11. Ma, S.H.; He, Y.; Hui, L.; Xu, L. Effects of hygrothermal and thermal aging on the low-velocity impact properties of carbon fiber composites. *Adv. Compos. Mater.* **2020**, *29*, 55–72. [CrossRef]
12. Zhu, B.L.; Wang, J.; Zheng, H.; Ma, J.; Wu, J.; Wu, R. Investigation of thermal conductivity and dielectric properties of LDPE-matrix composites filled with hybrid filler of hollow glass microspheres and nitride particles. *Compos. Part B Eng.* **2015**, *69*, 496–506. [CrossRef]

13. Xian, G.; Guo, R.; Li, C.; Wang, Y. Mechanical performance evolution and life prediction of prestressed CFRP plate exposed to hygrothermal and freeze-thaw environments. *Compos. Struct.* **2022**, *293*, 115719. [CrossRef]
14. Lu, Z.; Li, J.; Xie, J.; Huang, P.; Xue, L. Durability of flexurally strengthened RC beams with prestressed CFRP sheet under wete-dry cycling in a chloride-containing environment. *Compos. Struct.* **2020**, *255*, 112869. [CrossRef]
15. Ouyang, S.; Li, Y.B.; Ouyang, D.G.; Li, S.J.; Xu, N.N.; Xiang, R.F. Microstructural evolution of carbon fibers by silicon vapor deposition and its effect on mullite-corundum castables. *Ceram. Int.* **2021**, *47*, 7824–7830. [CrossRef]
16. Huang, Z.; Zhang, W.; Qian, X.; Su, Z.; Pham, D.-C.; Sridhar, N. Fatigue behaviour and life prediction of fifilament wound CFRP pipes based on coupon tests. *Mar. Struct.* **2020**, *72*, 102756. [CrossRef]
17. Zhao, X.; Zhang, Z.Y.; Pang, J.Y.; Su, L. Preparation of carbon fibre-reinforced composite panels from epoxy resin matrix of nano lignin polyol particles. *J. Clean. Prod.* **2023**, *428*, 139170. [CrossRef]
18. Sun, Z.; Xiao, J.; Tao, L.; Wei, Y.; Wang, S.; Zhang, H.; Zhu, S.; Yu, M. Preparation of high-performance carbon fiber-reinforced epoxy composites by compression resin transfer molding. *Materials.* **2019**, *12*, 13. [CrossRef] [PubMed]
19. Xie, H.B.; Zhang, J.L.; Li, F.L.; Yuan, G.Q.; Zhu, Q.; Jia, Q.L.; Zhang, H.J.; Zhang, S.W. Selective laser melting of SiCp/Al composites: Densification, microstructure, and mechanical and tribological properties. *Ceram. Int.* **2021**, *47*, 30826–30837. [CrossRef]
20. Liu, S.; Yuan, Q.; Gong, Y.; Xu, G.; Qiao, W. Relationship between microstructure and dry wear behavior of compo-cast nano-$SiC_{(p)}$+micro-$Gr_{(p)}$/Zn-35Al-1.2Mg-0.2Sr composite under different chilling conditions. *Kovove Mater.* **2020**, *58*, 49–57. [CrossRef]
21. Abakah, R.R.; Huang, F.; Hu, Q.; Wang, Y.; Liu, J. Comparative study of corrosion properties of different graphene nanoplate/epoxy composite coatings for enhanced surface barrier protection. *Coatings* **2021**, *11*, 285. [CrossRef]
22. Liu, H.B.; Falzon, G.B.; Tan, W. Experimental and numerical studies on the impact response of damage-tolerant hybrid unidirectional/woven carbon-fibre reinforced composite laminates. *Compos. Part B Eng.* **2018**, *136*, 101–108. [CrossRef]
23. Kumar, R.A.; Devaraju, A. A comparative investigation on cast and aging (T6) response on mechanical and dry sliding wear behavior of Al7075/SiCp metal matrix composite. *Surf. Rev. Lett.* **2021**, *28*, 2150044. [CrossRef]
24. Sun, G.Y.; Yu, H.W.; Wang, Z.; Xiao, Z.; Li, Q. Energy absorption mechanics and design optimization of CFRP/aluminum hybrid structures for transverse loading. *Int. J. Mech. Sci.* **2019**, *150*, 767–783. [CrossRef]
25. Song, Z.G.; Zhang, L.W.; Liew, K.M. Dynamic responses of CNT reinforced composite plates subjected to impact loading. *Compos. Part B Eng.* **2016**, *99*, 154–161. [CrossRef]
26. Gu, G.X.; Takaffoli, M.; Hsieh, A.J.; Markus, M.J. Biomimetic additive manufactured polymer composites for improved impact resistance. *Extreme Mech. Lett.* **2016**, *9*, 317–323. [CrossRef]
27. Li, G.; Cui, S.S. Grain modeling and finite element simulation of damage evolution for AA5182-O aluminum alloy sheet. *J. Mater. Res. Technol.* **2020**, *9*, 10559–10575. [CrossRef]
28. Lei, Z.X.; Ma, J.; Sun, W.K.; Yin, B.B.; Liew, K.M. Low-velocity impact and compression-after-impact behaviors of twill woven carbon fiber/glass fiber hybrid composite laminates with flame retardant epoxy resin. *Compos. Struct.* **2023**, *321*, 117253. [CrossRef]
29. Zhang, J.Y.; Xie, J.; Zhao, X.Z.; Chen, J.L.; Li, Z.Y. Influence of void defects on impact properties of CFRP laminates based on multi-scale simulation method. *Int. J. Impact Eng.* **2023**, *180*, 104706. [CrossRef]
30. Hu, S.D.; Jiang, Y.N.; Zhou, C.; Li, L.X.; Wang, X.Y.; Wang, C. Prediction and prevention of cracks in free-cutting stainless steel bar forming. *Metall. Mater. Trans B.* **2020**, *51*, 1687–1696. [CrossRef]
31. Mendoza, I.; Lamberson, L. Damage tolerancing in carbon fiber-reinforced polymer (CFRP) laminates under combined impact fatigue and environmental conditioning. *Compos. Part A-Appl. S.* **2024**, *180*, 108062. [CrossRef]
32. Ren, Y.Y.; Zhang, L. Impact characterization of fiber reinforced epoxy resin composites. *Plast. Sci. Technol.* **2023**, *10*, 67–70.
33. Jia, Y.X.; Ao, Q.Y.; Zhang, W.Z.; Wang, T.; Liu, Y.Y.; Tian, Y. Analysis of low-velocity impact damage performance of carbon fiber composite laminate. *Arms Mater. Sci. Eng.* **2022**, *45*, 170–174.
34. Zhong, Y.; Xu, M.T.; Wang, P.; Li, Y.Y. Low-velocity impact properties and failure mechanism of carbon fiber-UHMWPE fiber hybrid reinforced epoxy resin composites. *Acta Mater. Compos. Sin.* **2022**, *39*, 3202–3211.
35. Yan, S.L.; Li, X.Y.; Liu, Y. Preparation of low-density polyethylene/carbon fiber composites and their properties. *Plast. Sci. Technol.* **2023**, *51*, 72–75.
36. Yu, M.M.; Zhu, X.L.; Liu, X.Q.; Fang, L.; Xie, W.; Ren, M.S.; Sun, J.L. Failure mechanism and assessment of residual strength of carbon fiber/epoxy resin matrix composite laminates under multiple impacts at low velocities. *Acta Mater. Compos. Sin.* **2023**, *40*, 5359–5370.
37. Sun, X.C.; Hallett, S.R. Failure mechanisms and damage evolution of laminated composites under compression after impact (CAI): Experimental and numerical study. *Compos. Part. A-Appl. S.* **2018**, *104*, 41–59. [CrossRef]
38. Seamone, A.; Davidson, P.; Waas, A.M.; Ranatunga, V. Low velocity impact and compressive response after impact of thin carbon fiber composite panels. *Int. J. Solids. Struct.* **2022**, *257*, 111604. [CrossRef]
39. Yang, G.D.; Zhang, J.L.; Xie, H.B.; Li, F.L.; Huang, Z.; Yuan, G.Q.; Zhang, J.Z.; Jia, Q.L.; Zhang, H.J.; Yeprem, H.A.; et al. Preparation of B_4Cp/Al composites via selective laser melting and their tribological properties. *Materials* **2022**, *15*, 8340. [CrossRef]
40. Pawlik, M.; Lu, Y.L. Effects of the graphene nanoplatelets reinforced interphase on mechanical properties of carbon fibre reinforced polymer–A multiscale modelling study. *Compos. Part B Eng.* **2019**, *177*, 107097. [CrossRef]

41. Shao, Y.Z.; Okubo, K.; Fujii, T.; Shibata, O.; Fujita, Y. Effect of matrix properties on the fatigue damage initiation and its growth in plain woven carbon fabric vinylester composites. *Compos. Sci. Technol.* **2014**, *104*, 125–135. [CrossRef]
42. *ASTM D7136/D7136M-20*; Standard Test Method for Measuring the Damage Resistance of a Fiber-Reinforced Polymer Matrix Composite to a Drop-Weight Impact Event. ASTM: West Conshohocken, PA, USA, 2020.
43. Wang, H.H.; Qin, Z.P.; Wan, X.L.; Wei, R.; Wu, K.M. Continuous cooling transformation behavior and impact toughness in heat-affected zone of Nb-containing fire-resistant steel. *Met. Mater. Int.* **2017**, *23*, 848–854. [CrossRef]
44. Siegfried, M.; Tola, C.; Claes, M.; Lomov, S.V.; Verpoest, I.; Gorbatikh, L. Impact and residual after impact properties of carbon fiber/epoxy composites modified with carbon nanotubes. *Compos. Struct.* **2014**, *111*, 488–496. [CrossRef]
45. Ismail, K.I.; Sultan, M.T.H.; Shah, A.U.M.; Jawaid, M.; Safri, S.N.A. Low velocity impact and compression after impact properties of hybrid bio-composites modified with multi- walled carbon nanotubes. *Compos. B Eng.* **2019**, *163*, 455–463. [CrossRef]
46. Tuo, H.L.; Lu, Z.X.; Ma, X.P.; Xing, J.; Zhang, C. Damage and failure mechanism of thin composite laminates under low-velocity impact and compression-after-impact loading conditions. *Compos. B Eng.* **2019**, *163*, 642–654. [CrossRef]
47. Liang, W.; Zhang, Y.F.; Zhang, L.F.; Wu, R. Precipitation behavior and strengthening mechanism of Ti micro-alloyed steel via CSP process. *Mater. Res. Express.* **2019**, *6*, 116533. [CrossRef]
48. Ge, X.X.; Zhang, P.; Zhao, F.; Liu, M.; Liu, J.; Cheng, Y.S. Experimental and numerical investigations on the dynamic response of woven carbon fiber reinforced thick composite laminates under low-velocity impact. *Compos. Struct.* **2022**, *279*, 114792. [CrossRef]
49. Wang, A.N.; Liu, X.G.; Yue, Q.R.; Xian, G.J. Effect of volume ratio and hybrid mode on low-velocity impact properties of unidirectional flax/carbon fiber hybrid reinforced polymer composites. *Thin-Walled Struct.* **2023**, *187*, 110764. [CrossRef]
50. Damghani, M.; Ersoy, N.; Piorkowski, M.; Murphy, A. Experimental evaluation of residual tensile strength of hybrid composite aerospace materials after low velocity impact. *Compos. Part B Eng.* **2019**, *179*, 107537. [CrossRef]
51. Gao, W.Q.; Zhang, C.L.; Yang, M.X.; Zhang, S.Q.; Juul Jensen, D.; Godfrey, A. Strain distribution and lattice rotations during in-situ tension of aluminum with a transmodal grain structure. *Mater. Sci. Eng. A.* **2021**, *828*, 142010. [CrossRef]
52. Zhou, S.B.; Hu, F.; Wang, K.; Hu, C.Y.; Zhou, W.; Yershov, S.; Wu, K.M.; Zhang, Z.C.; Pan, X.M. Nanomechanics of retained austenite in medium-carbon low-temperature bainitic steel: A critical analysis of a one-step versus a two-step treatment. *Materials* **2022**, *15*, 5996. [CrossRef] [PubMed]

materials

MDPI

Article

Energy Absorption and Failure Modes of Different Composite Open-Section Crush Elements under Axial Crushing Loading

Xulong Xi [1,2], Pu Xue [1], Xiaochuan Liu [2], Chunyu Bai [2], Xinyue Zhang [2], Xiaocheng Li [2], Chao Zhang [1] and Xianfeng Yang [3,*]

1 School of Aeronautics, Northwestern Polytechnical University, Xi'an 710072, China; 2020160673@mail.nwpu.edu.cn (X.X.); xuepu@nwpu.edu.cn (P.X.)
2 National Key Laboratory of Strength and Structural Integrity, Aircraft Strength Research Institute of China, Xi'an 710065, China
3 National Key Laboratory of Strength and Structural Integrity, School of Aeronautic Science and Engineering, Beihang University, Beijing 100191, China
* Correspondence: yangxf@buaa.edu.cn

Abstract: In order to study the energy absorption characteristics of the open-section thin-walled composite structures with different cross-sections, axial compression tests were carried out at loading speeds of 0.01 m/s, 0.1 m/s, and 1 m/s. Finite element models were built to predict the crushing response and energy absorption behaviors of these open-section structures. The effects of the cross-section's shape, cross-section aspect ratio, trigger mechanism, and loading speed on the energy absorption characteristics of the composite structures were analyzed. The results show that the average crushing loads of the hat-shaped and Ω-shaped open-section structures are 14.1% and 14.6% higher than those of C-shaped open-section structures, and the specific energy absorption (SEA) values are 14.3% and 14.8% higher than that of C-shaped open-section structures, respectively. For the C-shaped open-section structures, a 45° chamfer trigger is more effective in reducing the initial peak load, while a 15° steeple trigger is more appropriate for the hat-shaped open-section structures. The average crushing loads and SEA of C-shaped, hat-shaped, and Ω-shaped open-section structures are reduced when the loading speed is increased from 0.01 m/s to 1 m/s. The increase in loading speed leads to the splashing of debris and thus reduces the loading area and material utilization of open-section structures, leading to a decrease in energy absorption efficiency.

Keywords: composite; failure mode; energy absorption; numerical simulation

1. Introduction

Composite materials have become the overwhelming choice for the structural components of aircraft in recent years due to their excellent advantages in strength, weight reduction, and energy absorption behaviors [1–3]. For example, the number of composite materials used in Airbus A350 and Boeing 787 has exceeded 50% of their total structural weights. However, aircraft structures are inevitably subjected to impact loads in the event of a crash. In order to protect occupants from injury, the structural deformation and material failure of aircraft components may be involved to absorb the impact's kinetic energy [4,5]. It is a critical point that aerospace composite materials possess the same crashworthiness as metal materials during an emergency landing. Thus, the energy absorption design of aerospace composite structures has always remained a top priority to prevent catastrophic structural failure and significant casualties.

Composite tubes have been increasingly used in aerospace applications. Existing studies on crashworthiness have mainly focused on evaluating and improving the crushing behavior and energy absorption capacity of various composite thin-walled structures, such as circular carbon fiber-reinforced plastic (CFRP) tubes [6,7], squared CFRP tubes [8,9], and double-hat-shaped CFRP tubes [10,11]. It was found that circular composite tubes exhibit the best specific energy absorption characteristics among all the various composite tubes with different sectional profiles, but they have the same wall thickness [12]. A large number of approaches have been proposed to enhance crashworthiness behaviors, such as filling cellular materials in composite tubes. Lightweight foams are usually used as a core material of composite tubes to improve buckling strength, stiffness, and impact resistance while maintaining minimal mass [13,14]. However, the usage of cellular materials may decrease weight efficiency, leading to undesirable specific energy absorption compared with hollow composite tubes.

Composite materials also have disadvantages such as high cost and brittle fracture behaviors. The hybrid material design strategy provides a good solution to meet the requirements of lightweight property and energy absorption [15–17]. The SEA of the metal–composite hybrid structure can be improved by 37% compared to pure aluminum tubes. The external inversion mode of the metal–clad composite hybrid circular tube is the main mechanism for energy absorption [18]. In addition, it is able to avoid the excessive loads transferred to the entire structure and improve energy absorption behaviors by using the trigger mechanism [19]. Traditional composite tubes usually exhibit Euler buckling, shell buckling, and a brittle fracture mode. However, these composite tubes can collapse with a progressive crushing mode after introducing a chamfer trigger mechanism [20,21]. The crashworthiness investigation of the double- and triple-coupling triggered composite tubes reveals that the triple-coupled triggers improve the energy absorption capacity while double-coupled triggers can weaken the triggered zones to decrease the initial peak force [22]. In addition, multi-scale impact fracture mechanisms were investigated in hybrid materials [23].

Open-section energy-absorbing structures are extensively used as the subfloor structure of modern airplanes. Thus, a number of studies have been conducted to investigate the crushing behavior and energy absorption of open-section structures [24,25]. Bolukbasi and Laananen [26] investigated the energy-absorbing characteristics of the composite flat plates, angles, and channel sections under axial compression loading, and a semi-empirical analysis methodology was developed to predict the energy absorption capability. Riccio et al. [27] investigated the structural behavior of a channel section composite component subjected to quasi-static compression and dynamic loads. The interfaces between plies oriented according to the impact loading direction have been found more susceptible to inter-laminar damage. Jackson et al. [28] studied the crushing response of the carbon fiber/epoxy crush elements. The impact testing results indicate a 6–15% reduction in SEA compared to the quasi-static crushing case. Waimer et al. [29] investigated the dynamic failure behavior of the CFRP components assembled using a CFRP half-tube and an L-shaped strut. The assembly of the half-tube and L-strut provides an Improved section modulus, which leads to a progressive crushing mode.

The existing studies on open-section energy-absorbing structures have focused on the crushing response with the impact velocity arranged from 1.8 m/s to 10 m/s, while it still lacks the crushing data of the composite channel section structures with different cross-sections when the impact velocity is arranged from 0.01 m/s to 1 m/s. In addition, the crushing mechanism and energy absorption performance of different composite thin-walled open-section structures have not been revealed. Accordingly, this paper aims to study the energy absorption characteristics of different open-section thin-walled composite structures using a high-speed hydraulic servo testing system. Then, a finite element model was built to further study the crushing response and energy absorption performance of open-section thin-walled composite structures. Finally, the influences of the cross-section's shape, cross-section aspect ratio, trigger mechanism, and impact velocity on the failure modes and energy absorption characteristics of thin-walled composite structures were analyzed.

2. Experimental Testing

2.1. Specimen Description

The testing specimens are fabricated via the hot-press molding process based on CCF300/8552A high-temperature cured epoxy/carbon fiber prepregs (AVIC Composite Co., Ltd., Beijing, China). The volume fraction of the carbon fibers is approximately 67% ± 2%, and the 0° tensile strength of a single-layer prepreg is 1500 Mpa. The material density of the testing specimens is 1.6 g/cm^3. The testing specimens are laid at angles of $[45°/0°/90°/-45°/45°/0°/90°/-45°]_s$, and the subscript "s" means the symmetric lay-up method. The axial compression direction aligns with the 0° direction of the carbon fibers, and the specimen's thickness is 2 mm. To investigate the effect of the cross-sectional configuration, cross-section aspect ratio, and triggering method on the energy absorption characteristics of composite thin-walled structures, three cross-sectional configurations are considered, including the C-type, hat-type, and Ω-type configurations, as shown in Figure 1. These geometric configurations have three different section aspect ratios, namely 1.65, 1.06, and 2.31, and the three types of cross-section aspect ratios are denoted as C1, C2, and C3, respectively. In addition, two different triggering methods, including the 45° chamfer trigger mechanisms and the 15° steeple trigger mechanisms, have been adopted to improve crushing responses. Figure 2 exhibits the testing specimens of the open-section crushing elements. All specimens have a height of 100 mm and the same cross-sectional area. In order to guarantee that these specimens are fixed in the material's testing machine stably, a composite base along the horizontal plane is added at the bottom of the testing specimens. The composite base consists of a metal shell and a resin material with dimensions of 80 mm × 60 mm × 20 mm. The testing specimens are partially inserted into the composite base during crushing experiments.

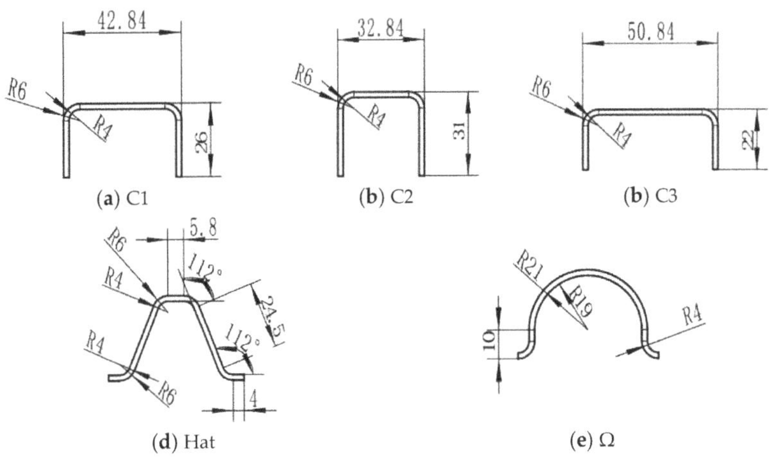

Figure 1. The cross-section dimensions of the testing specimens.

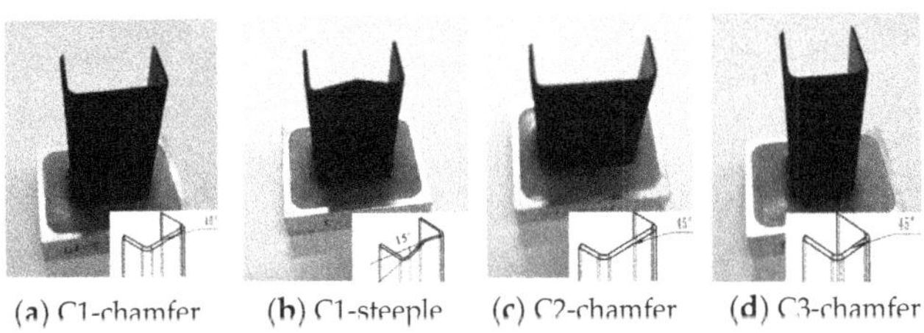

Figure 2. *Cont.*

(**e**) Hat-chamfer (**f**) Hat-steeple (**g**) Ω-chamfer

Figure 2. The testing specimens of the open-section crushing elements.

2.2. *Testing Setup and Procedure*

The axial compression tests of composite open-section thin-walled structures were conducted using an Instron VHS 160/100-20 high-speed hydraulic servo testing machine (Instron Inc., Norwood, MA, USA), as shown in Figure 3a. This testing machine consists of a hydraulic system, water cooling system, frame, and control system. The constant dynamic loading process is realized by the loading energy provided by the combination of a hydraulic actuator and a gas accumulator. The maximum loading speed of the testing machine is 20 m/s, and the maximum impact dynamic load is 100 kN. The testing load and displacement are, respectively, measured by the force sensors and displacement sensors installed in the testing machine. The impact velocity was set at 0.01 m/s, 0.1 m/s, and 1 m/s, with a maximum compression stroke of 54 mm. The testing equipment and the specimen clamping method are illustrated in Figure 3b. The testing specimen is fixed to the base of the testing machine using two fixed blocks and bolted joints. The support can guarantee the stability of the loading device and prevent the head of the testing machine from becoming unstable during the loading process. Table 1 presents all experimental conditions of the composite open-section thin-walled structures. During the loading process, a Photron SA-X high-speed camera (Photron Inc., Tokyo, Japan), as shown in Figure 3c, was used to record the deformation and failure behaviors of the composite open-section thin-walled structures.

Figure 3. Experimental setup for the dynamic crushing tests: (**a**) INSTRON VHS 160 high-speed servo-hydraulic testing machine; (**b**) installation and schematic of the specimens; (**c**) Photron SA-X high-speed camera.

2.3. *Crashworthiness Evaluation Criterion*

Several crashworthiness evaluation criteria are selected to evaluate and compare the energy absorption performance of various structures [30]. Figure 4 shows a typical force-displacement curve of energy-absorbing materials during a progressive crushing process. The force–displacement curve can be classified into three typical stages: the initial crushing stage (Stage I), the stable crushing stage (Stage II), and the densification stage (Stage III). During Stage I, the crushing force shows an approximate linearity increase first and then a certain degree of decrease after reaching the peak force. During Stage II, the crushing force maintains a relatively stable interval and fluctuates around the mean crushing load. The following evaluation criteria are adopted to evaluate the energy absorption capacity of open-section crushing elements.

Table 1. Experimental conditions of composite open-section thin-walled structures.

Configuration	Triggering Mechanism	Impact Velocity (m/s)	Mass (g)
C1	45° Chamfer trigger	1.00	14.680
Hat	45° Chamfer trigger	1.00	14.678
Ω	45° Chamfer trigger	1.00	14.677
C2	45° Chamfer trigger	1.00	14.680
C3	45° Chamfer trigger	1.00	14.680
C1	15° steeple trigger	1.00	13.507
Hat	15° steeple trigger	1.00	14.029
C1	45° Chamfer trigger	0.01	14.680
C1	45° Chamfer trigger	0.10	14.680
Hat	45° Chamfer trigger	0.01	14.678
Hat	45° Chamfer trigger	0.10	14.678
Ω	45° Chamfer trigger	0.01	14.677
Ω	45° Chamfer trigger	0.10	14.677

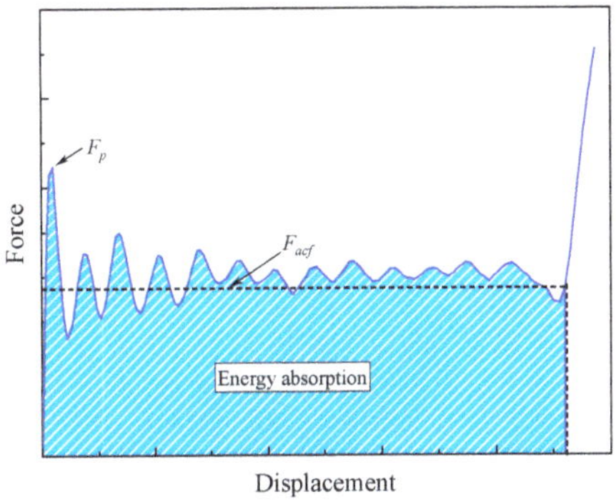

Figure 4. A typical load–displacement curve of an energy-absorbing material.

(1) Initial peak force F_p: The maximum force during the crushing process, and F_p should be controlled within an allowable range to avoid transmitting excessive loads to passengers.
(2) Average crushing force F_{acf}: The average crushing force during Stage II can be expressed as

$$F_{mcf} = \frac{\int F dl}{l} \tag{1}$$

where F is the instantaneous crushing force during the crushing process, and l is the effective compression stroke when an energy-absorbing material is fully compacted.

(3) Crushing force efficiency CFE: The crushing force efficiency is defined as the ratio of the averaged load during the plateau stage to the initial peak force, which represents the load uniformity of an energy-absorbing material. It can be given by

$$CLE = \frac{F_{mcf}}{F_p} \tag{2}$$

(4) Specific energy absorption SEA: It is the most significant evaluation criterion to compare the energy absorption capacity of different energy-absorbing materials,

which is defined as the energy absorbed per unit mass of a structure, and it can be given by

$$SEA = \frac{\int F dl}{m} \quad (3)$$

where m is the mass of the energy-absorbing material. In this work, the displacement when the head of the material testing machine contacts the open-section crushing elements is taken as the starting point for energy absorption.

3. Numerical Modeling

3.1. Finite Element Modeling

Numerical simulations can provide more detailed insights into failure mechanisms and deformation evolution, which are not accessible via physical testing methods. This section focuses on developing a finite element model to simulate the axial crushing response of composite open-section thin-walled structures. Figure 5 shows the finite element model of composite open-section thin-walled structures under axial crushing loads. The composite laminated material is simulated using continuous shell elements (SC8R) with a mesh size of 1.5 mm. These composite specimens are composed of sixteen layers, and a two-layer unidirectional strip is set for each layer. In addition, cohesive elements are established to simulate delamination failure between the adjacent layers. A schematic diagram of the lay-up configuration and the cohesive elements is shown in Figure 6. The interlayer adhesive layer is modeled using cohesive elements (C3H8) with a thickness of 0.01 mm. The indenter is regarded as the rigid body in the finite element model. The clamping portion at the bottom of the test specimens is neglected in the numerical model. The nodes located at the bottom of the test specimens remain fixed by restricting all degrees of freedom. In order to facilitate loading and obtain the reaction forces, a reference point RP-1 is established and coupled with the indenter. The indenter moves downwards at a prescribed velocity. In addition, a general automatic contact algorithm is adopted to simulate the contact interface, the possible friction interactions between the different parts and the self-contact of the specimen are defined in the tangential contact behavior. The modeling of friction contact is based on the general contact algorithm, and its friction coefficient is set to 0.5. The hard contact relationship is adopted to define the normal contact behaviors of the model. The traction–separation model is adopted to define the cohesive contact behavior between the adjacent layers.

(**a**) The C-shaped open-section element.

Figure 5. *Cont.*

(**b**) The hat-shaped open-section element.

(**c**) The Ω-shaped open-section element.

Figure 5. Finite element models of the open-section crushing elements.

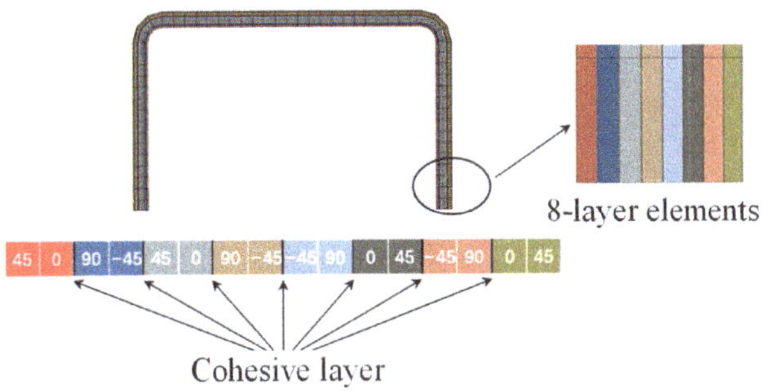

(**a**) Lay-up configuration

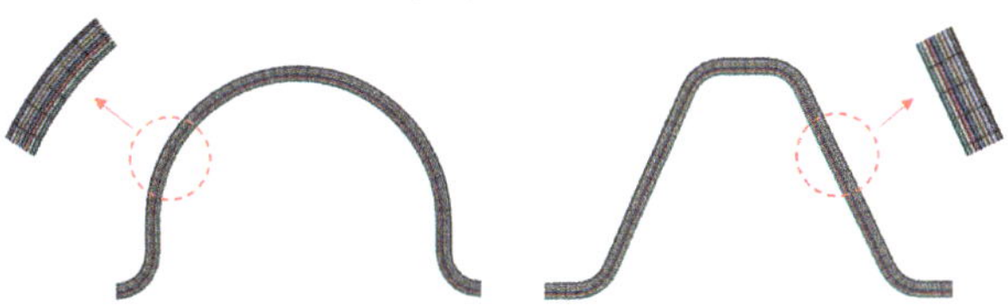

(**b**) Cohesive elements

Figure 6. Detailed modeling method for the open-section crushing elements.

3.2. Material Damage Modeling

(1) Intralaminar properties

In this study, the 2D Hashin failure criterion is employed to determine the failure initiation for open-section crushing elements. The Hashin criterion defines four types of failure modes, including fiber tension, fiber compression, matrix tension, and matrix compression. The expressions for these failure modes are as follows:

Fiber tension failure ($\sigma_{11} \geq 0$):

$$F_{ft} = \left(\frac{\sigma_{11}}{X_T}\right)^2 + \alpha\left(\frac{\sigma_{12}}{S_{12}}\right)^2 = 1 \quad (4)$$

Fiber compression failure ($\sigma_{11} < 0$):

$$F_{fC} = \left(\frac{\sigma_{11}}{X_C}\right)^2 = 1 \quad (5)$$

Matrix tension failure ($\sigma_{22} \geq 0$):

$$F_{mt} = \left(\frac{\sigma_{22}}{Y_T}\right)^2 + \beta\left(\frac{\sigma_{12}}{S_{12}}\right)^2 = 1 \quad (6)$$

Matrix compression failure ($\sigma_{22} < 0$):

$$F_{mc} = \left(\frac{\sigma_{22}}{Y_C}\right)^2 + \gamma\left(\frac{\sigma_{12}}{S_{12}}\right)^2 = 1 \quad (7)$$

Here, σ_{ij} ($i, j = 1, 2$) denotes the stress components. X_T is the axial tension strength, X_C is the axial compression strength, Y_T is the transverse tension strength, Y_C is the transverse compression strength, and S_{12} is the shear strength. α, β, and γ are scale coefficients. When F_g ($g = f, m, t, c$) reaches 1, this indicates the onset of material damage in the open-section crushing elements. Here, after the initiation of failure, the material stiffness gradually degrades, and it enters the damage evolution stage, as shown in Figure 7. Point A represents the failure initiation point that satisfies the Hashin criterion, followed by linear stiffness degradation until the material failure occurs.

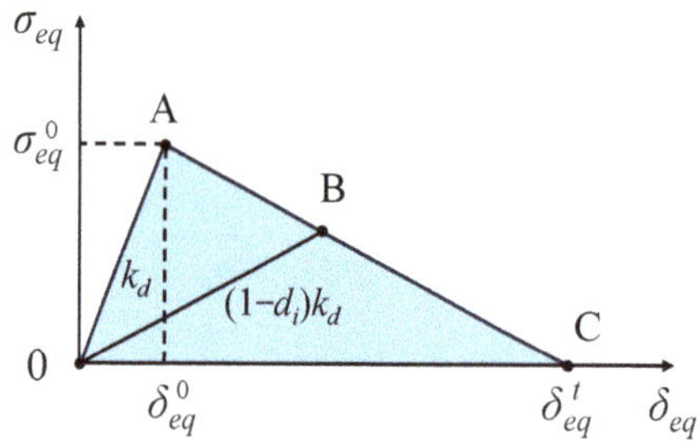

Figure 7. Constitutive model and damage evolution model.

Once the damage initiation criterion is satisfied, a damage evolution method is required to describe its subsequent development. A linear progressive damage evolution is selected to represent the evolution process of intralaminar damage, and the damage variable d_i can be given by

$$d_i = \frac{\delta_{eq}^t(\delta_{eq} - \delta_{eq}^0)}{\delta_{eq}(\delta_{eq}^t - \delta_{eq}^0)} \quad (8)$$

where δ_{eq} represents the equivalent displacement corresponding to the failure mode, δ_{eq}^0 denotes the equivalent displacement at the initiation of the failure, while δ_{eq}^t represents the equivalent displacement when the material is fully damaged. δ_{eq}^0 and δ_{eq}^t can be expressed as

$$\delta_{eq}^0 = \frac{\delta_{eq}}{\sqrt{\varphi}} \tag{9}$$

$$\delta_{eq}^t = \frac{2G^C}{\sigma_{eq}/\sqrt{\varphi}} \tag{10}$$

where φ is the variable related to the failure mode, G^C is the fracture energy density, and σ_{eq} is the equivalent stress corresponding to the failure mode. In addition, the calculation methods of the equivalent displacement and stress for every failure mode are shown in Table 2.

Table 2. Equivalent displacement and stress for every failure mode.

Failure Mode	Equivalent Stress	Equivalent Displacement
Fiber tension	$\sigma_{ft,eq} = \frac{L_c(\langle\sigma_{11}\rangle\langle\varepsilon_{11}\rangle+\sigma_{12}\varepsilon_{12})}{\delta_{ft,eq}}$	$\delta_{ft,eq} = L_c\sqrt{\langle\varepsilon_{11}\rangle^2+(\varepsilon_{12})^2}$
Fiber compression	$\sigma_{fc,eq} = \frac{L_c\langle-\sigma_{11}\rangle\langle-\varepsilon_{11}\rangle}{\delta_{fc,eq}}$	$\delta_{fc,eq} = L_c\langle-\varepsilon_{11}\rangle$
Matrix tension	$\sigma_{mt,eq} = \frac{L_c(\langle\sigma_{22}\rangle\langle\varepsilon_{22}\rangle+\sigma_{12}\varepsilon_{12})}{\delta_{mt,eq}}$	$\delta_{mt,eq} = L_c\sqrt{\langle\varepsilon_{22}\rangle^2+(\varepsilon_{12})^2}$
Matrix compression	$\sigma_{mc,eq} = \frac{L_c(\langle-\sigma_{22}\rangle\langle-\varepsilon_{22}\rangle+\sigma_{12}\varepsilon_{12})}{\delta_{mc,eq}}$	$\delta_{mc,eq} = L_c\sqrt{\langle-\varepsilon_{22}\rangle^2+(\varepsilon_{12})^2}$

In Table 2, L_c represents the element's characteristic length; the symbol "< >" is the Macaulay operator, and it can be expressed as

$$< x >= \begin{cases} x, \ x > 0 \\ 0, \ x \leq 0 \end{cases} \tag{11}$$

The elasticity constants and damage initiation coefficients of the composite materials are exhibited in Table 3.

Table 3. Elasticity constants and damage initiation coefficients for the intra-laminar model [31].

Description	Variable	Value
Longitudinal Young modulus	E_1	171,420 MPa
Transversal Young modulus	E_2	9080 MPa
Principal Poisson's ratio	v_{12}	0.32
Shear modulus	G_{12}	5290 MPa
Longitudinal tensile strength	X_T	1773 MPa
Longitudinal compressive strength	X_C	1264 MPa
Transversal tensile strength	Y_T	62.3 MPa
Transversal compressive strength	Y_C	199.8 MPa
In-plane shear strength	S_{12}	92.3 MPa
Longitudinal tensile fracture energy	G_{ft}^c	120 N/mm
Longitudinal compressive fracture energy	G_{fc}^c	100 N/mm
Transverse tension fracture energy	G_{mt}^c	2 N/mm
Transverse compression fracture energy	G_{mc}^c	5 N/mm

(2) Interlaminar properties

To simulate the delamination failure of the composite open-section thin-walled structures during the crushing process, the cohesive elements based on the traction–separation law are used between the layers that are prone to failure. The elastic constitutive relationships are described as follows:

$$\begin{bmatrix} \sigma_n \\ \sigma_s \\ \sigma_t \end{bmatrix} = \begin{bmatrix} K_n & 0 & 0 \\ 0 & K_s & 0 \\ 0 & 0 & K_t \end{bmatrix} \begin{bmatrix} \delta_n \\ \delta_s \\ \delta_t \end{bmatrix} \quad (12)$$

where σ_k ($k = n, s, t$) and K_k ($k = n, s, t$) denote the traction stress and stiffness in the normal direction and two shear directions, and δ_k is the separation displacement. The quadratic nominal stress criterion is adopted to determine the initiation of the delamination failure, and it can be given by

$$\left(\frac{\langle \sigma_n \rangle}{\sigma_n^c}\right)^2 + \left(\frac{\sigma_s}{\sigma_s^c}\right)^2 + \left(\frac{\sigma_t}{\sigma_t^c}\right)^2 = 1 \quad (13)$$

where σ_n^c, σ_s^c, and σ_t^c are the corresponding interface strengths of the interlaminar layer.

A power law fracture criterion is employed to predict the damage evolution of the interlaminar properties. It indicates that the delamination failure under mixed-mode conditions is governed by the power law interaction of the energies required to cause failure in the individual modes:

$$\left(\frac{G_I}{G_{IC}}\right)^2 + \left(\frac{G_{II}}{G_{IIC}}\right)^2 + \left(\frac{G_{III}}{G_{IIIC}}\right)^2 = 1 \quad (14)$$

where G_I, G_{II}, and G_{III} are the interface energies; and G_{IC}, G_{IIC}, and G_{IIIC} are the critical fracture energy values required to cause failure in the different modes. In addition, the material parameters of the cohesive element are given in Table 4.

Table 4. Material parameters of the cohesive elements [32].

Description	Variable	Value
The stiffness in the normal direction	K_n	1×10^6 N/mm^3
The stiffness in the first shear direction	K_s	5×10^5 N/mm^3
The stiffness in the second shear direction	K_t	5×10^5 N/mm^3
The interface strength in the normal direction	σ_n^c	60 MPa
The interface strength in the first shear direction	σ_s^c	110 MPa
The interface strength in the second shear direction	σ_t^c	110 MPa
Mode-I fracture toughness	G_{IC}	0.2 N/mm
Mode-II fracture toughness	G_{IIC}, G_{IIIC}	5 N/mm

4. Results

To validate the crushing response of the testing specimens, the C1-shaped and hat-shaped specimens were selected to establish the finite element models at a loading velocity of 1 m/s. The load–displacement curves obtained from the experimental and simulation results are shown in Figure 8. The comparisons between the peak load and the average crushing load of the testing specimens are presented in Table 5. It can be concluded that the finite element models can effectively capture the crushing response of the open-section crushing elements during the compression process. Furthermore, convergence analysis is conducted with mesh sizes of 1 mm, 1.5 mm, 2mm, and 3 mm. The simulation results shown in Figure 8c prove that 1.5 mm is an appropriate mesh size with acceptable accuracy and computational costs. In addition, the oscillatory processes are noticeable for the force–displacement curves of the specimens during the crushing process. The crushing force decreases when composite thin-walled open-section structure fractures appear. Different failure mechanisms, such as matrix cracking, fiber failure, and delamination, may be involved during the impact process. In addition, local buckling may occur, which leads to a folding sequence for composite thin-walled open-section structures; this folding phenomenon can also result in the oscillatory process.

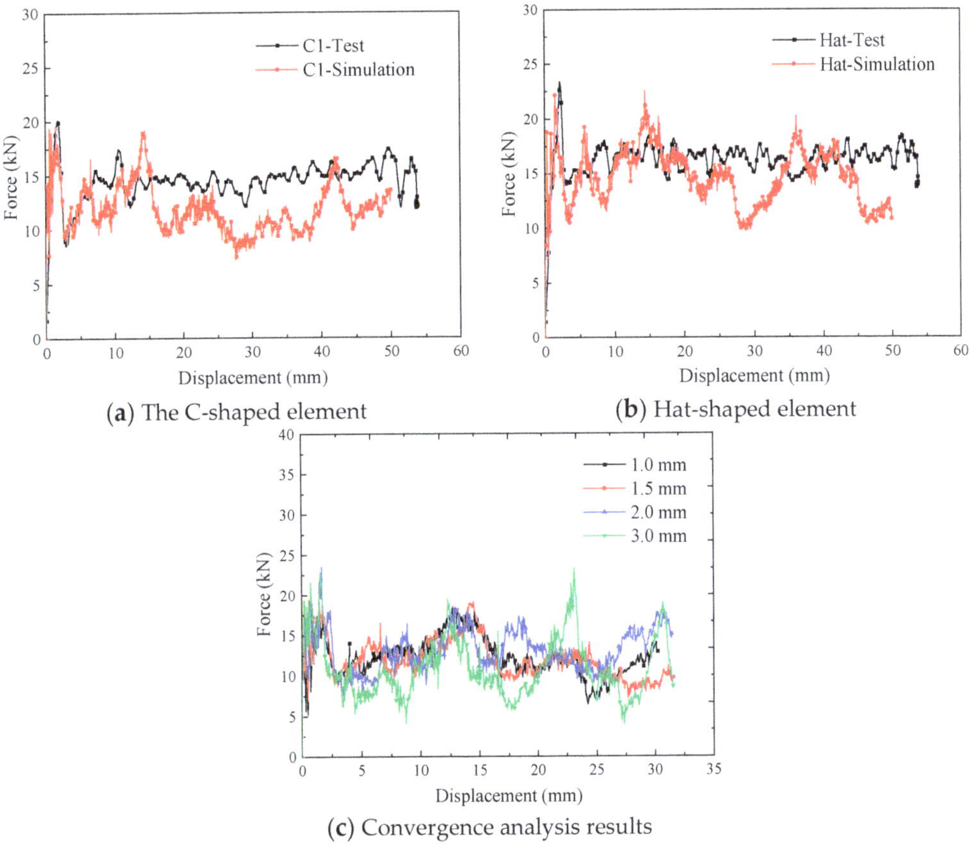

Figure 8. Comparison of simulation and experimental results for the crushing process.

Table 5. Initial peak forces and mean crushing forces of different testing specimens.

Description	F_P (Kn)	F_{MCF} (Kn)
C1-Test	20.14	14.36
C1-Simulation	19.53	11.90
Error	3.01%	17.10%
Hat-Test	23.38	16.13
Hat-Simulation	22.57	14.65
Error	3.46%	9.18%

Figure 9 exhibits the failure modes of C1-shaped and hat-shaped testing specimens with a crushing displacement of 10 mm. It is evident that both C1-shaped and hat-shaped specimens primarily experience in-plane shear deformation and interlaminar delamination during the compression failure process. For C1-shaped open-section specimens, the stress concentration at the corners leads to the formation of a relatively long axial matrix tensile failure zone. However, no significant axial matrix tensile failure zone is observed at the corners of the hat-shaped open-section specimens.

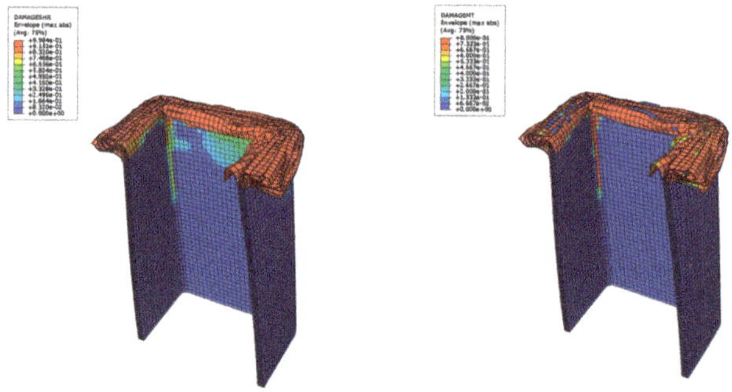

(**a**) In-plane shear failure and matrix compression failure of the C-shaped element

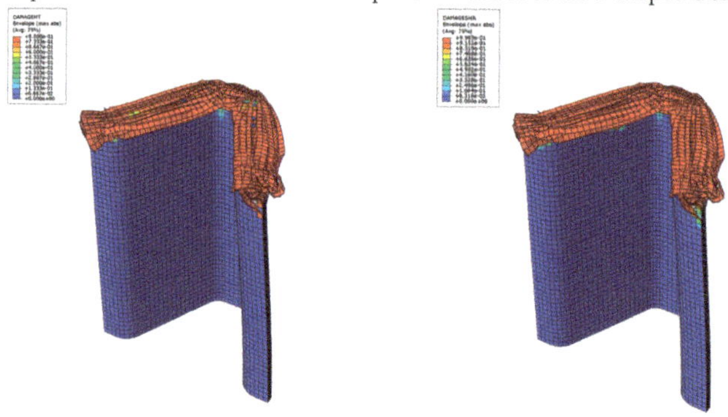

(**b**) In-plane shear failure and matrix compression failure of the hat-shaped element

Figure 9. Simulation results of the failure modes of different open-section crushing elements.

5. Discussions

5.1. The Effect of the Cross-Section Configuration

To investigate the effect of the cross-section configuration on the energy absorption characteristics of composite thin-walled structures, three geometric configurations, including the C1-shape, hat-shape, and Ω-shape, were selected. The triggering mode for all the configurations was a 45° chamfer triggering mechanism, and the loading speed was 1 m/s. Figure 10 shows the typical load–displacement curves of the open-section elements with three different cross-section configurations during the crushing process. It can be found that all three specimens exhibit a progressive crushing failure mode. The C1-shaped specimen presents a more pronounced load decrease after reaching the peak force compared with the hat-shaped and Ω-shaped specimens. In addition, the crushing force of the C1-shaped specimen at the stable crushing stage is lower than that of the other two configurations. The initial peak force, the average crushing force, and specific energy absorption of the hat-shaped and Ω-shaped specimens are almost the same with respect to the triggering mode and loading speed. The average crushing forces of the hat-shaped and Ω-shaped specimens are 14.1% and 14.6% higher than that of the C1-shaped specimen. The specific energy absorption increases by 14.3% and 14.8% for the hat-shaped and Ω-shaped specimens compared with the C1-shaped specimen, as shown in Figure 11. The relatively higher initial peak load of the hat-shaped and Ω-shaped open-section specimens can be attributed to the combined influence of the cross-section configuration and triggering mode.

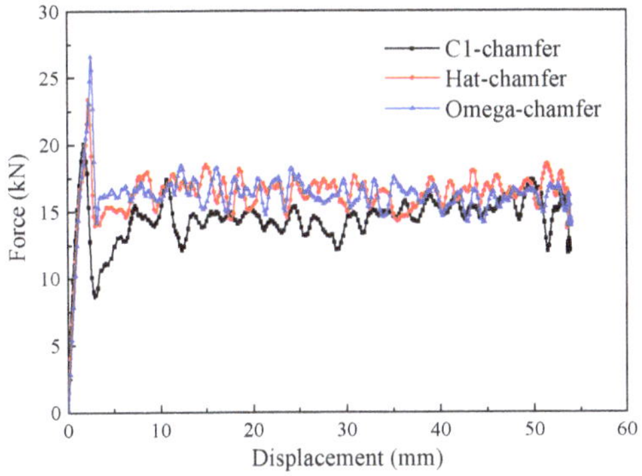

Figure 10. Typical force–displacement curves of the testing specimens with different cross-section shapes.

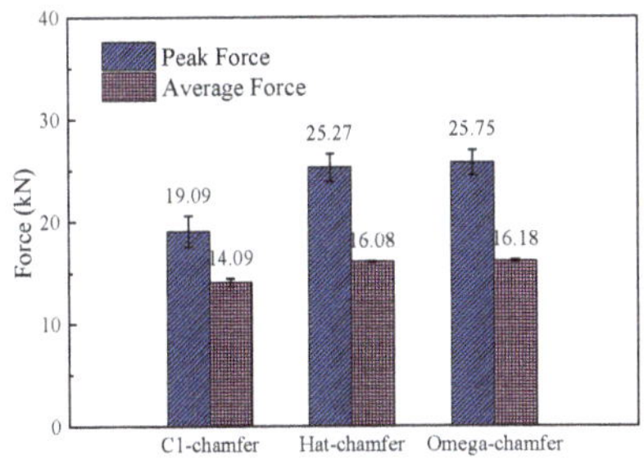

(**a**) Peak force and average force

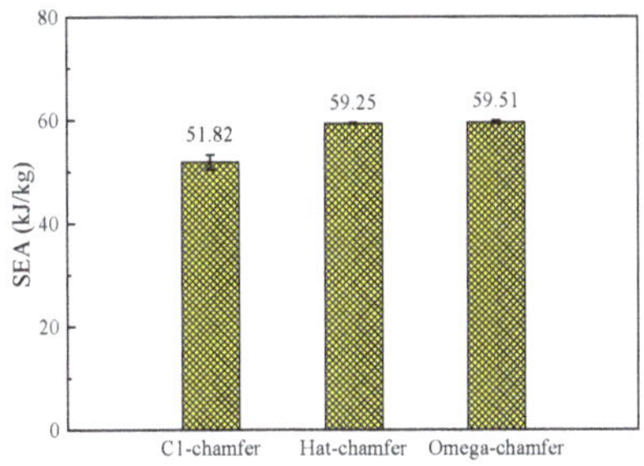

(**b**) Specific energy absorption

Figure 11. Comparison of the energy absorption characteristics of the testing specimens with different cross-section shapes.

During the crushing process of the testing specimens, material deformation primarily occurs through bending, delamination failure, and shear failure. The top of the testing specimen experiences bending due to interlaminar cracking, causing the inner carbon fiber fabric to bend inward and the outer carbon fiber fabric to bend outward. Simultaneously, numerous short intralaminar cracks form during the crushing process, and shear failure occurs at the root of these short intralaminar cracks, leading to the formation of numerous fragments, as shown in Figure 12.

Figure 13 exhibits the failure modes of the testing specimens with different cross-section configurations. The C1-shaped testing specimen has the largest fragment size, while the Ω-shaped testing specimen has the smallest fragment size. For the C1-shaped and hat-shaped specimens, a portion of the outer carbon fiber fabric forms a longer axial tearing region due to stress concentration at the corners. Furthermore, a portion of the inner carbon fiber fabric bends inward, and some relatively intact carbon fiber fabric remains after these specimens are fully crushed. The larger size of the remaining fragments and the presence of large intact carbon fiber fabrics in the C1-shaped specimen indicate insufficient damage during the crushing process, which is averse to energy absorption. Consequently, the C1-shaped specimen exhibits a lower average crushing force and specific energy absorption. However, no evident stress concentration zone occurs for the Ω-shaped open-section specimen during the crushing process, resulting in more complete material failure. Accordingly, the Ω-shaped specimen shows a higher average crushing force and better specific energy absorption. In addition, the stress concentration is relieved in the hat-shaped specimen due to the smoother corners; thus, these materials are fully damaged, leading to a considerable average crushing load and specific energy absorption compared to the Ω-shaped specimen.

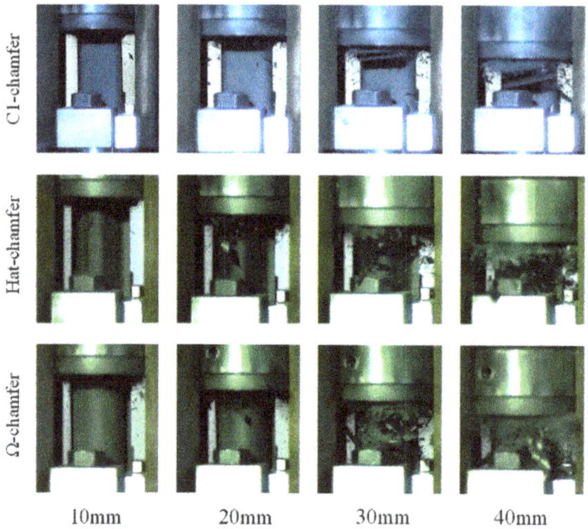

Figure 12. Crushing process of the testing specimens with different cross-section configurations.

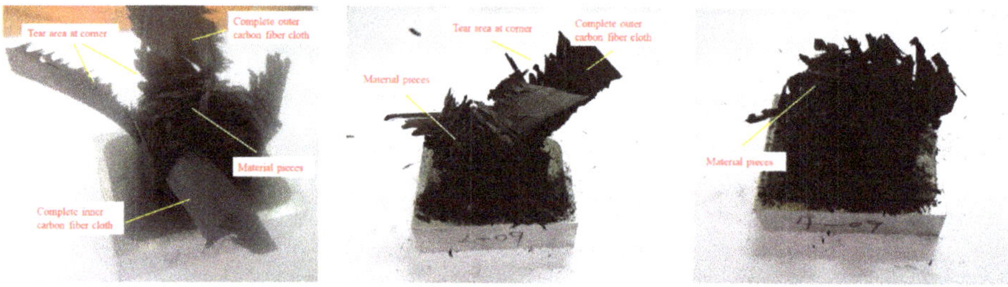

(**a**) C1-chamfer (**b**) Hat-chamfer (**c**) Ω-chamfer

Figure 13. Failure modes of the testing specimens with different cross-section configurations.

5.2. The Effect of the Cross-Section Aspect Ratio

To investigate the effect of the cross-section aspect ratio on the energy absorption characteristics of thin-walled composite material structures, the crushing responses of the C1, C2, and C3 specimens were analyzed. The triggering mode of all specimens was a 45° chamfer trigger. Figure 14 shows the typical load–displacement curves for the C-shaped open-section elements with three different cross-section aspect ratios. It can be observed that the load–displacement curves for the three C-shaped open-section elements are quite similar. Figure 15 shows a comparison of the energy absorption characteristics of the testing specimens with different cross-section aspect ratios. The average crushing load of the C1 specimen is simply 6.4% higher than that of the C2 specimen and 5.1% higher than that of the C3 specimen. In addition, the SEA of the C1 specimen is 6.4% and 5.0% higher than that of the C2 and C3 specimens. The C1, C2, and C3 specimens almost have the same failure mode in that axial tearing occurs at the corners of the carbon fiber fabric. The width of the residual fragments is comparable to the specimen's thickness, as shown in Figure 16. However, the length of the residual fragments during the crushing process is related to the specimen's dimensions, resulting in slight differences in energy absorption characteristics for the specimens with three aspect ratios.

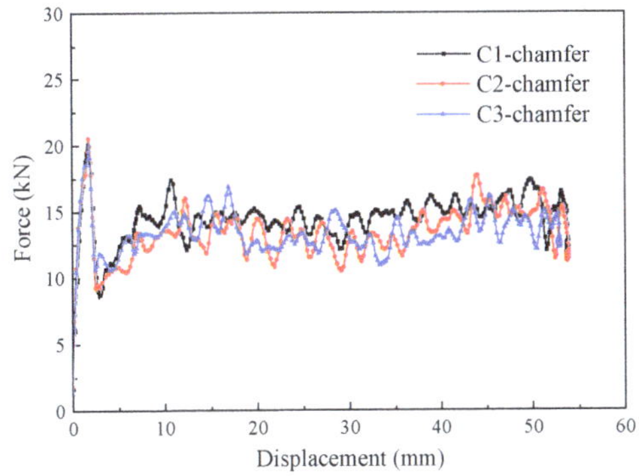

Figure 14. Typical force–displacement curves of the testing specimens with different cross-section aspect ratios.

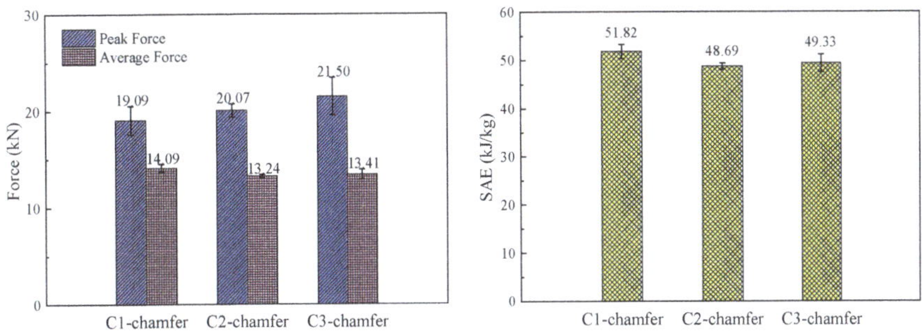

(**a**) Peak force and average force (**b**) Specific energy absorption

Figure 15. Comparison of the energy absorption characteristics of the testing specimens with different cross-section aspect ratios.

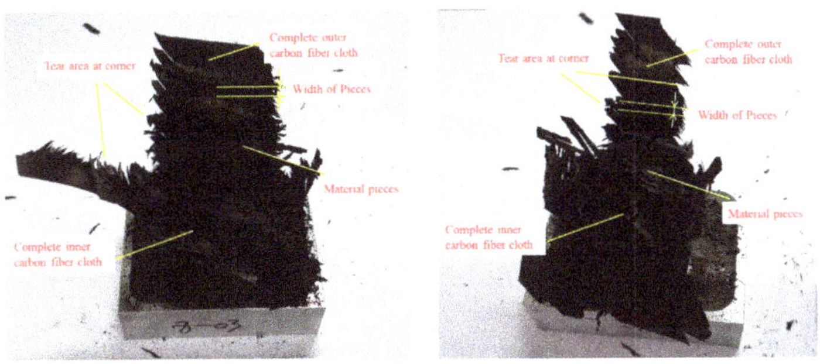

(**a**) C2 (**b**) C3

Figure 16. Failure modes of the testing specimens with different cross-section aspect ratios.

5.3. Trigger Mechanism

To investigate the influence of trigger mechanisms on the energy absorption characteristics of composite thin-walled open-section structures, the crushing responses of the C1-shaped and hat-shaped open-section specimens with the 45° chamfer trigger and the 15° steeple trigger are analyzed. Figure 16 shows the typical load–displacement curves of the C1-shaped and hat-shaped specimens with different triggering modes under axial crushing loading. Figure 17 presents a comparison of the energy absorption characteristics of the testing specimens with different trigger methods. It can be observed that the time needed to reach the peak force significantly increases for testing specimens with the 15° steeple trigger at the initial stage of the crushing process. However, no evident differences can be found for the load–displacement curves of the testing specimens with the two triggering mechanisms during the stable crushing stage. For the C1-shaped open-section specimens, the 45° chamfer trigger is more effective in reducing the peak force during the crushing process. For the testing specimens with the 15° steeple trigger, the load–displacement curve exhibits a small plateau when crushing displacement reaches approximately 4 mm. Subsequently, the crushing force continues to rise, and the slope of the load–displacement curve also increases. This behavior is attributed to the excessive weakening of the testing specimen above the corner by the 15° steeple trigger, while the weakening below the corner is insufficient, resulting in the inadequate induction of the progressive failure.

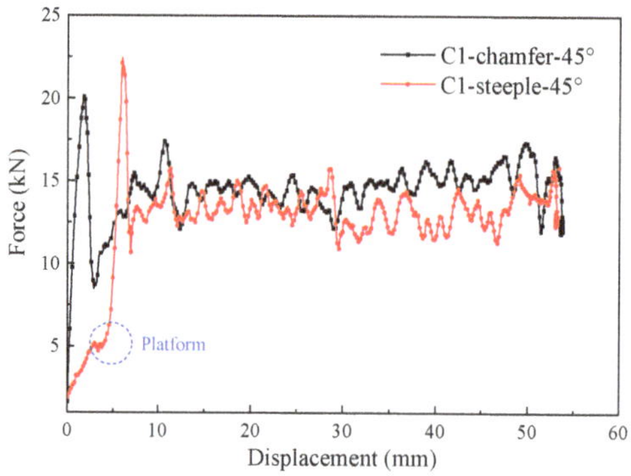

(**a**) C1

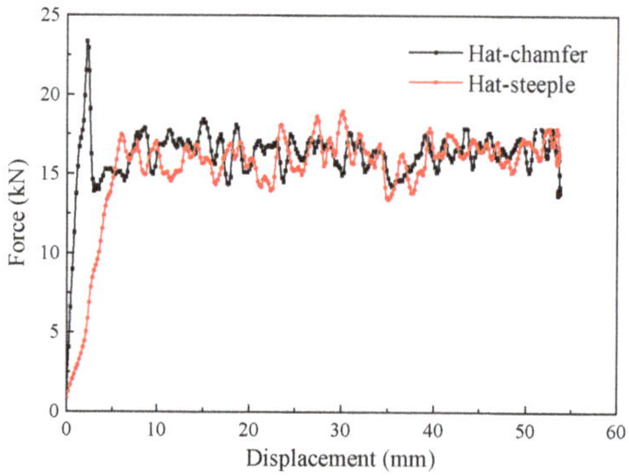

(**b**) Hat

Figure 16. Typical force–displacement curves of the testing specimens with different trigger mechanisms.

For the hat-shaped open-section specimens, the initial peak force can be eliminated by using the 15° steeple trigger, and the load efficiency can reach 89.50%, indicating that the hat-shaped open-section specimen is well matched with the corresponding steeple trigger. Figure 18 exhibits the failure modes of the testing specimens with the steeple trigger method. It can be found that numerous material fragments occur during the crushing process of the hat-shaped specimen with the 15° steeple trigger, which indicates that more thorough material damage appears at the top of the testing specimen. In Section 5.1, the initial peak forces of the hat-shaped and Ω-shaped open-section specimens are relatively high. The main reason is that the optimal triggering mechanism may be different compared to composite thin-walled open-section structures with different cross-section configurations. The 45° chamfer trigger does not sufficiently weaken the hat-shaped and Ω-shaped specimens, which is not the optimal triggering mode for this loading condition.

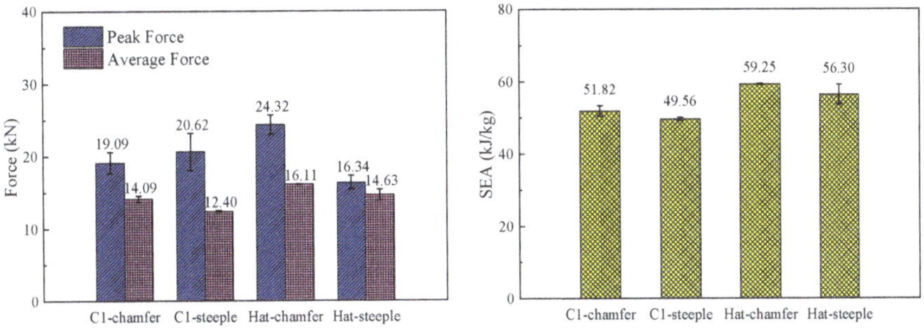

(**a**) Peak force and average force (**b**) Specific energy absorption

Figure 17. Comparison of the energy absorption characteristics of the testing specimens with different trigger mechanisms.

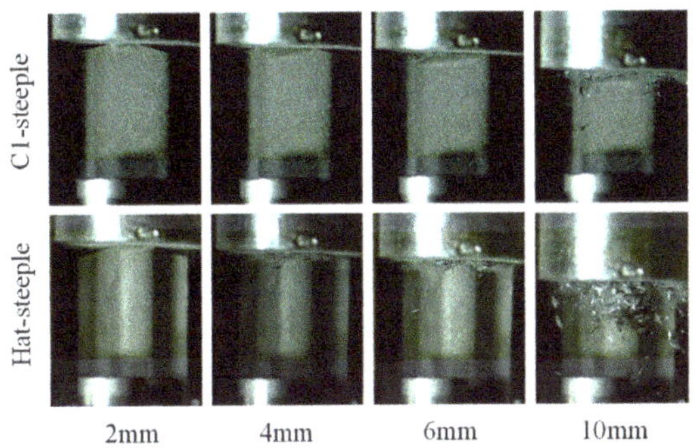

Figure 18. Failure modes of the testing specimens with steeple triggers.

5.4. Impact Velocity

To study the effect of the loading rate on the energy absorption performance, the crushing responses of C1-shaped, hat-shaped, and Ω-shaped open-section specimens at the impact velocities of 0.01 m/s, 0.1 m/s, and 1 m/s are investigated in this section. Figure 19 shows the typical load–displacement curves for the three types of testing specimens with different impact velocities. Figure 20 presents a comparison of the energy absorption characteristics of testing specimens with different impact velocities. The initial peak force, average crushing force, and specific energy absorption of three types of the testing specimen decrease as the impact velocity increases. When the loading speed is increased from 0.01 m/s to 1 m/s, the average crushing force of C1-shaped, hat-shaped, and Ω-shaped open-section specimens decreases by 6.1%, 10.9%, and 6.1%, respectively. However, the specific energy absorption of these testing specimens decreases by 6.2%, 11.0%, and 6.2%, respectively. Figure 21 shows the variation in the specific energy absorption of these testing specimens with different impact velocities. It can be concluded that the hat-shaped specimen experiences a more pronounced decline in specific energy absorption with an increase in impact velocity compared with the C1-shaped and Ω-shaped open-section specimens.

Figure 22 shows the failure modes of the testing specimens with a loading speed of 0.01 m/s, which were observed by comparing the failure behaviors of the testing specimens at loading speeds of 0.01 m/s and 1 m/s. A few material fragments from the testing specimens with a loading rate of 0.01 m/s appear during the crushing process, indicating that a thorough material failure occurs for this loading condition. The fracture morphology of the testing specimens reveals that the end section of the crushed zone retains more material fragments and exhibits more chaotic and twisted morphology, as shown in Figure 23. The crushed zone experiences adequate friction and compaction with the indenter, which can contribute to an increased structural load-bearing area. However, the testing specimens with the loading rates of 1 m/s ejected a large number of material fragments during the crushing process, reducing the load-bearing area and material utilization. Furthermore, frictional energy absorption was observed between the indenter and the crushed zone, as well as between the layers and fragments. Thus, the initial peak load, the average crushing load, and the specific energy absorption of the specimens also decrease with an increase in impact velocity.

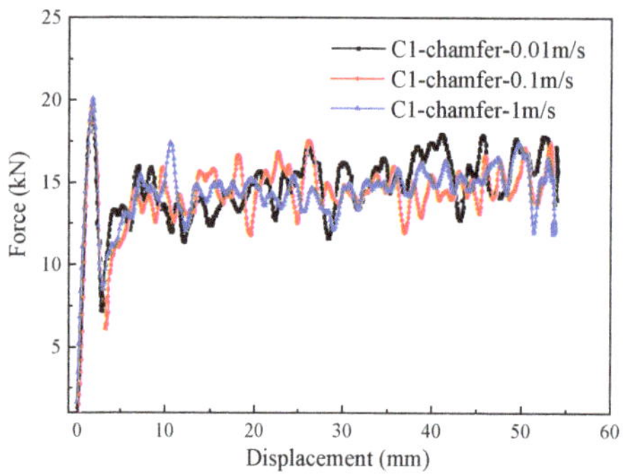

(**a**) C1-chamfer

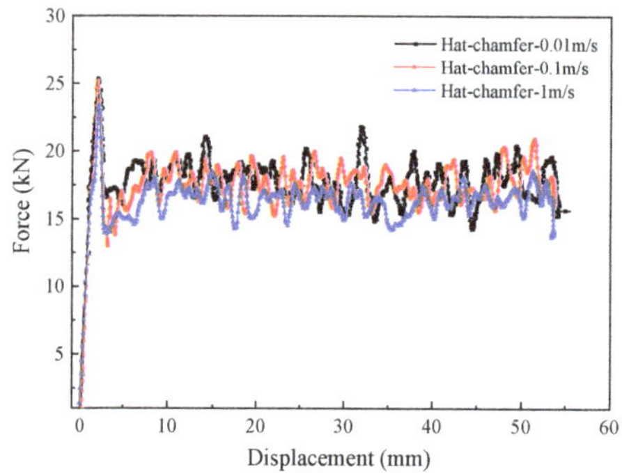

(**b**) Hat-chamfer

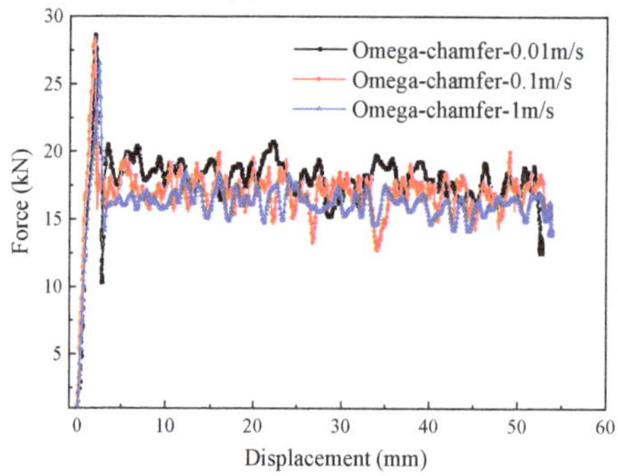

(**c**) Omega-chamfer

Figure 19. Typical force–displacement curves of the testing specimens with different loading rates.

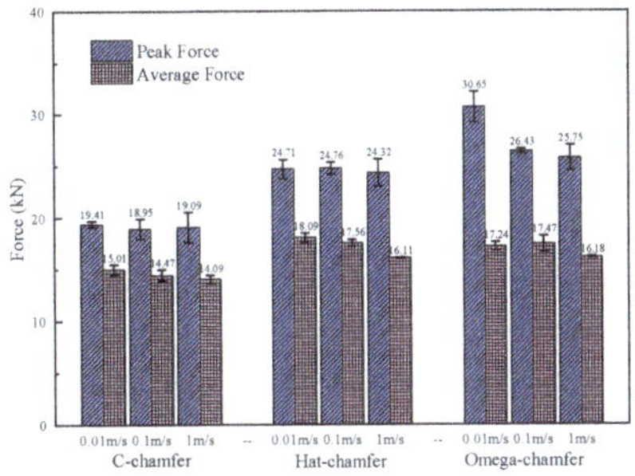

(**a**) Peak force and average force

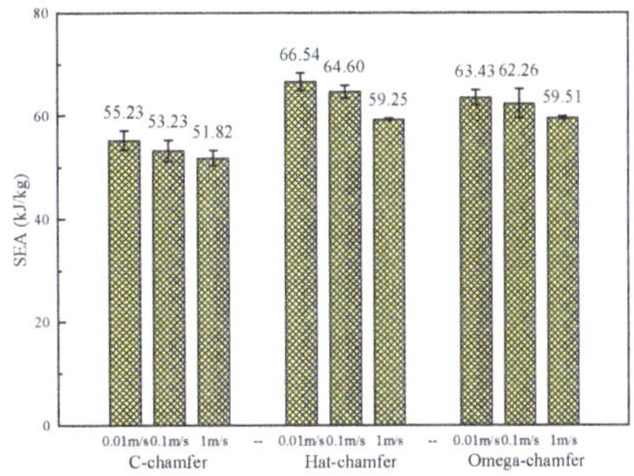

(**b**) Specific energy absorption

Figure 20. Comparison of the energy absorption characteristics of testing specimens with different loading rates.

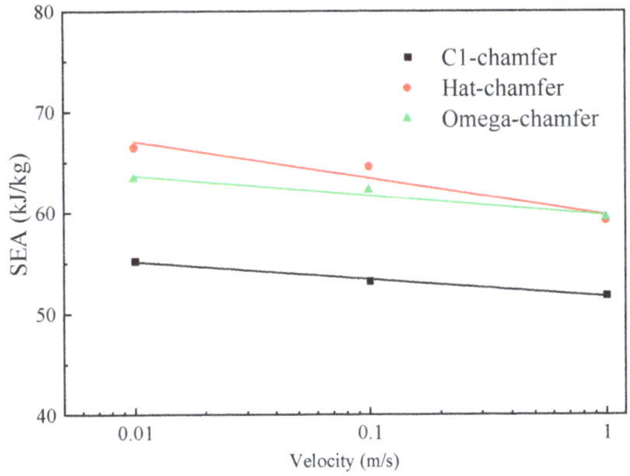

Figure 21. Variation in the SEA of testing specimens with different loading rates.

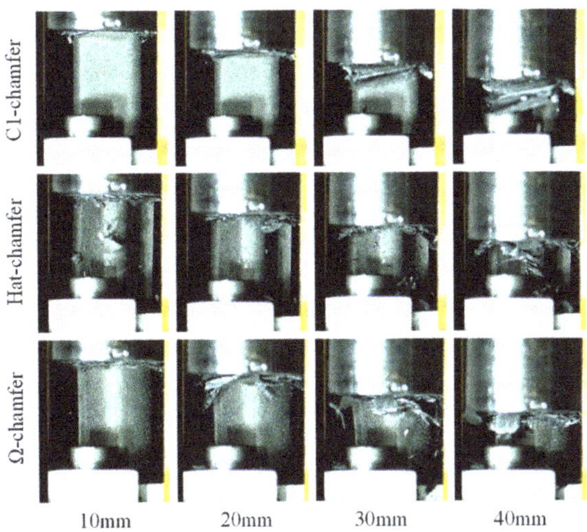

Figure 22. Failure modes of the testing specimens with a loading speed of 0.01 m/s.

(**a**) C1-chamfer (**b**) Hat-chamfer (**c**) Ω-chamfer

Figure 23. Failure modes of the testing specimens with a loading speed of 0.01 m/s.

6. Conclusions

This work mainly studies the crushing response and energy absorption characteristics of composite open-section thin-walled structures with different geometric configurations and loading conditions. The main conclusions obtained are as follows:

(1) For the composite thin-walled open-section structures with different geometric configurations, a brittle failure mode can be observed under different loading rates. The impact kinetic energy is mainly absorbed via material bending, delamination failure, shear failure, and friction between the crushing zone during the crushing.
(2) The cross-section configuration significantly influences the energy absorption characteristics of composite thin-walled open-section structures. The insufficient material damage caused by stress concentration is the main reason for the low energy absorption efficiency of C1-shaped specimens compared to the hat-shaped and Ω-shaped specimens.
(3) Different triggering mechanisms primarily affect the initial crushing stage of the composite structures, while it has little influence on the stable crushing stage. For C-shaped specimens, a 45° chamfer trigger yields better energy absorption. However, the 15° steeple trigger is the optimal triggering mode for the hat-shaped structures.
(4) The average crushing force and specific energy absorption of composite thin-walled structures decrease with an increasing in loading rates. More material fragments can be ejected under higher loading rates, which reduces the structural load-bearing area, material utilization, and frictional energy absorption in the crushing zone.

Author Contributions: Conceptualization, X.Y. and P.X.; methodology, X.X. and C.Z.; software, X.X. and X.L. (Xiaochuan Liu); validation, X.X. and X.L. (Xiaochuan Liu); formal analysis, X.X.; investigation, X.X., X.Z. and X.L. (Xiaocheng Li); resources, C.B.; data curation, X.X.; writing—original draft preparation, X.X., X.Z., X.L. (Xiaocheng Li) and C.B.; writing—review and editing, X.Y.; visualization, X.Y.; supervision, X.Y. All authors have read and agreed to the published version of the manuscript.

Funding: The work described in this paper is financially supported by the National Natural Science Foundation of China (Grant No. 12002027, 12272319) and the Beijing Nova Program (Grant No. 20230484360). The authors would like to gratefully acknowledge their support.

Institutional Review Board Statement: Not applicable.

Informed Consent Statement: Not applicable.

Data Availability Statement: The original contributions presented in the study are included in the article, further inquiries can be directed to the corresponding author.

Conflicts of Interest: The authors declare no conflicts of interest.

References

1. Bhong, M.; Khan, T.K.H.; Devade, K.; Krishna, B.V.; Sura, S.; Eftikhaar, H.K.; Thethi, H.P.; Gupta, N. Review of composite materials and applications. *Mater. Today Proc.* **2023**, *in press*. [CrossRef]
2. Mania, R.; Kowal-Michalska, K. Behaviour of composite columns of closed cross-section under in-plane compressive pulse loading. *Thin-Walled Struct.* **2007**, *45*, 902–905. [CrossRef]
3. Krolak, M.; Kowal-Michalska, K.; Mania, R.; Swiniarski, J. Experimental tests of stability and load carrying capacity of compressed thin-walled multi-cell columns of triangular cross-section. *Thin-Walled Struct.* **2007**, *45*, 883–887. [CrossRef]
4. Yang, X.F.; Ma, J.X.; Wen, D.S.; Yang, J.L. Crashworthy design and energy absorption mechanisms for helicopter structures: A systematic literature review. *Prog. Aerosp. Sci.* **2020**, *114*, 100618. [CrossRef]
5. Ha, N.S.; Lu, G. Thin-walled corrugated structures: A review of crashworthiness designs and energy absorption characteristics. *Thin-Walled Struct.* **2020**, *157*, 106995. [CrossRef]
6. Sebaey, T.A. Crashworthiness of GFRP composite tubes after aggressive environmental aging in seawater and soil. *Compos. Struct.* **2022**, *284*, 115105. [CrossRef]
7. Chen, D.; Liu, Y.; Meng, M.; Li, B.; Sun, X.; Yang, B.; Xiao, S.; Wang, T. Dynamic axial crushing behaviors of circular composite tubes with different reinforcing fibers and triggers. *Int. J. Mech. Sci.* **2023**, *244*, 108083. [CrossRef]
8. Batuwitage, C.; Fawzia, S.; Thambiratnam, D.; Liu, X.; Al-Mahaidi, R.; Elchalakani, M. Impact behaviour of carbon fibre reinforced polymer (CFRP) strengthened square hollow steel tubes: A numerical simulation. *Thin-Walled Struct.* **2018**, *131*, 245–257. [CrossRef]
9. Mamalis, A.G.; Manolakos, D.E.; Ioannidis, M.B.; Papapostolou, D.P. Crashworthy characteristics of axially statically compressed thin-walled square CFRP composite tubes: Experimental. *Compos. Struct.* **2004**, *63*, 347–360. [CrossRef]
10. Chen, D.; Sun, G.; Jin, X.; Li, Q. Quasi-static bending and transverse crushing behaviors for hat-shaped composite tubes made of CFRP, GFRP and their hybrid structures. *Compos. Struct.* **2020**, *239*, 111842. [CrossRef]
11. Chen, D.; Sun, X.; Xiao, S.; Yang, G.; Yang, B.; Zhu, T.; Wang, M. On axial crushing behavior of double hat-shaped CFRP and GFRP structures. *Compos. Struct.* **2023**, *319*, 117117. [CrossRef]
12. Zhu, G.; Sun, G.; Li, G.; Cheng, A.; Li, Q. Modeling for CFRP structures subjected to quasi-static crushing. *Compos. Struct.* **2018**, *184*, 41–55. [CrossRef]
13. Bao, F.; Wang, J.; Wang, J.; Zeng, S.; Guo, X. Static and impact responses of syntactic foam composites reinforced by multi-walled carbon nanotubes. *J. Mater. Res. Technol.* **2020**, *9*, 12391–12403. [CrossRef]
14. Yang, H.; Lei, H.; Lu, G. Crashworthiness of circular fiber reinforced plastic tubes filled with composite skeletons/aluminum foam under drop-weight impact loading. *Thin-Walled Struct.* **2021**, *160*, 107380. [CrossRef]
15. Chen, D.; Luo, Q.; Meng, M.; Li, Q.; Sun, G. Low velocity impact behavior of interlayer hybrid composite laminates with carbon/glass/basalt fibres. *Compos. Part B Eng.* **2019**, *176*, 107191. [CrossRef]
16. Yang, H.; Lei, H.; Lu, G.; Zhang, Z.; Li, X.; Liu, Y. Energy absorption and failure pattern of hybrid composite tubes under quasi-static axial compression. *Compos. Part B Eng.* **2020**, *198*, 108217. [CrossRef]
17. Zhu, G.; Liao, J.; Sun, G.; Li, Q. Comparative study on metal/CFRP hybrid structures under static and dynamic loading. *Int. J. Impact Eng.* **2020**, *141*, 103509. [CrossRef]
18. Reuter, C.; Tröster, T. Crashworthiness and numerical simulation of hybrid aluminium-CFRP tubes under axial impact. *Thin-Walled Struct.* **2017**, *117*, 1–9. [CrossRef]
19. Siromani, D.; Henderson, G.; Mikita, D.; Mirarchi, K.; Park, R.; Smolko, J.; Awerbuch, J.; Tan, T.-M. An experimental study on the effect of failure trigger mechanisms on the energy absorption capability of CFRP tubes under axial compression. *Compos. Part A Appl. Sci. Manuf.* **2014**, *64*, 25–35. [CrossRef]
20. Sigalas, I.; Kumosa, M.; Hull, D. Trigger mechanisms in energy-absorbing glass cloth/epoxy tubes. *Compos. Sci. Technol.* **1991**, *40*, 265–287. [CrossRef]
21. Hull, D. A unified approach to progressive crushing of fibre-reinforced composite tubes. *Compos. Sci. Technol.* **1991**, *40*, 377–421. [CrossRef]
22. Ren, Y.; Jiang, H.; Liu, Z. Evaluation of double- and triple-coupled triggering mechanisms to improve crashworthiness of composite tubes. *Int. J. Mech. Sci.* **2019**, *157–158*, 1–12. [CrossRef]
23. Stukhlyak, P.D.; Buketov, A.V.; Panin, S.V.; Maruschak, P.O.; Moroz, K.M.; Poltaranin, M.A.; Vukherer, T.; Kornienko, L.A.; Lyukshin, B.A. Structural fracture scales in shock-loaded epoxy composites. *Phys. Mesomech.* **2015**, *18*, 58–74. [CrossRef]
24. Joosten, M.W.; Dutton, S.; Kelly, D.; Thomson, R. Experimental and numerical investigation of the crushing response of an open section composite energy absorbing element. *Compos. Struct.* **2011**, *93*, 682–689. [CrossRef]
25. Lv, R.; Ren, Y.; Liu, Z.; Jiang, H. Energy Absorption Characteristics and Failure Mechanism of Fabric Composite Channel Section Structure under Axial Crushing Loading. *J. Aerosp. Eng.* **2022**, *35*, 04022046. [CrossRef]
26. Bolukbasi, A.O.; Laananen, D.H. Energy absorption in composite stiffeners. *Composites* **1995**, *26*, 291–301. [CrossRef]
27. Riccio, A.; Raimondo, A.; Di Caprio, F.; Fusco, M.; Sanità, P. Experimental and numerical investigation on the crashworthiness of a composite fuselage sub-floor support system. *Compos. Part B Eng.* **2018**, *150*, 93–103. [CrossRef]
28. Jackson, A.; Dutton, S.; Gunnion, A.J.; Kelly, D. Investigation into laminate design of open carbon–fibre/epoxy sections by quasi-static and dynamic crushing. *Compos. Struct.* **2011**, *93*, 2646–2651. [CrossRef]

29. Waimer, M.; Kohlgrüber, D.; Hachenberg, D.; Voggenreiter, H. Experimental study of CFRP components subjected to dynamic crash loads. *Compos. Struct.* **2013**, *105*, 288–299. [CrossRef]
30. Lu, G.T.; Yu, X. *Energy Absorption of Structures and Materials*; Elsevier: Amsterdam, The Netherlands, 2003.
31. Zhang, D.; Zheng, X.; Wu, T. Damage characteristics of open-hole laminated composites subjected to longitudinal loads. *Compos. Struct.* **2019**, *230*, 111474. [CrossRef]
32. Jiang, H.; Ren, Y. Crashworthiness and failure analysis of steeple-triggered hat-shaped composite structure under the axial and oblique crushing load. *Compos. Struct.* **2019**, *229*, 111375. [CrossRef]

materials

MDPI

Article

Study on Dynamic Mechanical Properties of Sandwich Beam with Stepwise Gradient Polymethacrylimide (PMI) Foam Core under Low-Velocity Impact

Mousab Mahgoub, Cong Liu and Zhuhua Tan *

School of Mechanical Engineering, Hebei University of Technology, Tianjin 300401, China; mousaba71@gmail.com (M.M.)
* Correspondence: zhtan@hebut.edu.cn

Abstract: Different PMI foam materials of 52, 110, and 200 kg/m^3 were used to design stepwise gradient cores to improve the impact resistance of the sandwich beam. The stepwise gradient core consists of three layers arranged in positive gradient, negative gradient, and sandwich-core (e.g., 200/52/200). These sandwich beams were subjected to the impact of a steel projectile under impact momentum of 10 to 20 kg·m/s, corresponding to impact energy in the range of 12.5 to 50 J. During the test, the impact force was recorded by an accelerometer, and the different failure modes were also obtained. Subsequently, the influence of the layer arrangement on the energy absorption and load transfer mechanism between the different layers was analyzed. The results showed that the top layer with a large density can improve the impact force, but the middle/bottom layer with a low density promoted specific energy absorption. Thus, based on these two points, the negative gradient core (200/110/52) had an excellent specific energy absorption because it can transfer and expand the area to bear the load layer by layer, which improved the energy absorption in each layer. Combined with the failure modes, the load transfer and deformation mechanisms between the layers were also discussed. The present work provided a valuable method to design an efficient lightweight sandwich structure in the protection field.

Keywords: stepwise gradient; sandwich structure; low-velocity impact; failure mode; energy absorption

1. Introduction

The sandwich structure consists of two face sheets and a core, which has been widely used in the fields of aerospace, automotive, marine vehicle, and other industry applications due to their attractive properties, such as high specific strength, high specific stiffness, and excellent energy absorption [1–6].

Over the past decades, a great deal of work has been conducted on the quasi-static and dynamic behavior of the different sandwich structures (e.g., beams [7], plates [8], shells [9], and tubes [10]), and the corresponding failure modes and mechanisms were extensively studied, such as indentation [11], core shear failure [12], tension failure [13], and debonding [14]. It is well known that the core is a key component of the sandwich structure, which not only suffers from the shear load but also dissipates energy through deformation or collapse failure. Different lightweight materials and structures have been used to design the core, such as aluminum foam [15], polymeric foam [16], honeycomb core [17], lattice core [18], corrugated core [19], etc. To some degree, most of these research studies focused on the core with uniform or monolithic materials or structures. Thus, the development of an efficient core is still a challenge in the field of sandwich structures.

Recently, the primary concept of density-graded foams was developed to improve the load-bearing capacity and energy absorption of the sandwich structures, which consisted of several layers with different densities. Lots of work has been conducted on the compressive mechanical properties of density-graded foam under quasi-static and dynamic loading [20–24], such as deformation and failure mechanisms [20], stress wave propagation [21], stress–strain response [22], constitutive model relationship [23], and energy absorption characteristics [24]. These works revealed a noticeable difference in the deformation mechanism between quasi-static and dynamic loading due to the inertia efficiency of the impact loading. Moreover, the arrangement of the density-graded foam layers significantly influenced deformation and energy absorption [20,21,24–35]. Koohbor et al. [20] demonstrated that gradient greatly influences the propagation of stress waves in foams with density gradients. Furthermore, the overall ability of the structure to absorb energy corresponds to the stability of the shock waves through the gradient foam. Kazemi et al. [25,26] studied the sandwich beam/panel with stepwise density-graded cores under quasi-static three-point bending and punch loading. The results showed that the arrangement of the density layer and the number of layering have the most significant influences on the bending resistance, contact peak force, and energy absorption. Nie et al. [27] reported the effects of the density-graded changes of foam core sandwich panel subjected to high-velocity impact, and they concluded that the sandwich panel with a stepwise negative gradient core outperforms to resist the perforation and energy dissipation compared to a positive gradient and uniform-density core sandwich panel. In addition, the failure modes and deformation mechanisms of stepwise density gradient sandwich structures under different load cases were also investigated, such as flexural bending [28,29], blast loading [30,31], low-velocity impact [32,33], and high-velocity impact [34,35].

However, the choice of materials with high specific strength, stiffness, and excellent energy absorption is another significant way to improve the mechanical performance of the sandwich structure. Polymethacrylimide (PMI) material is lightweight and has high energy absorption, which has attracted extensive attention in recent years. The sandwich structure with PMI core has been applied in transportation, aerospace, and military protection equipment [36–39]. The mechanical properties of the sandwich structure with PMI core were also focused under different load cases, such as indentation load [36], three-point bending load [37], damage and fatigue [38], and energy absorption [39]. Most of these works focused on increasing the strength and stiffness of sandwich structures by increasing the density and thickness of the PMI core [40,41] and filling the honeycomb core with PMI foam materials [42].

Most of the previous work focused on sandwich structures with a single-layer PMI foam core, and the work on the sandwich beams with stepwise density-graded PMI foam core is seldom performed. Therefore, in the present paper, the PMI foam materials with different densities of 52, 110, and 200 kg/m^3 were used to design the stepwise gradient core of the sandwich beam. The dynamic mechanical properties of the sandwich beam were studied under the impact of a steel projectile. The impact force was recorded using an accelerometer during the test, and different failure modes were obtained. Based on the experimental results, the effects of the arrangement of the layers on the energy absorption and specific energy absorption were analyzed. The corresponding deformation and failure mechanisms were also discussed.

2. Experimental Procedures

2.1. Materials and Specimens

PMI foam materials (purchased from Suzhou Zhongbao Composite Material Co., Ltd., Suzhou, China) [43] with different densities of 52, 110, and 200 kg/m^3 were used to design a stepwise gradient core for the sandwich beam. Aluminum alloy AL-1050 (purchased from Lutai Co., Ltd., Suzhou, China) [44] was used to design the face sheet of the sandwich beam. The dimensions of the sandwich beam were 260 mm in length and 40 mm in width. The thickness of the face sheet was 0.5 mm. The core consisted of three layers of PMI foams with different densities of 52, 110, and 200 kg/m^3, and each layer was 5 mm thick. The face sheets and PMI foam layers were glued together in different arrangements using epoxy resin (produced by 3M Scotch-Weld Co., Ltd., St. Paul, MN, USA), and the fabrication steps of the specimen are shown in Figure 1a. Five types of stepwise gradient cores were designed as follows: (1) sandwich-core 52/X/52, which means the top and bottom layers of the sandwich-core are 52 kg/m^3 in density, and the middle layer varied from 52 kg/m^3 to 200 kg/m^3 and is denoted as X; (2) sandwich-core: 110/X/110; (3) sandwich-core: 200/X/200; (4) positive gradient core: 52/110/200; (5) negative gradient core: 200/110/50. More details are listed in Table 1.

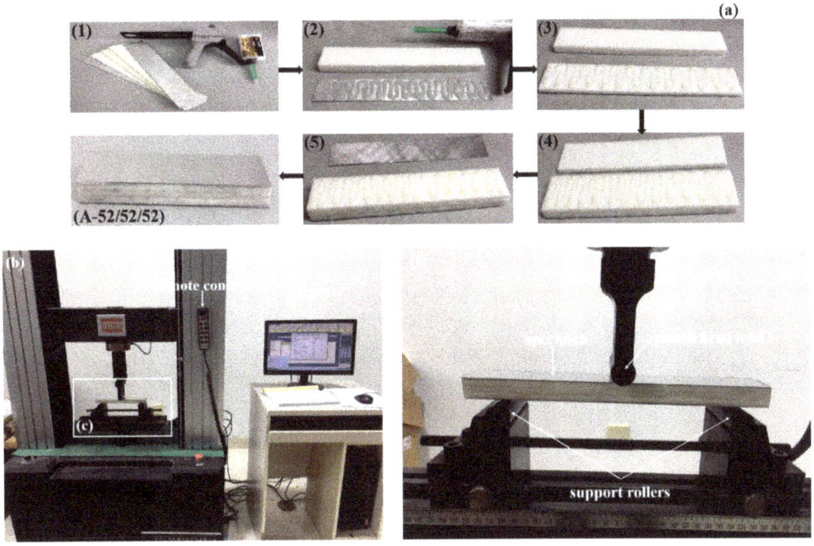

Figure 1. Specimen and the three-point bending test. (**a**) The fabrication steps of the sandwich beam with stepwise gradient core (step 1 is the cutting of the face sheets and foam core from the large plates with specific dimensions of the sandwich beam while steps 2–5 are the gluing the lower face sheet with the bottom layer foam core and consequentially gluing other layers core and upper face sheet); (**b**) the three-point bending test; (**c**) enlarged view of the specimen and support rollers.

Table 1. Details of the sandwich beam specimens.

No.	Specimen Code	Thickness of Each Layer (mm)	Density of the Core Layer (kg/m^3)	Average Density of the Core (kg/m^3)
1	A	0.5/5/5/5/0.5	52/52/52	52
2	B	0.5/5/5/5/0.5	52/110/52	71
3	C	0.5/5/5/5/0.5	52/200/52	101
4	D	0.5/5/5/5/0.5	110/52/110	91
5	E	0.5/5/5/5/0.5	110/110/110	110
6	F	0.5/5/5/5/0.5	110/200/110	140
7	G	0.5/5/5/5/0.5	200/52/200	151
8	H	0.5/5/5/5/0.5	200/110/200	170
9	I	0.5/5/5/5/0.5	200/200/200	200
10	P	0.5/5/5/5/0.5	52/110/200	121
11	N	0.5/5/5/5/0.5	200/110/52	121

2.2. Experimental Apparatus

The quasi-static three-point bending tests were performed by a universal material testing machine SUNS WAW000 (product of Shenzhen SUNS Technology Stock Co. Ltd., Shenzhen, China) according to the ASTM C-393 test standard [45], as shown in Figure 1b,c. Two rollers supported the specimen, and two ends of the specimen could not move in the vertical direction. The compression load was applied on the mid-span of the specimen by a steel compression head. During the tests, the friction force between the bottom face sheet of the sandwich beam and the two support rollers is little, which can be neglected [45]. The effective length was 200 mm between the two support rollers [46]. The load rate was 2 mm/min, and a computer was connected to the machine to record the load and displacement data.

For the impact tests, a steel projectile was launched by a gas gun and impacted the clamped sandwich beam at the mid-span position, as shown in Figure 2. The mass of the steel striker bar was 8 kg, and the dimensions were 40 mm in diameter and 800 mm in length. Two ends of the sandwich beam were fixed in a frame, and the effective span was 200 mm. The projectile's velocity was controlled by gas pressure and measured using a laser velocimeter. Moreover, a 352C04 PCB accelerometer (produced by PCB Company, Depew, NY, USA), with a sensitivity of 2.51 mV/g, was fixed on the projectile to measure the impact force, as shown in Figure 2b. The sampling rate was 5 MHz during the tests, and the accelerometer signal was amplified by a PCB 482C05 signal conditioner (produced by PCB Company, Depew, NY, USA) and saved in the Tektronix MDO3034 oscilloscope (produced by Tektronix, Beaverton, OR, USA).

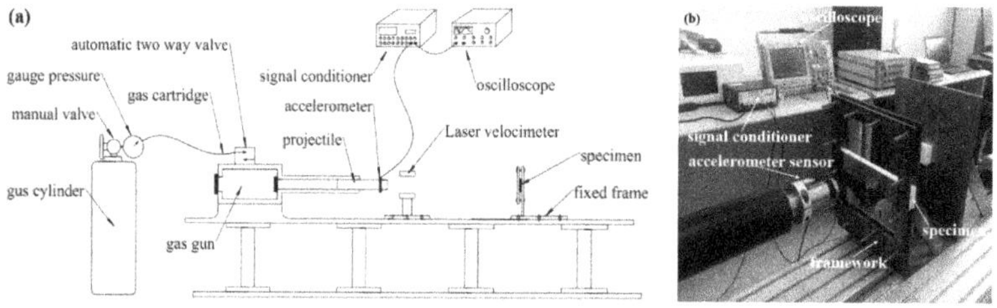

Figure 2. The impact apparatus for the dynamic experiments: (**a**) schematic graph; (**b**) experimental apparatus.

3. Results and Discussions

3.1. *Quasi-Static Load Case*

3.1.1. Compressive Force–Displacement Curves

Figure 3 illustrates the sandwich beams' force–displacement curves and failure modes with different cores in the three-point bending tests. It can be found that the arrangement of the layers in the core has a significant influence on the peak force. The peak force increased from 450 N to 1288 N for the different sandwich beams. Moreover, there is a sudden change in the curve of a sandwich beam of 52/110/52 and 52/200/52 in Figure 3a, and the rest of the sandwich beams had a similar tendency in the force–displacement curves.

For sandwich-core of 52/X/52, the top layer with low strength easily fails in indentation failure mode, and if the middle layer is also 52 kg/m^3 in density, all the layers would fail in the same mode; if the middle layer is with a density of 110 or 200 kg/m^3, the strength of the middle layer increased, which would improve the load bearing capacity and prevent the drop of the curves. This is also another reason for the sudden curve changes in Figure 3a.

For sandwich-cores of 110/X/110 and 200/X/200, the curves ascended as the density of the middle layer increased, as shown in Figure 3b. The bearing load is because the top and bottom layers have a larger strength and can suffer the external load, which determines and transfers the load to the following middle layer. Though the stiffness and strength increased, the deformation decreased, which is much less than those of 52/X/52.

For the positive gradient core, the arrangement of the layers is 52, 110, and 200 kg/m^3 in the sequence. The top layer of 52 kg/m^3 in density has a low strength, which would undergo a similar deformation process and failure mechanism of 52/X/52; the top layer would fail at the load exertion point, then the load keeps the concentration state, which cannot be expanded effectively. However, for the negative gradient core, on the contrary, the arrangement of the layers is 200, 110, and 52 kg/m^3 in the sequence for the negative stepwise gradient core; the top layer of 200 kg/m^3 has enough strength to suffer the external load and distribute the concentrated load, then the distributed load was transferred to the next layer. The area of load exertion on the following layers would expand layer by layer. Thus, the middle and bottom layers can sustain the load and absorb the energy efficiently.

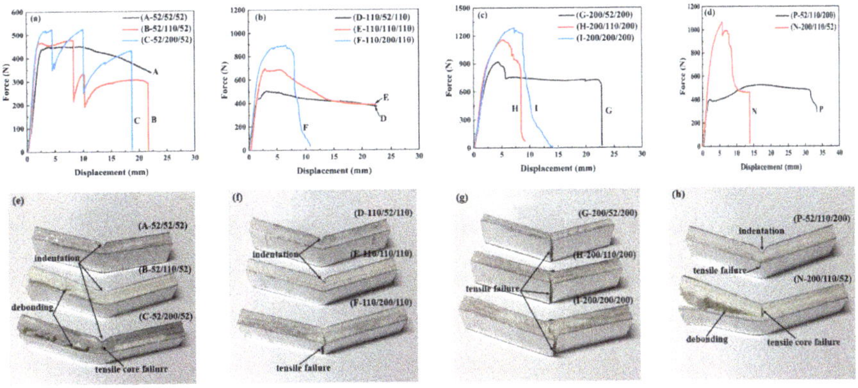

Figure 3. Force–displacement curves and failure modes of the sandwich beams with different cores under the three-point bending tests: (**a**,**e**) for 52/X/52; (**b**,**f**) for 110/X/110; (**c**,**g**) for 200/X/200; (**d**,**h**) for 52/110/200 and 200/110/52, respectively.

3.1.2. Energy Absorption of the Sandwich Beam under Three-Point Bending Tests

The energy absorption (*EA*) of the sandwich beam was calculated by using the force-displacement curve, and the specific energy absorption (*SEA*) was obtained by dividing the energy absorption by the average density of the core. The corresponding equations are as follows [40]:

$$EA = \int F d\delta \quad (1)$$

$$SEA = \frac{\int F d\delta}{\overline{\rho}} \quad (2)$$

where F and δ are the corresponding force and displacement, EA represents energy absorption, SEA is the specific energy absorption, and $\overline{\rho}$ denotes the average density of the core.

Figure 4 illustrates the energy absorption (*EA*) and specific energy absorption (*SEA*) of the sandwich beams with different cores. During the calculation of the *EA* and *SEA*, the influence of the face sheet on the EA and SEA was neglected. Because the thickness of the face sheet is just 0.5 mm, the stiffness and the corresponding load-bearing capacity are small. For the sandwich-core of 52/X/52, both EA and SEA of the sandwich beam increased with the increasing density in the middle layer. The increasing density in the middle layer can improve the stiffness of the sandwich beam, which can bear a larger load and absorb more energy. However, for the sandwich-cores of 110/X/110 and 200/X/200, both EA and SEA decreased with the increasing density in the middle layer. Though the top layers of 110/X/110 and 200/X/200 can suffer from large external loads, only a slight deformation occurs.

Furthermore, the larger the density of the middle layer, the lower the EA and SEA of the sandwich beam. So, the SEA of 110/X/110 and 200/X/200 are less than those of 52/X/52. Moreover, the arrangement of layers in the core varied from a positive to negative gradient; both EA and SEA increased due to the larger strength of the top layer, which can suffer a larger load. Furthermore, the following layers with low strength can absorb the energy transferred from the top layer. Thus, as shown in Figure 4, the negative gradient core of 200/110/52 has the largest SEA.

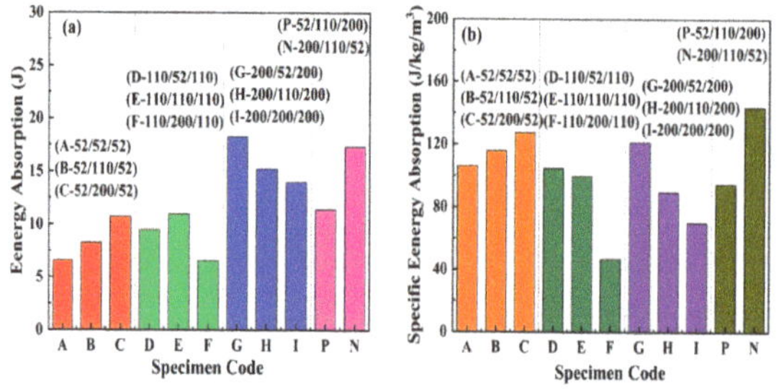

Figure 4. (**a**) Energy absorption and (**b**) specific energy absorption of the sandwich beam with different cores.

3.2. Dynamic Load Cases

The sandwich beams with different cores were subjected to the impact of the projectile, and the impact momentum is in the range of 10 to 20 kg·m/s, corresponding to the impact energy in the range of 12.5 to 50 J. And projectile impacted the top layer side of the sandwich-core, which was marked as the impact side. During the tests, the impact force was recorded by an accelerometer, and the corresponding failure modes were also obtained, which were used to analyze the load transfer mechanism between the core layers.

3.2.1. The Impact Force–Time Curves of the Sandwich-Core 52/X/52

Figure 5 illustrates that the force curves and failure modes of the sandwich-core 52/X/52 under impact loading differed from the quasi-static load case. The peak force increased with the increasing impact momentum/energy and middle-layer density. In particular, when the density of the middle layer increased from 52 kg/m^3 to 200 kg/m^3, the maximum peak forces were 1190 and 2250 N, respectively. It is well known that the maximum shear stress is located at the center of the beam. The increased density in the middle layer can improve the load-sustaining capacity efficiently. Moreover, the density of the top and bottom layers is only 52 kg/m^3, which is easy to fail due to its low strength. So, the corresponding failure mode is mainly in shear and debonding modes, which is a nearly disintegrated failure mode.

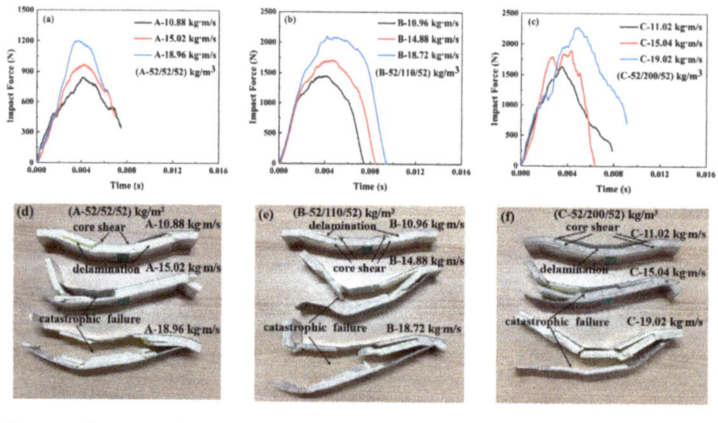

Figure 5. The impact force–time curves and corresponding failure modes of the sandwich beam with different cores: (**a**,**d**) for A-52/52/52; (**b**,**e**) for B-52/110/52; (**c**,**f**) for C-52/200/52.

3.2.2. The Impact Force–Time Curves of Sandwich-Core 110/X/110

The variations of the impact force with time of the sandwich beams with sandwich-core 110/X/110 are shown in Figure 6. The impact force increased with the density in the middle layer, which is larger than the results in Figure 5. As the density of the middle layer increased from 52 kg/m^3 to 200 kg/m^3, the maximum peak forces were 1570 and 3250 N, respectively. However, the failure modes are pretty different from those in Figure 5; most of the failure modes are in shear failure, and there is nearly no debonding or disintegrated failure. The reason can be explained as follows: the top layer with 110 kg/m^3 has larger strength and stiffness than that of 52 kg/m^3, which can sustain a large load and deform in bending deformation, expanding the load exertion area on the next layer. A more effective area of the next layer can be involved in load bearing. So, the impact load of 110/X/110 is larger than that of 52/X/52, and the degree of failure is less than that of 52/X/52.

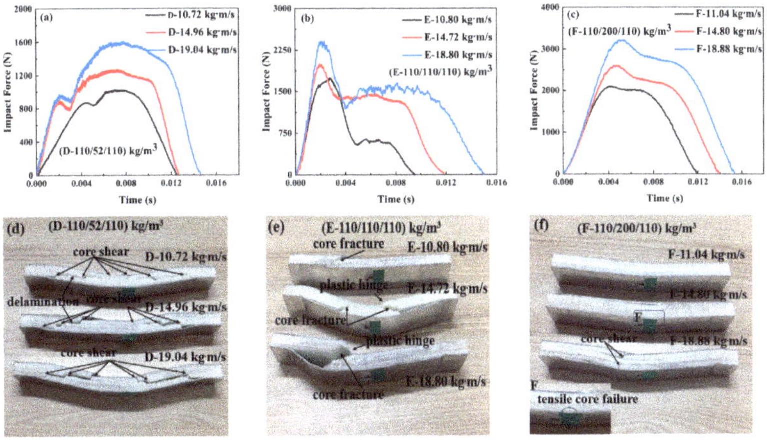

Figure 6. The impact force–time curves and corresponding failure modes of the sandwich beam with different cores: (**a**,**d**) for D-110/52/110; (**b**,**e**) for E-110/110/110; (**c**,**f**) for F-110/200/110.

3.2.3. The Impact Force–Time Curves of Sandwich-Core 200/X/200

Figure 7 illustrates the impact force curves and failure modes of the sandwich beams with a sandwich-core of 200/X/200. The variation tendency in the impact force of 200/X/200 in Figure 7 was similar to those in Figures 6 and 7. As the density of the middle layer increased from 52 kg/m^3 to 200 kg/m^3, the maximum peak forces were 1750 N and 4550 N, respectively. Moreover, it can be observed that the sandwich-core 200/52/200 was in shear failure; however, both 200/110/200 and 200/200/200 failed in tensile failure. The top layer of the 200/X/200 has enough stiffness and strength to suffer the impact of loading and then transfer the load to the middle layer. If the middle layer is 52 kg/m^3 in density, it would fail in shear failure mode and absorb the energy, which results in a large reduction of the load transferring to the bottom layer, as shown in Figure 7d. However, the middle layer, which is 110 or 200 kg/m^3 in density, also has enough stiffness and strength, which makes all the layers deform in bending and tensile failure, as shown in Figure 7e,f.

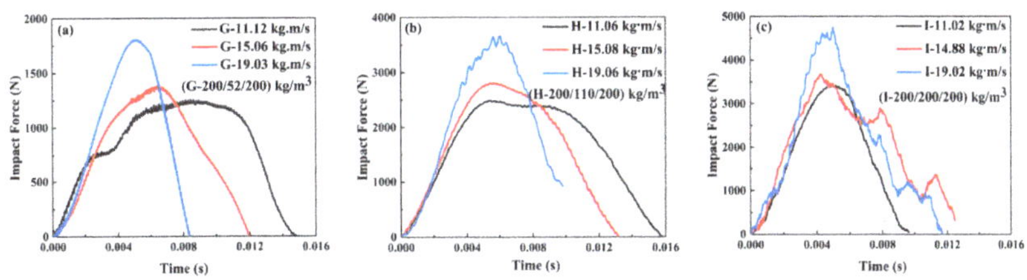

Figure 7. *Cont.*

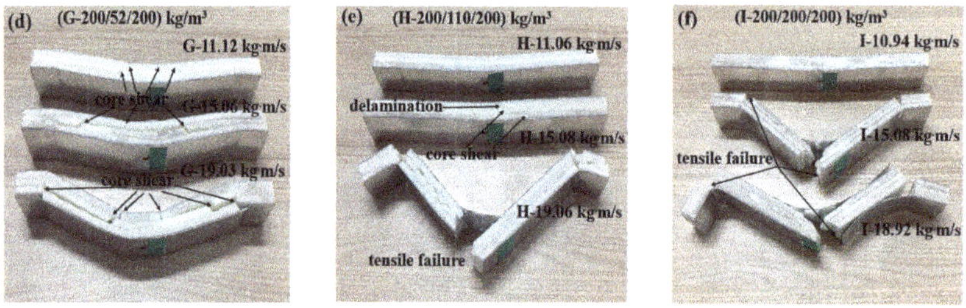

Figure 7. The impact force–time curves and failure modes of the sandwich beam with different cores: (**a**,**d**) G-200/52/200; (**b**,**e**) H-200/110/200; (**c**,**f**) I-200/200/200.

3.2.4. The Impact Force–Time Curves of the Positive and Negative Gradient Cores 52/110/200 and 200/110/52

The variations of the impact force and failure modes of the sandwich beams with positive and negative gradient cores are shown in Figure 8. For the positive gradient core, the projectile impacted on the top layer core of 52 kg/m^3, and the bottom layer is 200 kg/m^3 in density. However, the projectile impacted the top layer core of 200 kg/m^3 for the negative gradient core, and the bottom layer is 52 kg/m^3 in density. Both positive and negative gradient cores have a similar variation tendency in the impact force, which increases with the impact momentum/energy. And the impact forces for both cases are at the same level. However, the corresponding duration time of the negative gradient core is larger than that of the positive gradient core at the different impact momentum/energy. It can be observed that the duration time is 0.012 s, 0.014 s, and 0.0145 s for the case of negative gradient core, which is more than 0.009 s, 0.0013 s, and 0.004 s for the positive gradient case, respectively.

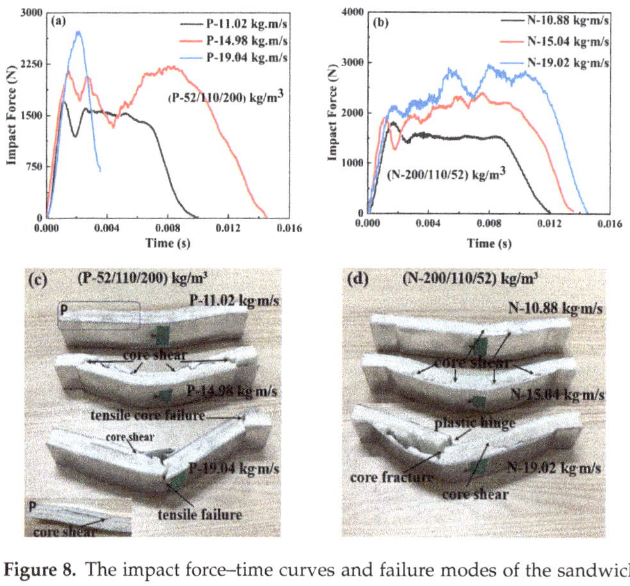

Figure 8. The impact force–time curves and failure modes of the sandwich beams with different cores: (**a**,**c**) positive gradient core (52/110/200); (**b**,**d**) negative gradient core (200/110/52).

Moreover, there is a noticeable difference in the failure modes shown in Figure 8. The failure modes of the positive gradient cores varied from core shear failure to tensile failure; however, all negative gradient cores failed in core shear failure mode. It can be explained as follows: for the negative gradient case, the top layer is 200 kg/m^3 in density, which can sustain the impact load and then transfer the load to the next layer, but the middle and bottom layers with densities of 110 and 52 kg/m^3 have low strength, which would fail and collapse under the load transferred from the top layer, and the load and the energy were also dissipated. Thus, the failure mode is a core shear failure. For the positive gradient core, though the top layer of 52 kg/m^3 failed, the load and energy experienced little loss due to the inertial effect. The concentrated impact load did not spread transversally. Most energy of the impact load still acts on the next layer as a concentrated impact loading, which results in a tensile failure in the bottom layer and face sheet, as shown in Figure 8.

3.3. *Energy Absorption of the Sandwich Beam with Different Cores under Impact Loading*

During calculation of the energy absorption of the sandwich beam with different cores under impact loading, the corresponding force and displacement of the projectile were derived based on the data of the accelerometer, which can also be found in Ref. [47]. The variation in the velocity of the projectile $\Delta V(t)$ can be calculated as follows:

$$\Delta V(t) = \int_0^t a_0(t)d(t) \tag{3}$$

Subsequently, the instant velocity of the projectile can be obtained:

$$V(t) = V_0 - \int_0^t a_0(t)d(t) \tag{4}$$

Hence, the displacement of the projectile during the impact process can be calculated as follows:

$$U(t) = \int_0^t V(t)d(t) \tag{5}$$

Moreover, the corresponding impact force is as follows:

$$F(t) = ma(t) \tag{6}$$

where $a(t)$, m, and $U(t)$ are the de-acceleration, mass, and displacement of the projectile, respectively; $\Delta V(t)$ and $V(t)$ are the changes in velocity and the initial velocity of the projectile, respectively; $F(t)$ is the impact force. Then, the corresponding energy absorption can be obtained based on the impact force and displacement using Equations (1)–(3).

Based on Equations (1)–(6), the energy absorption (EA) and specific energy absorption (SEA) of the sandwich beams with different cores under impact loading were calculated and shown in Figure 9. Generally, for the sandwich-core, EA increased with the increasing density of the top layer. However, the SEA of 200/X/200 was less than that of 52/X/52 and 110/X/110, which is due to the slight deformation that occurred in the sandwich beam with core 200/X/200, and the sandwich beams with cores 52/X/52 and 110/X/110 absorbed more energy through the deformation and failure of the core.

Furthermore, compared to the sandwich-core, the negative gradient core (200/110/52) has an excellent capacity for both EA and SEA. The top layer of 200 kg/m^3 can sustain the impact load and transform the concentrated impact load into a distributed load through bending deformation. Then, the following layers absorbed the energy with 110 and 52 kg/m^3.

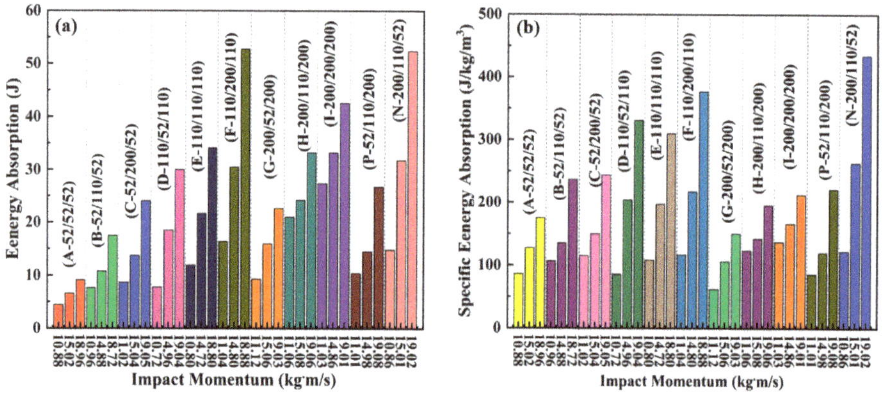

Figure 9. (**a**) Energy absorption and (**b**) specific energy absorption of the sandwich beam with different cores under impact loading.

3.4. Deformation and Failure Mechanism

3.4.1. Deformation and Failure Mechanism of Quasi-Static Load Case

Most of the failure modes of the sandwich beams are indentation, core shear, and face yield, which can be mapped by the Gibson model [46]. The failure mode map is divided into three regions, each one corresponding to a predominant failure mode, e.g., face yield, indentation, and core shear. And the lines between these failure modes are expressed as follows [48]:

$$\frac{c}{L} = \frac{1}{2}\sqrt{\frac{\sigma_{yc}}{\sigma_{yf}}} - \frac{t}{L} \tag{7}$$

$$\frac{c}{L} = \frac{1}{2}\left[\left(1 + \frac{H}{L}\right)\frac{\tau_{yc}}{\sigma_{yf}} - \frac{2t}{L}\right]^{-1}\left(\frac{t}{L}\right)^{2} \tag{8}$$

$$\frac{c}{L} = \frac{L}{L+H}\frac{\sigma_{yf}}{\tau_{yc}}\left[\left(\frac{\sigma_{yc}}{\sigma_{yf}}\right)^{1/2}\frac{t}{L} - \frac{3}{2}\left(\frac{t}{L}\right)^{2}\right] \tag{9}$$

where L is the effective span length, b is the width of the sandwich beam, and H is the free end distance outer roller supports. The thickness of the face sheet and each layer foam core are denoted as t and c, respectively, as shown in Figure 10.

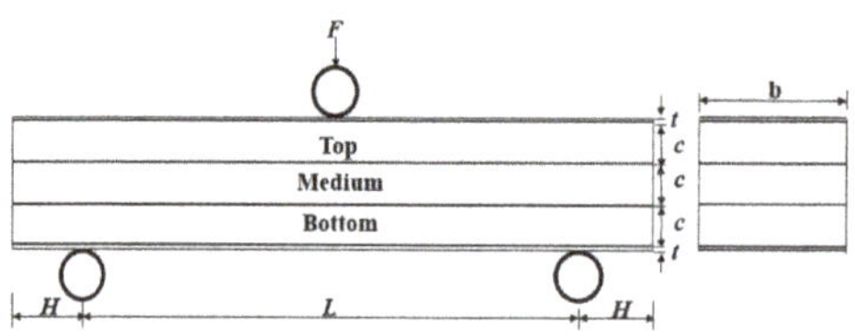

Figure 10. A schematic graph of the sandwich beam with stepwise gradient core under three-point bending.

Moreover, the predicted failure load for different modes can be calculated as follows [44]: The failure load of face yield failure mode is:

$$F_{fy} = \frac{8\sigma_f bt(c+t)}{L} + \frac{\sigma_f bt^2}{L} + \frac{4\sigma_c bc(t+c)}{L} \tag{10}$$

The failure load of indentation failure mode is:

$$F_{in} = 2bt\sqrt{\sigma_f \sigma_c} \tag{11}$$

The failure load of core shear failure mode is:

$$F_{cs} = \frac{7bt^2}{2L}\sigma_f + 4\tau_c bc\left(1 + \frac{2H}{L}\right) \tag{12}$$

In the present paper, the material parameters of the PMI foam are as follows: the yield strength σ_f and elastic modulus E_f of the face sheet are 134 MPa and 68.9 GPa, respectively. And σ_c and E_c are the yield strength and elastic modulus of PMI. For the PMI foam with densities of 52, 110, and 200 kg/m^3, the yield strength values σ_c are 1 MPa, 3.5 MPa, and 7.5 MPa, and elastic modulus values E_c are 74 MPa, 200 MPa, and 443 MPa, respectively. The shear strength $\tau_c \approx \frac{2}{3}\sigma_{Yc}$.

Figure 11 shows the predicted failure mode maps of the sandwich beams with different cores under the three-point bending loading, and the three-point bending experimental results also agreed well with the predicted failure mode map, which demonstrated the reliability of the failure mode map. As shown in Figure 11a–c, when the density of the middle layer of the sandwich-core increased from 52 kg/m^3 to 200 kg/m^3, the region of the face yield mode expanded, whereas the region of the indentation and core shear modes shrunk. The reason can be explained by the fact that the middle layer with a large density can improve the stiffness of the sandwich beam and resist the external load. Moreover, as the density of the top layer increased, similar changes in the regions of the failure modes also occurred. Because the top layer with a large density has enough strength and can suffer the load, the sandwich beam would not be in indentation failure. A similar tendency can also be observed in Figure 11d.

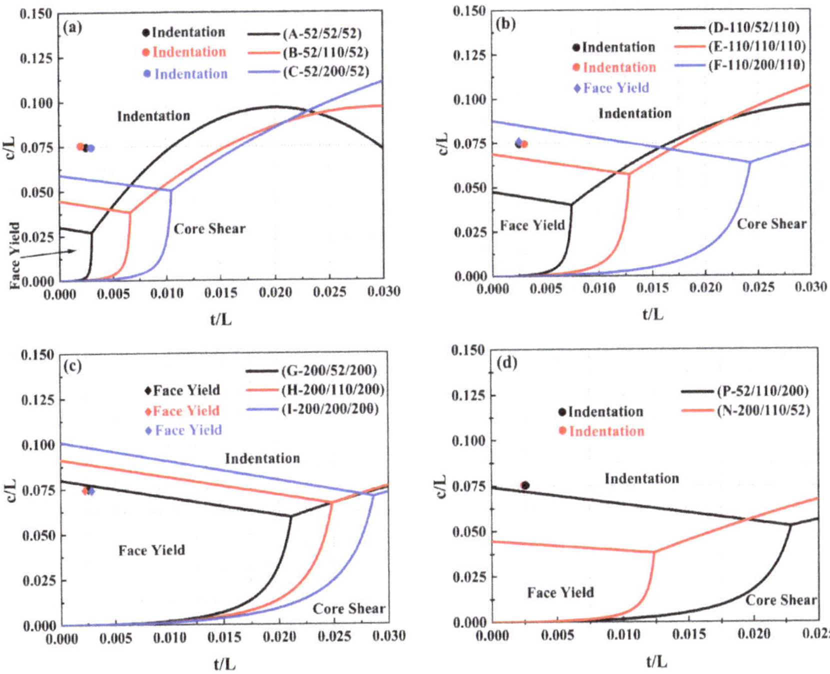

Figure 11. Failure mode map of sandwich beam with different cores: (**a**) 52/X/52; (**b**) 110/X/110; (**c**) 200/X/200; (**d**) 52/110/200 and 200/110/52.

3.4.2. Deformation and Failure Mechanisms under Impact Loading

Combined with the impact force curves and failure modes, the deformation and failure mechanisms of the sandwich beam with different cores were analyzed. Generally, the failure modes are sensitive to the layer density (strength), arrangement, and impact momentum/energy, significantly influencing the load-transferring mechanism between the core layers. According to the core type, it can be divided into three cases, including sandwich-core, positive gradient core, and negative gradient core. The details are as follows:

Case 1: sandwich-core of 52/X/52, 110/X/110, and 200/X/200

For the sandwich-cores of 52/X/52 and 110/X/110, the strength of the top layer was low, and the top layer failed in shear failure mode under the projectile's impact. Then, the impact loading would continue to act on the middle layer, but the load exertion area has little change. If the middle layer was still in low density (52 or 110 kg/m^3), it would also fail in core shear failure and transfer the impact load to the bottom layer, as shown in Figure 12a; if the middle layer was in large density (200 kg/m^3), it would have enough strength to sustain the load and deform in bending, which can spread the concentrated impact loading into a large area on the bottom layers. Due to the low density of the bottom layer, it would also fail in shear failure mode, as observed in Figure 12b. Still, more energy can be absorbed by its deformation and failure because a more effective part of the bottom layer was involved in energy absorption.

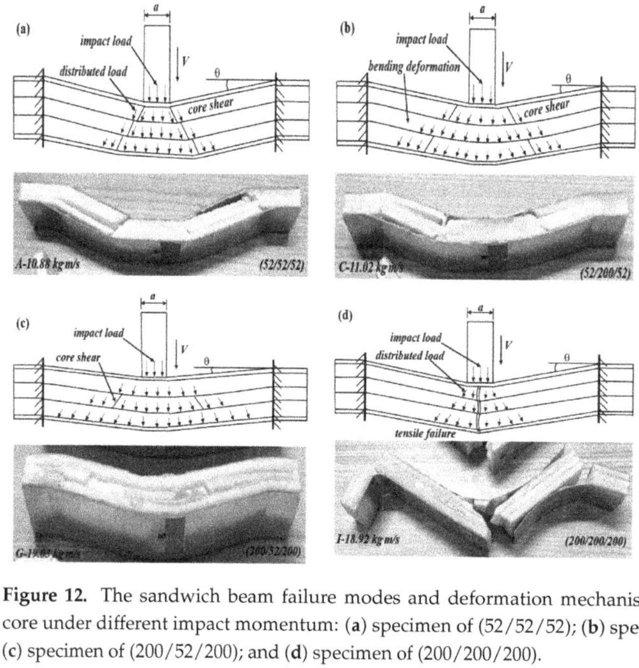

Figure 12. The sandwich beam failure modes and deformation mechanisms with a sandwich-core under different impact momentum: (**a**) specimen of (52/52/52); (**b**) specimen of (52/200/52); (**c**) specimen of (200/52/200); and (**d**) specimen of (200/200/200).

For the sandwich-core of 200/X/200, the top layer can sustain the impact loading and deform in bending, and then the concentrated impact load was distributed and transferred to the middle layer. If the middle layer has a low density, the energy would be absorbed by the collapse deformation and failure of the middle layer, and the load would be further distributed and weakened. When the distributed and weakened load is transferred to the bottom layer, it will deform in bending or tensile failure, as shown in Figure 12c. If the middle layer has a large density, it can sustain the load and deform in bending, further distributing the concentrated load. Still, there is little reduction in energy and load. So, the load applied on the core layers would result in a tensile failure, as shown in Figure 12d.

Case 2: positive gradient core (52/110/200)

For the positive gradient core of 52/110/200, the top layer of 52 kg/m^3 failed in shear failure under the impact of the projectile, and the concentrated impact load was distributed and weakened. Then, the load was sustained by the middle layer of 110 kg/m^3; it would deform in bending and transfer to the bottom layer of 200 kg/m^3. Thus, both the middle and bottom layers deformed in bending. When the impact momentum/energy is low, the core will fail in shear failure, as shown in Figure 13a. The sandwich beam will fail in tensile failure mode if the momentum is large enough, as observed in Figure 13b.

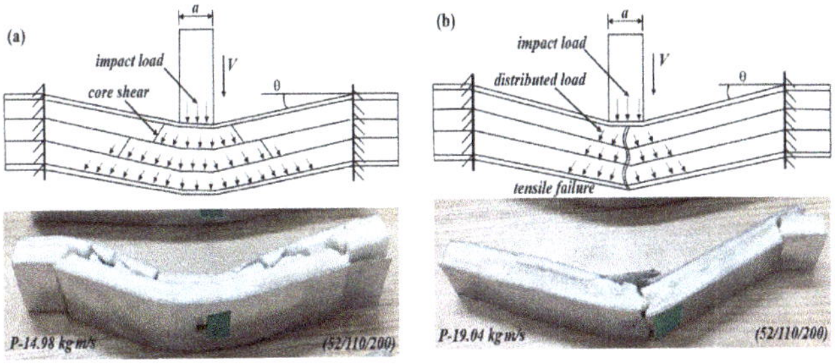

Figure 13. The sandwich beam failure modes and deformation mechanisms with positive stepwise gradient core: (**a**) specimen of (52/110/200); and (**b**) specimen of (200/110/52).

Case 3: negative gradient core (200/110/52)

For the negative gradient core of 200/110/52, the top layer has enough strength to sustain the load and deform in bending under the impact load, as shown in Figure 14. Then, the concentrated impact load was also distributed, but the magnitude of the load was a little weakened due to the lower energy absorption. When the distributed load is transferred to the middle layer of 110 kg/m^3, the middle layer will fail in shear failure mode. The reason is that the strength of 110 kg/m^3 is much lower than that of 200 kg/m^3; the middle layer cannot sustain the load in a bending way. After the energy absorption and load spread by the middle layer, the load acted on the 52 kg/m^3 bottom layer in a large area with a decreased magnitude. Though the 52 kg/m^3 bottom layer still failed in core shear, most of the energy was absorbed by the middle and bottom layers. The load action area increases layer by layer, and more effective areas of the layers can sustain the load and absorb the energy. The above deformation process and mechanism agreed with the maximum SEA of the sandwich beam with a negative core (200/110/52) in Figure 9.

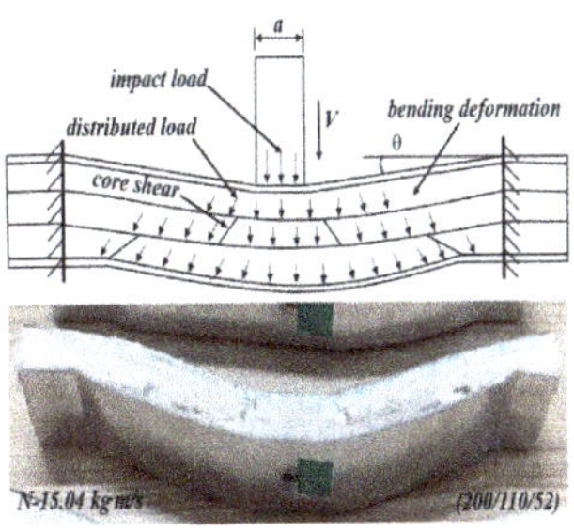

Figure 14. The deformation mechanism of the sandwich beams with a negative gradient core under different impact momentum.

4. Conclusions

The PMI foams with 52, 110, and 200 kg/m^3 densities were used to design the stepwise gradient cores for the sandwich beams. The dynamic response and failure mechanisms of the different sandwich beams were investigated under three-point bending and impact loading. Combining with the experimental results, the load transferring and failure mechanisms of the sandwich beams were analyzed. The results showed that the arrangement of the layers in the core has a significant influence on the load-bearing capacity and failure modes of the sandwich beam under three-point bending and impact loading. The top layer with a large density can improve the load-bearing capacity, but the middle/bottom layer with a low density promotes the specific energy absorption. For instance, the top layer with low strength (for 52 or 110 kg/m^3) would fail in indentation (or shear) mode in quasi-static (or impact) load case, respectively, because the top layer cannot dissipate the concentrated load and transfer it to the following layers; however, the top layer with high strength (for 200 kg/m^3) would deform in bending and transfer the concentrated load into distributed load, which expanded the area to bear the load layer by layer and also improved the energy absorption in each layer. Thus, the 200/110/52 core satisfied the large strength in the top layer and low strength in the middle/bottom layers, resulting in the largest specific energy absorption.

In the future, more work can focus on the performance of stepwise gradient sandwich structures with different materials, such as honeycomb, metal foam, lattice materials, etc. Moreover, different loads can also be considered, such as penetration and blasting loads. The present work provided a valuable method to design an efficient lightweight sandwich structure in the protection fields.

Author Contributions: Conceptualization, Z.T.; Formal analysis, M.M., C.L. and Z.T.; Investigation, M.M. and C.L.; Writing—original draft, M.M.; Writing—review & editing, Z.T. All authors have read and agreed to the published version of the manuscript.

Funding: This work was supported by the National Natural Science Foundation of China (No. 12272121).

Institutional Review Board Statement: Not applicable.

Informed Consent Statement: Not applicable.

Data Availability Statement: The raw data needed to reproduce these findings cannot be shared at this time, as the data will be used in ongoing research.

Acknowledgments: This work was supported by the National Natural Science Foundation of China (No. 12272121), and it is gratefully acknowledged.

Conflicts of Interest: The authors declare no conflict of interest.

References

1. Li, J.; Gao, G.; Yu, Y.; Zhuo, T. Experimental and numerical study on the lightweight design of load-bearing energy absorption structure for subway train. *Thin-Walled Struct.* **2024**, *197*, 111542. [CrossRef]
2. Ghanbari, J.; Panirani, P.N. A hybrid bio-inspired sandwich structures for high strain rate energy absorption applications. *Sci. Rep.* **2024**, *14*, 2865. [CrossRef] [PubMed]
3. Xiao, W.; Hu, Y.; Li, Y. Ice impact response and energy dissipation characteristics of PVC foam core sandwich plates: Experimental and numerical study. *Mar. Struct.* **2023**, *89*, 103407. [CrossRef]
4. Zhang, J.; Huang, W.; Yuan, H.; Wu, X. Failure behavior of a sandwich beam with GLARE face-sheets and aluminum foam core under three-point bending. *Thin-Walled Struct.* **2023**, *183*, 110438. [CrossRef]
5. Zhao, X.; Wei, L.; Wen, D.; Zhu, G.; Yu, Q.; Ma, Z.D. Bending response and energy absorption of sandwich beams with novel auxetic honeycomb core. *Eng. Struct.* **2021**, *247*, 113204. [CrossRef]
6. Khan, F.; Hossain, N.; Mim, J.J.; Rahman, S.M.; Iqbal, M.J.; Billah, M.; Chowdhury, M.A. Advances of Composite Materials in Automobile Applications—A Review. *J. Eng. Res.* **2024**. [CrossRef]
7. Zhang, J.; Huang, W.; Miao, F.; Du, J.; Sun, H. Plastic behavior of sandwich beams with fiber metal laminate face-sheets and metal foam core: Combined local denting and overall deformation. *Thin-Walled Struct.* **2022**, *173*, 108973. [CrossRef]
8. Zhu, Y.; Sun, Y. Dynamic response of foam core sandwich panel with composite face sheets during low-velocity impact and penetration. *Int. J. Impact Eng.* **2020**, *139*, 103508. [CrossRef]

9. Li, Z.; Gao, Y.; Wang, Y.; Xue, P.; Gong, C.; Wang, W.; Wei, X.; Xiong, J. Failure mechanisms and acoustic emission pattern recognition of all-CFRP cylindrical honeycomb sandwich shell under three-point bending. *Compos. Sci. Technol.* **2023**, *237*, 110003. [CrossRef]
10. Ge, J.; Luo, T.; Qiu, J. Experimental Investigation of the Dynamic Responses of Thin-Walled and Foam-Filled Steel Tubes Subjected to Repeated Impacts. *Materials* **2024**, *17*, 1018. [CrossRef]
11. Lu, H.; Wang, X.; Chen, T. Quasi-static bending response and energy absorption of a novel sandwich beam with a reinforced auxetic core under the fixed boundary at both ends. *Thin-Walled Struct.* **2023**, *191*, 111011. [CrossRef]
12. Xie, H.; Li, W.; Fang, H.; Zhang, S.; Yang, Z.; Fang, Y.; Yu, F. Flexural behavior evaluation of a foam core curved sandwich beam. *Compos. Struct.* **2024**, *328*, 117729. [CrossRef]
13. Zhang, W.; Li, J.; Wang, Z.; Li, K.; Bai, C.; Qin, Q. The influence of asymmetric faces on low-velocity impact failure of CFRP/aluminum foam composite sandwich beams. *Eng. Struct.* **2023**, *292*, 116574. [CrossRef]
14. Mei, J.; Chen, Y.; Liu, Z.; Liu, J.; Huang, W. Experimental and theoretical study on the response of the X-frame CFRP sandwich beam under the local impulsive loading. *Int. J. Impact Eng.* **2024**, *186*, 104885. [CrossRef]
15. Tan, Z.H.; Luo, H.; Long, W.; Han, X. Dynamic response of clamped sandwich beam with aluminium alloy foam core subjected to impact loading. *Compos. Part B Eng.* **2013**, *46*, 39–45. [CrossRef]
16. Chen, C.; Airoldi, A.; Caporale, A.; Sala, M.; Yin, G.; Xiao, J. Impact response of composite energy absorbers based on foam-filled metallic and polymeric auxetic frames. *Compos. Struct.* **2024**, *331*, 117916. [CrossRef]
17. Acanfora, V.; Zarrelli, M.; Riccio, A. Experimental and numerical assessment of the impact behaviour of a composite sandwich panel with a polymeric honeycomb core. *Int. J. Impact Eng.* **2023**, *171*, 104392. [CrossRef]
18. Li, J.; Zhang, W.; Wang, Z.; Wang, Q.; Wu, T.; Qin, Q. Dynamic response and failure of CFRP Kagome lattice core sandwich panels subjected to low-velocity impact. *Int. J. Impact Eng.* **2023**, *181*, 104737. [CrossRef]
19. Sun, H.; Yuan, H.; Zhang, J.; Zhang, J.; Du, J.; Huang, W. Dynamic response of multilayer sandwich beams with foam-filled trapezoidal corrugated and foam cores under low-velocity impact. *J. Eng. Struct.* **2023**, *286*, 116080. [CrossRef]
20. Koohbor, B.; Ravindran, S.; Kidane, A. In situ deformation characterization of density-graded foams in quasi-static and impact loading conditions. *Int. J. Impact Eng.* **2021**, *150*, 103820. [CrossRef]
21. Liu, H.; Ng, B.F. Dynamic response of density-graded foam subjected to soft impact. *Compos. Struct.* **2022**, *284*, 115145. [CrossRef]
22. Astafurov, S.V.; Maier, G.G.; Melnikov, E.V.; Moskvina, V.A.; Panchenko, M.Y.; Astafurova, E.G. The strain-rate dependence of the Hall-Petch effect in two austenitic stainless steels with different stacking fault energies. *Mater. Sci. Eng. A* **2019**, *756*, 365–372. [CrossRef]
23. Liu, X.; Wang, Y.; He, X.; Liu, H.; Cao, S. Deformation failure mechanism and constitutive model of gradient aluminum foam under impact loading. *Compos. Struct.* **2024**, *327*, 117684. [CrossRef]
24. Yang, L.; Li, X.; Yang, L.; Lu, J.; Wang, Z.; Yang, J. Experimental and numerical analysis of dynamic response of graded PVC foam sandwich panel under impact load. *Mech. Adv. Mater. Struct.* **2022**, *29*, 5567–5578. [CrossRef]
25. Kazemi, M. Experimental analysis of sandwich composite beams under three-point bending with an emphasis on the layering effects of foam core. *Structures* **2021**, *29*, 383–391. [CrossRef]
26. Kazemi, M. Experimental investigation on the energy absorption characteristics of sandwich panels with layering of foam core under quasi-static punch loading. *J Mech. Adv. Mater. Struct.* **2022**, *29*, 3067–3075. [CrossRef]
27. Nia, A.A.; Kazemi, M. Experimental study of ballistic resistance of sandwich targets with aluminum face-sheet and graded foam core. *I.J. Sndwich Struct. Mater.* **2020**, *22*, 461–479. [CrossRef]
28. Dhaliwal, G.S.; Newaz, G.M. Flexural response of degraded polyurethane foam core sandwich beam with initial crack between face sheet and core. *Materials* **2020**, *13*, 5399. [CrossRef]
29. Wang, F. Static plastic analysis of metallic sandwich beam with functionally graded core. *Int. Eur. J. Mech. A/Solids* **2021**, *90*, 104370. [CrossRef]
30. Jing, L.; Su, X.; Chen, D.; Yang, F.; Zhao, L. Experimental and numerical study of sandwich beams with layered-gradient foam cores under low-velocity impact. *Thin-Walled Struct.* **2019**, *135*, 227–244. [CrossRef]
31. Zhang, W.; Qin, Q.; Li, K.; Li, J.; Wang, Q. Effect of stepwise gradient on dynamic failure of composite sandwich beams with metal foam core subject to low-velocity impact. *Int. J. Solids Struct.* **2021**, *228*, 111125. [CrossRef]
32. Fang, B.; Huang, W.; Xu, H.; Jiang, C.; Liu, J. High-velocity impact resistance of stepwise gradient sandwich beams with metal foam cores. *Thin-Walled Struct.* **2022**, *181*, 110054. [CrossRef]
33. Zhao, Z.; Jing, L. The response of clamped sandwich panels with layered-gradient aluminum foam cores to foam projectile impact. *Mech. Adv. Mater. Struct.* **2020**, *27*, 744–753. [CrossRef]
34. Khaire, N.; Gupta, M.; Tiwari, G. Blast resistance of graded aluminium foam core sandwich structure against blast loading. *Mater. Today Proc.* **2023**, *87*, 159–163. [CrossRef]
35. Khondabi, R.; Khodarahmi, H.; Hosseini, R.; Ziya-Shamami, M. Dynamic plastic response of sandwich structures with graded polyurethane foam cores and metallic face sheets exposed to uniform blast loading: Experimental study and numerical simulation. *J. Braz. Soc. Mech. Sci. Eng.* **2023**, *45*, 526. [CrossRef]
36. Flores-Johnson, E.; Li, Q. Experimental study of the indentation of sandwich panels with carbon fibre-reinforced polymer face sheets and polymeric foam core. *Compos. Part B Eng.* **2011**, *42*, 1212–1219. [CrossRef]

37. Cheon, S.; Yu, S.; Kim, K.-Y.; Lim, D.Y.; Lee, J.-C. Improvement of Interfacial Bonding Force between PMI Foam and CFRP in PMI Foam-Cored CFRP Sandwich Composites. *Fibers Polym.* **2021**, *22*, 2281–2284. [CrossRef]
38. Huo, X.; Jiang, Z.; Luo, Q.; Li, Q.; Sun, G. Mechanical characterization and numerical modeling on the yield and fracture behaviors of polymethacrylimide (PMI) foam materials. *Int. J. Mech. Sci.* **2022**, *218*, 107033. [CrossRef]
39. Song, S.; Xiong, C.; Zheng, J.; Yin, J.; Zou, Y.; Zhu, X. Compression, bending, energy absorption properties, and failure modes of composite Kagome honeycomb sandwich structure reinforced by PMI foams. *Compos. Struct.* **2021**, *277*, 114611. [CrossRef]
40. Mahgoub, M.; Zhang, Y.; Yang, C.; Tan, Z.H. Dynamic Responses of Sandwich Beams with Polymethacrylimide (PMI) Foam Cores When Subjected to Impact Loading. *Materials* **2023**, *16*, 1108. [CrossRef]
41. Zhou, H.; Liu, R.; Hu, Y.; Song, P.; Guo, R. Quasi-static compressive strength of polymethacrylimide foam-filled square carbon fiber reinforced composite honeycombs. *J. Sandw. Struct. Mater.* **2021**, *23*, 2358–2374. [CrossRef]
42. Wang, F.; Ming, Y.; Zhao, Y.; Yang, F.; Lou, J.; Zhu, Y.; Duan, Y.; Wang, B.; Xiao, H. Fabrication of a novel continuous fiber 3D printed thermoset all-composite honeycomb sandwich structure with polymethacrylimide foam reinforcement. *J. Compos. Commun.* **2024**, *45*, 101794. [CrossRef]
43. Suzhou Zhongbao Composite Material Co., Ltd., Suzhou, China. Available online: http://www.szzbmf.com (accessed on 6 June 2023).
44. Lutai Co., Ltd., Suzhou, China. Available online: http://www.lttc.com.cn (accessed on 6 June 2023).
45. Kaboglu, C.; Yu, L.; Mohagheghian, I.; Blackman, B.R.; Kinloch, A.J.; Dear, J.P. Effects of the core density on the quasi-static flexural and ballistic performance of fibre-composite skin/foam-core sandwich structures. *J. Mater. Sci.* **2018**, *53*, 16393–16414. [CrossRef]
46. Yuan, H.; Zhang, J.; Sun, H. The failure behavior of double-layer metal foam sandwich beams under three-point bending. *Thin-Walled Struct.* **2022**, *180*, 109801. [CrossRef]
47. Zhao, Y.; Yang, Z.; Yu, T.; Xin, D. Mechanical properties and energy absorption capabilities of aluminium foam sandwich structure subjected to low-velocity impact. *Constr. Build. Mater.* **2021**, *273*, 121996. [CrossRef]
48. Shunmugasamy, V.C.; Mansoor, B. Aluminum foam sandwich with density-graded open-cell core: Compressive and flexural response. *Mater. Sci. Eng. A* **2018**, *731*, 220–230. [CrossRef]

materials

MDPI

Review

Auxetics and FEA: Modern Materials Driven by Modern Simulation Methods

Russell Galea Mifsud [1], Grace Anne Muscat [1], James N. Grima-Cornish [1], Krzysztof K. Dudek [2], Maria A. Cardona [1], Daphne Attard [1], Pierre-Sandre Farrugia [1], Ruben Gatt [1,3], Kenneth E. Evans [4] and Joseph N. Grima [1,5,*]

1 Metamaterials Unit, Faculty of Science, University of Malta, MSD 2080 Msida, Malta; russell.galea.15@um.edu.mt (R.G.M.); grace.muscat.16@um.edu.mt (G.A.M.); james.grima.16@um.edu.mt (J.N.G.-C.); maria.cardona@um.edu.mt (M.A.C.); daphne.attard@um.edu.mt (D.A.); pierre-sandre.farrugia@um.edu.mt (P.-S.F.); ruben.gatt@um.edu.mt (R.G.)
2 Institute of Physics, University of Zielona Gora, ul. Szafrana 4a, 65-069 Zielona Gora, Poland; k.dudek@if.uz.zgora.pl
3 Centre for Molecular Medicine and Biobanking, University of Malta, MSD 2080 Msida, Malta
4 Department of Engineering, Faculty of Environment, Science and Economy, University of Exeter, North Park Road, Exeter EX4 4QF, UK; k.e.evans@exeter.ac.uk
5 Department of Chemistry, University of Malta, MSD 2080 Msida, Malta
* Correspondence: joseph.grima@um.edu.mt

Abstract: Auxetics are materials, metamaterials or structures which expand laterally in at least one cross-sectional plane when uniaxially stretched, that is, have a negative Poisson's ratio. Over these last decades, these systems have been studied through various methods, including simulations through finite elements analysis (FEA). This simulation tool is playing an increasingly significant role in the study of materials and structures as a result of the availability of more advanced and user-friendly commercially available software and higher computational power at more reachable costs. This review shows how, in the last three decades, FEA proved to be an essential key tool for studying auxetics, their properties, potential uses and applications. It focuses on the use of FEA in recent years for the design and optimisation of auxetic systems, for the simulation of how they behave when subjected to uniaxial stretching or compression, typically with a focus on identifying the deformation mechanism which leads to auxetic behaviour, and/or, for the simulation of their characteristics and behaviour under different circumstances such as impacts.

Keywords: auxetics; negative Poisson's ratio; metamaterials; finite elements

1. Introduction

The last three decades have seen an extensive growth in our knowledge about various new materials and metamaterials, more known as auxetics [1], which exhibit the rather unusual mechanical property of a negative Poisson's ratio [2,3]. Auxetic materials defy common expectations by expanding laterally when uniaxially stretched, rather than becoming thinner. Conversely, these materials become thinner, rather than thicker, when uniaxially compressed. In fact, the Poisson's ratio measures the magnitude of this effect [4] and is defined for a particular cross-section of a material in the plane Ox_i-Ox_j in terms of the applied strain ε_i in the direction Ox_i and the transverse strain ε_j in the orthogonal transverse direction Ox_j as

$$\nu_{ij} = -\frac{\varepsilon_j}{\varepsilon_i} \tag{1}$$

In general, for an isotropic three-dimensional (3D) material, the Poisson's ratio can assume any positive or negative value. However, for isotropic materials, this Poisson's ratio is restricted to have values $-1 \leq \nu \leq +0.5$ ($-1 \leq \nu \leq +1$ for two-dimensional (2D) orthotropic) [5]. Materials which are auxetic in all planes for loading in any direction have been termed as "complete auxetics" whilst materials which are only auxetic in specific planes for loading in specific directions (common in crystalline materials and cubic metals [6,7]) have been termed "partially auxetic" [8–10].

Auxetics have associated with them a number of benefits and enhanced properties that are not that commonly encountered in most everyday materials [11]. They are highly desirable for their exceptional properties, including high indentation resistance [12–20], high fracture toughness [20–23], shear resistance [18,24,25], energy absorption [26,27] and other enhanced dynamic characteristics [28–31]. These unique attributes enable auxetics to find diverse applications across various fields such as in the medical field [32–37], sportswear applications [38–42], military defence equipment [43–46], as well as in the automotive [47–49] and aerospace [50–52] industries. For example, personal protective equipment (PPE) would benefit from being resistant to indentations. However, traditional materials tend to displace material away from the point of impact, resulting in reduced density upon impact. Auxetics, upon impact, behave in a manner that the material flows towards the point of impact, increasing their density and resisting indentation.

As a result of these enhanced properties and possible applications, scientists have availed themselves of various research tools to characterize, optimize and design de novo metamaterials which exhibit auxeticity. In this field of research, pure experimental research is not always deemed feasible. This may be attributed, at least in part, to the complex geometries that auxetic structures have, the lack of availability of "off the shelf" auxetic materials on a large scale, issues relating to sustainability, practical drawbacks of physical prototyping (prototypes are expensive, unsustainable, difficult to manufacture, and just a general burden to test) as well as ethical issues which may arise in certain experimental testing in real applications where auxetics could be truly needed (e.g., stents [53,54]). As a consequence, researchers have resorted to rapidly developing research protocols based on computer simulation and numerical analysis, often using experimental results and analytical studies to verify the numerical results obtained [9,55–69].

One of the numerical approaches which researchers have used to make significant inroads in the field of auxetics is the finite element method (FEM), a technique that provides an approximate solution indicating how a system behaves when subjected to specific physical constraints. The method provides easy solutions when dealing with problems involving complex geometries and/or nonhomogeneous domains having different properties in different regions. It involves the subdivision of the original domain into smaller parts called elements. The subdomains are easier to treat as they can be chosen so that their governing equations are much simpler than those for the whole domain. These elements are then connected through the nodes that they have in common. Nodes are locations in space. In general, in the case of one-dimensional (1D) elements, they would represent points on a line, while if the elements are 2D or 3D, they represent corners [70]. The shape of the elements is then determined by joining the nodes via straight lines. (Here it should be noted that sometimes elements might also have internal nodes.) Once divided into these subregions, the network of nodes is referred to as a mesh or grid. FEM then involves solving the governing equations locally at the nodes to determine the physical variables at these points. Putting together the local solutions provides a piecewise solution that represents an approximation to the actual one for the whole domain [71]. The accuracy of the solution depends on the size of the elements and increases with the number of nodes used. At the same time, the computational time increases with the number of nodes. Hence, a balance needs to be found between accuracy and processing time. Most frequently, this is established by requiring that the solution that is used does not vary by more than a small percentage (many times taken to be 1%) from that obtained using more nodes. At this point, the solution is considered to be mesh-independent. Once the FEM is applied and the solution obtained, the subsequent analysis is referred to as the finite element analysis (FEA). This method dates back to the 1940s, having been developed to overcome the mathematical difficulties when applying the theory of elasticity [72,73]. In 1956, FEA was adopted by the aerospace industry as it allowed the modelling of complex geometries and provided instant analysis of their mechanical properties [74]. The method provided a solution to many problems of material analysis by calculating the stress within a structure [75]. The FEA method is well-known for its reliability in determining the location, magnitude, and direction of forces, as well as in assigning stresses and deformations. Therefore, within a few decades, the method was adopted by several research fields, ranging from dentistry [76–80] to biomechanics [81–84] amongst others [85–88].

A major advantage of using FEA as an integral part of a research protocol to study auxetics during the design stages is that it reduces the need for prototypes, by providing a quick, non-invasive and repeatable analysis [89–91]. Moreover, apart from the obvious sustainability advantages FEA brings with it when used for the optimization of material design whilst reducing experimental waste, FEA offers the advantage that it is an excellent tool to simulate and animate the "deformation mechanism" of auxetics, thereby providing researchers with essential information on how the deformations are leading to auxetic behaviour. This also applies to more complex systems such as 3D negative Poisson's ratio mechanical metamaterials, that are not only usually elastically stable under large compressive deformations but are also capable of exhibiting diverse properties by changing the feature size, making them potential candidates for both functional and structural applications [92]. This statement was equally relevant before the advent of 3D printing, as it is now when the current expectations regarding the performance capabilities of materials, metamaterials and structures have grown tremendously. Moreover, FEA offers the advantage that it is not limited to the conventional test parameters and can accurately and reliably simulate real-life scenarios even with unconventional parameters, such as an isotropic negative Poisson's ratio. This is particularly useful in the field of auxetics to study the behaviour of these unconventional materials in practical applications (e.g., shearing [93] or pressing [13]). Despite the lack of real materials with such properties, FEA enables us to anticipate their behaviour. Also, FEA can be used to investigate scenarios which would be difficult to achieve physically. One such study examined re-entrant hexagonal honeycombs and their post-yield behaviour under tension. Through FEA, the study revealed a plastic collapse mechanism and identified three stages of force–displacement curves [64].

FEA has been utilised in some of the earliest seminal papers on negative Poisson's ratio by Evans and his group who used the software ANSYS (Ansys Inc., Houston, TX, USA) to model two-dimensional re-entrant honeycombs as an embedded fibre in composites [94] (Figure 1a) and as a template for molecular level systems [95] (Figure 1b) to model 3D auxetic foams [96], to study continuous fibre-reinforced composites where either reinforcing or matrix constituents could have a negative Poisson's ratio [97] and to model auxetic microporous polymers [98]. Ole Sigmund [99] used a topology optimization protocol and a method based on a finite-element discretization of the base cells to propose and analyse various classical auxetic motifs. These included the double-V re-entrant system [100] and another motif [101], later referred to as the "anti-tetrachiral" motif [102–104] (Figure 1c). Nowadays, much more developed commercially available software such as ANSYS (Ansys Inc., Houston, TX, USA), MSC Marc (Hexagon AB, Newport Beach, CA, USA) and ABAQUS (Dassault Systèmes Simulia Corp., Johnston, RI, USA) combined with high computational power are essential tools when conducting research in the mechanical properties of materials and has, thus, been utilized by several research groups, as noted in Appendix A, in the study of auxetic metamaterials [63,105–113]. It is also most useful that various textbooks have been written to guide researchers through the background theory and application of FEA [70,71,114].

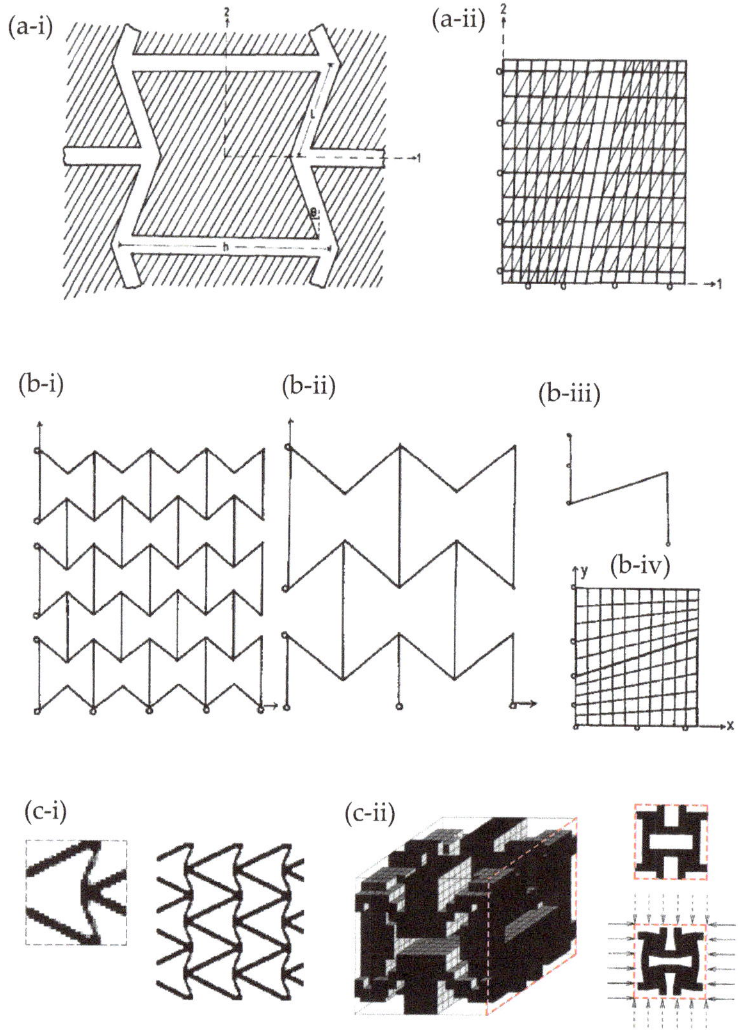

Figure 1. Some of the earliest uses of FEA to model auxetics: (**a**) Evans et al.'s (1992) matrix embedded re-entrant network-embedded fibre composite [94] with (**a-i**) showing the unit cell and (**a-ii**) showing a finite element grid used a typical re-entrant network composite with fibre re-enforcement being unshaded and matrix shaded; (**b**) various FEA representations of the analogues for the molecular level systems [95] where (**b-i**,**b-ii**,**b-iii**) correspond to systems with a different number of representative repeat units constructed from beam elements whilst (**b-iv**) represents a system where the honeycomb is embedded within an ultra-soft matrix having a near zero Young's modulus; (**c**) examples of the auxetics motifs generated by topology optimization by Sigmund, where (**c-i**) shows the double-V auxetic motif [99] and (**c-ii**) shows the motif which is now referred to as the anti-tetrachiral motif [101].

Recognizing the growth and development of auxetics, concurrently with finite element analysis, this review examines some of the major discoveries in the field of auxetics which have been made through FEA, and shows how this method of analysis contributes to the shaping of our current knowledge about negative Poisson's ratio.

2. Simulation of Auxetic Structures

The diverse range of enhanced properties associated with auxetics, together with their potential for use in several practical applications meant that, in the last four decades, considerable effort has been made to design and develop novel or improved auxetic structures. In fact, as clearly stated by [115], there is a constant "demand for new types of auxetic structures to achieve different design goals of their use (dynamic behaviour, fatigue resistance, manufacturability)".

If one had to look at the modus operandi which has been used to generate and optimise auxetics, one can clearly identify various approaches, including what may be referred to as the "systematic geometry-based approach" and the "topology optimisation approach". The former approach may be explained through the specific example of the 2D re-entrant geometry motif shown in Figure 1a, which has long been recognised for its potential as an auxetic system [1,116,117]. The systematic approach typically starts with the identification of geometric parameters which characterise the system, in this case, the length of the slanting ligament (l), the length of the ligaments connecting them (h) and the angle between them (θ). It then examines how each of these parameters, and/or combinations of them, affects Poisson's ratio. This can be achieved through, for example, the formulation of analytical models based on assumptions of how the systems deform (e.g., flexure, hinging and/or stretching of the ligaments) and/or running a series of FEA simulations where each parameter is varied whilst keeping the other parameters fixed. Both the analytical and simulation approaches offer distinct advantages and disadvantages. Analytical models lead to mathematical expressions for Poisson's ratio in terms of geometric parameters. On the other hand, the systematic FEA approach allows for a more realistic representation of systems, as discussed elsewhere [118]. The topology optimisation approach, first applied and pioneered in the field of auxetics in the 1990s by Sigmund and co-workers [100,119,120], has the capability of automatically achieving an optimal structure and material layout (geometry) with a negative Poisson's ratio subject to specific pre-defined constraints through the use of advanced algorithms. [115,121] Such constraints include the desired macroscopic properties and prescribed performance. For example, this approach can help identify optimal structural configurations that maximize the desired properties of auxetic materials, such as impact resistance or energy absorption, while maintaining or reducing material weight. Typically, the optimised structures obtained have their properties evaluated through FEA before potentially being manufactured as prototypes for experimental testing [122,123]. Various studies have applied the technique within the field of auxetics, including a study by Clausen et al. [124] who employed parameterized optimization to produce auxetic chiral topologies which maintain their desired Poisson's ratio over strains; a study by Bruggi et al. [125] who made use of SIMP-based topology optimization (i.e., solid isotropic material with penalization) to obtain auxetics based on micropolar materials; and a study by Wang et al. [126] who applied isogeometric topology optimization with the aim of reducing stress concentrations within star-shaped auxetic structures.

2.1. Two Dimensional Systems

A substantial number of studies on auxeticity were devoted to investigating the mechanical properties of two-dimensional motifs that are capable of exhibiting a negative Poisson's ratio. This focus on 2D systems is probably due to the fact that the Poisson's ratio ν_{ij} can essentially be regarded as a 2D property.

The re-entrant honeycomb geometry, shown in Figure 1, is one of the most studied auxetic motifs. This structure essentially represents a standard hexagonal honeycomb where the classical Y-shaped joints have been inverted to form arrow-shaped joints. Such systems are known to exhibit a negative Poisson's ratio when loaded on-axis, primarily due to flexure (or hinging) of the slanting ligaments. This motif, equally useful as a 2D system and as a model for particular cross-sections of 3D materials, was one of the earliest auxetic systems studied by FEA around three decades ago [94–96] (Figure 1a,b). Since then, numerous modified re-entrant structures were tested through FEA to diversify and/or optimise the geometry [127–129] and to analyse the resulting mechanical properties such as energy absorption capacity [127], synclastic behaviour [130,131] and impact resistance [132–134].

Numerous other studies have utilised FEA to investigate several variations based on the 2D re-entrant hexagonal honeycomb systems and Sigmund's double-V model (Figure 1c-i) to simulate their mechanical properties, including in-plane Poisson's ratio and Young's modulus under uniaxial loading [135–139]. Through FEA simulations, several studies were carried out on the shape optimization of the re-entrant honeycombs by analysing the effect of a number of variables on the physical properties. For instance, Lu et al. in 2016 [140] analysed a re-entrant honeycomb with an additional narrow rib in the unit cell, resulting in a cellular structure with significantly enhanced Young's modulus [140]. Bezazi et al. in 2005 [135], and the subsequent work by Harkati et al. in 2017 [128], investigated the system illustrated in Figure 2b. Most notably, Bezazi et al. reported that for certain internal angles, the proposed structure exhibits a decrease in the Poisson's ratio when compared to the conventional honeycomb. Also, the presence of edge corners in the proposed configuration gives rise to a cellular structure with enhanced flexibility compared to the classical centrosymmetric one [135]. In the follow-up work, Harkati et al. [128] investigated the shear and axial deformation using FEA, providing a better insight into the deformation mechanism for the auxetic behaviour and the geometric parameters governing them, specifically the cell wall thickness [128]. Moreover, Gohar et al. [127] used FEA to optimize and model various novel auxetic structures under compressive loading (Figure 2c), systems which they then constructed using 3D printing and tested physically. These included a set of novel re-entrant structures, termed mixed-star structures. These re-entrant structures exhibited a number of superior properties, including a high energy absorption capacity [127].

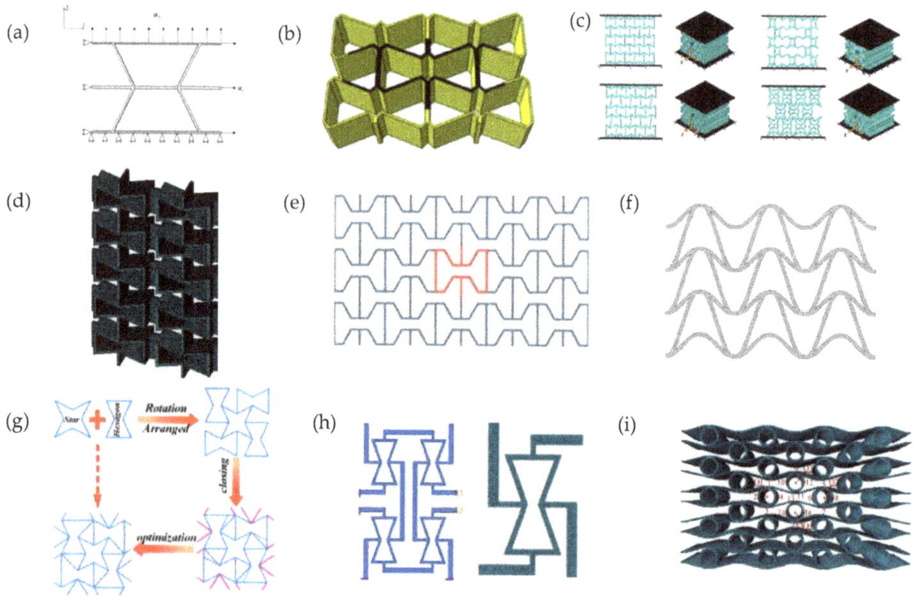

Figure 2. Examples of modified re-entrant structures, which can be seen as variations in the units in Figure 1, proposed by (**a**) Lu et al. (2016) [140]; (**b**) Bezazi et al. (2005) [135] and Harkati et al. (2017) [128]; (**c**) Gohar et al. (2021) [127]; (**d**) Huang et al. (2017) [141]; (**e**) Khan et al. (2019) [59] and Mustahsan et al. (2022) [142]; (**f**) Guo et al. (2020) [143]; (**g**) Wang et al. (2023) [144]; and (**h**) Zhang et al. (2021) [145] and (**i**) Li et al. (2022) [129].

Similarly, Huang et al. [141] investigated a new type of honeycomb design consisting of two distinct parts, a re-entrant hexagonal component, and a thin plate section, illustrated in Figure 2d. The authors developed theoretical models describing the in-plane uniaxial tensile modulus, shear modulus and Poisson's ratios, and verified them through FEA [141]. FEA was also instrumental in validating analytical and experimental methods, as demonstrated in the work by Khan et al. [59] and Mustahsan et al. [142] (Figure 2e). These used FEA in conjunction with experimental testing to validate the analytical model accounting for the bending, shearing and axial deformation of modified re-entrant honeycomb structures and showed a close agreement between the three methods (numerical, analytical and experimental). Other notable studies include the investigations by Guo et al. (2020) [143] on the double-U honeycomb structure shown in Figure 2f, in which the arrow-shaped system in Figure 1c-i proposed in the 1990s by Ole Sigmund [99] and Larsen et al. [100] was modified to have better load resistance and a higher energy absorbing capacity [143], and the WSH honeycombs composed of stars and hexagons studied by Wang et al. (2023) (Figure 2g) which exhibited excellent energy absorption capacity and enhanced anti-impact behaviour [144].

Another study carried out by Zhang et al. investigated two new hybrid metamaterial concepts combining a core unit cell of re-entrant or cross-chiral shape with lateral missing ribs (Figure 2h). FEA simulations optimised specific effective properties, while non-linear simulations were used to study the Poisson's ratio and stiffness of these metamaterials under large deformations [145]. Similarly, Li et al. proposed the composite auxetic structure consisting of corrugated sheets and tubes, shown in Figure 2i. This was studied using FEA, in conjunction with experiments and theoretical analysis, to investigate the variables affecting the Poisson's ratio and the mechanical properties of the structure, as well as shape optimization [129]. Other rather complex 2D systems studied include the hierarchical re-entrant honeycombs by Zhan et al. (2022), who used FEA to show that these systems exhibit enhanced mechanical properties under compression. Here, analysis of the deformation shows that when the system is compressed, the addition of a second order triangular hierarchy converts the deformation mechanism from bending-dominated to stretching-dominated, and revealed a combination of deformation mechanisms which contributed to significant improvement in the mechanical properties [146].

The FEA method has been employed to explore various other auxetic mechanisms including chiral systems and rotating rigid units. In particular, inspired by work the of Lakes [147,148] and Sigmund et al. [101], FEA was a key tool in studying 2D periodic systems characterised by highly ordered chiral sub-units [102–104,129,149–157] (such as the ones in Figure 3a,c), as well as systems with disordered or irregular chirals [158,159] (such as the system in Figure 3c).

Here, FEA was used to carry out parametric studies to optimise the geometries of the proposed chiral systems and to investigate their mechanical properties such as auxeticity, energy absorption capacity, shear resistance and much more. The unifying design aspect of these chiral systems is that they generally consist of rigid nodes to which thin flexible ligaments are attached in a manner that they form a chiral building block. As discussed by Alderson et al. [104], to model this system through a representative volume unit, it is essential to apply the appropriate boundary conditions and constraints (see Figure 3a-iii,a-iv as an illustration of typical constraints applied). Furthermore, as highlighted by Mizzi et al. [160], the correct application of the boundary conditions often dictates the success or otherwise of a simulation. Simulations of systems with a chiral building block such as the ones shown in Figure 3a–c demonstrate that auxeticity from the "chiral mechanism" occurs when the ligaments are flexible and the nodes are rigid. Generally, when the system is subjected to uniaxial compression, the nodes with the ligaments attached to it rotate, thus, constraining the ligaments to flex in synchrony (to some extent or another). This synchronized mode of deformation causes a lateral contraction which in turn results in auxeticity, something which is even evident in the irregular hexachirals simulated by FEA by Mizzi et al. (2018). [159] However, when the nodes are much less rigid than the ligaments, deformations occur predominantly in the nodes, as demonstrated by Attard et al. (see Figure 3c-ii) [150]. The authors referred to this mechanism as the "starchirals mechanism".

FEA was also one of the main tools used to explore how slits and perforations could generate a negative Poisson's ratio [161–167]. Notably, early studies on this idea (Figure 4a–c) highlight how strategically placed perforations could transform a regular sheet to an auxetic or zero-Poison's ratio system by making it mimic the "rotating squares" [168] and "rotating triangles" [169] auxetic mechanisms. This could be achieved when using appropriately placed diamond-shaped (Figure 4a) [161], slit [162] (Figure 4d), star-shaped [164] (Figure 4b) or triangular-shaped [164] (Figure 4c) perforations. Other similar studies investigated the mechanical properties and deformation mechanisms of "rotating rigid units"-mimicking systems. Amongst others, Wang et al. (2021) [170] and Atilla Yolcu et al. (2022) [171] explored such systems designed through the use of regularly and irregularly peanut-shaped perforations (depicted in Figure 4e); Acuna et al. [172] simulated rectangular perforations; whilst Mrozek and Strek [173] (Figure 4f) investigated a system of perforations mimicking the rotating squares model [64,174–176]. Afshar et al. utilised FEA to investigate non-porous perforated rotating rigid units having a soft inclusion in the perforations. It was shown that these inclusions still retained a negative Poisson's ratio but reduced the extent of auxeticity. This would be useful in applications of non-porous auxetic materials [105]. Similar conclusions were also drawn in an earlier FEA study by Mizzi et al. (2015) [177] which examined non-porous grooved single-material systems.

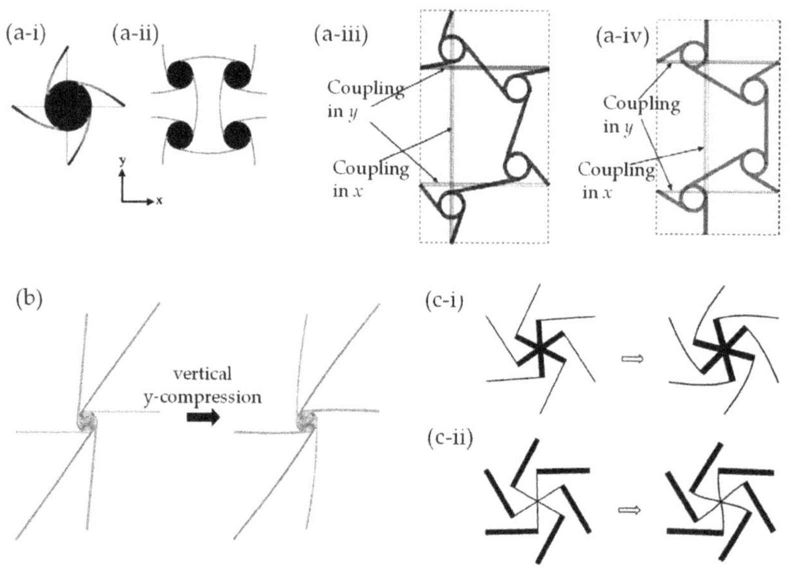

Figure 3. Examples of auxetics with chiral building blocks or based on rotating rigid units obtained via slits or perforations: (**a**) the repeat units in the regular (**a-i**) tetrachiral, (**a-ii**) anti-tetrachiral (**a-iii**) trichiral and (**a-iv**) anti-trichiral, as presented in the work by Alderson et al. (2010) [104]; (**b**) the repeat unit and the manner it deforms as predicted by FEA of an irregular hexachiral [159]; (**c**) FEA-simulated deformations of uniaxial compression in the vertical y-direction of (**c-i**) hexachirals, where the star-shaped node is much more rigid than the ligaments, and (**c-ii**) starchirals, where the ligaments are much more rigid than the nodes with the consequence that deformations occur within the nodes (the "starchiral mechanism") [150].

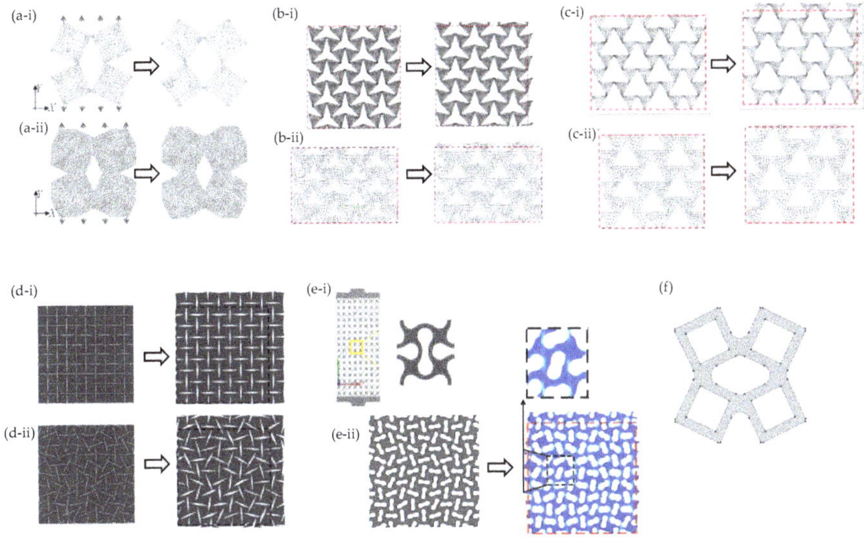

Figure 4. Examples of perforated systems which mimic rotating rigid units: (**a**) systems with diamond-shaped perforations where system in (**a-i**) has minimal overlap compared to system in (**a-ii**); (**b**) systems with star-shaped perforations where system in (**b-i**) has minimal overlap compared to system in (**b-ii**); (**c**) systems with triangle-shaped perforations, where system in (**c-i**) has minimal overlap compared to system in (**c-ii**); (**d**) systems with slits (regular (**d-i**) and randomly oriented (**d-ii**)); (**e**) system with (**e-i**) regularly and (**e-ii**) irregularly placed peanut-shaped perforations; and (**f**) implementation of a rotating squares model as proposed by Mrozek and Strek et al. (2022) [173]. Systems in (**a**,**d**,**e**,**f**) mimic the rotating squares model, whilst (**b**,**c**) mimic the rotating triangles model. Note that, if applicable, systems are being loaded vertically.

2.2. Three Dimensional Systems

3D auxetic materials and models, ranging from crystalline materials such as zeolites, silicates, etc. [178–180] to macroscopic systems such as the Hoberman sphere [181] are gaining importance due to their abundance, versatility, properties and potential applications. However, their added complexity compared to their 2D counterparts, can make the interpretation of their Poisson's ratio more challenging. In 3D auxetics, the overall deformations might result from multiple auxetic mechanisms acting simultaneously, as evident in some of the earliest seminal work in the field (Figure 1c-ii) [101]. FEA, thus, becomes an even more indispensable tool to elucidate their intricate behaviour. In fact, FEA has been extensively used to investigate a wide range of 3D auxetics including cellular materials [129,171,182], composites [183–187] and some rather novel constructs such as a surface auxetic structure (SAS) [188].

FEA has offered diverse insights into the effects of different geometric parameters on the mechanical properties and deformation mechanisms in 3D systems from as early as the 1990s. The diversity of systems studied can be appreciated from some cellular systems shown in Figure 5. Evans et al. [96] used the elongated dodecahedron as a model for auxetic foams (Figure 5a). More recently, Yang et al. [189] and Wang et. al. [190,191] studied a 3D re-entrant honeycomb (Figure 5b); Nasim and Etemadi [192] proposed a cellular structure (Figure 5c); Farrugia et al. (2019) [193] explored a novel 3D anti-tetrachiral honeycomb Figure 5d; and Wang et al. (2020) [194] investigated "3D cross-chiral auxetic materials", a system where some of its 2D projections bear a very close resemblance to the "rotating squares" [168] profile with the squares replaced by crosses [195] (Figure 5e). Later, in 2021, Photiou et al. worked on the so-called "tetra-petal auxetic [196] (Figure 5f), while in 2022, Grima-Cornish et al. [197] examined the crystalline material framework of boron arsenate as if it were a purely mechanical system (Figure 5g). Recently, it has also been proposed that 3D auxetic systems may be constructed by introducing continuous voids having constant cross-sectional areas at particular loci in different planes [198]. Collectively, these works highlight how FEA simulations have become the staple tool to bridge design and mathematical modelling with experimental work and physical testing. Moreover, for a given idealised design or concept, FEA has made it possible to obtain a glimpse into how real systems might behave, thereby guiding experimentalists towards the synthesis of systems with tailor-made auxetic properties. All this contributed significantly to the growth of the field of auxetics over the last decades. While most 3D cellular systems exhibit their auxetic behaviour through "re-entrant" or "rotating rigid units" (or its chiral variant) mechanisms, Su et. al. investigated a 3D-printable auxetic metamaterial operating through a sliding mechanism (Figure 5h). Their experimental and FEA simulations produced coherent results, and the proposed structure was claimed to exhibit superior performance to the 3D re-entrant honeycomb, due to higher compression resistance and more stable auxetic behaviour [199].

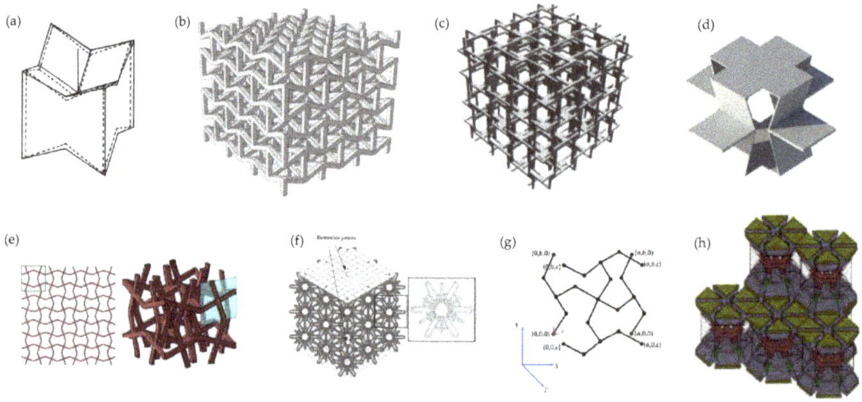

Figure 5. Examples of 3D cellular structures studied via FEA in the last three decades: (**a**) work by Evans, Nkansah and Hutchinson, 1994, on a cellular systems to simulate the microstructure of auxetic foams [96]; (**b**) work by Wang et al., 2017, on a re-entrant version of the classical re-entrant honeycomb [191]; (**c**) another 3D version of the re-entrant honeycomb as studied by Nasim and Etemadi, 2018 [192]; (**d**) the 3D chiral cellular structure as proposed and studied by Farrugia, Gatt and Grima, 2019 [193]; (**e**) a 3D cellular system as studied by Wang et al., 2020, which can be seen as a 3D render on the "rotating squares" [194]; (**f**) a 3D cellular structure proposed, modelled and tested experimentally by Photiou et al., 2021 [196]; (**g**) a mechanical version of a crystalline system modelled by Grima Cornish et al., 2022 [197]; and (**h**) a sliding system investigated by Su et. al. (2021) [199].

From a design and modelling perspective, the possibility to make use of the third dimension brings with it some very interesting possibilities on how to achieve auxeticity. FEA facilitates the transition from concept to an actual physical model. Thus, tubular structures were made by morphing novel or existing 2D auxetics into the shape of a tube for potential use as biomedical devices like artery stents. Wu et al. (2018) [62] (Figure 6a-i) followed the design principles proposed by Gatt et al. [200] (Figure 6a-ii) to explore the mechanical properties of the proposed artery stents. Understanding such properties is crucial for the mechanical integrity and biomechanical performance reliability of the stent–plaque–artery system. Wu et al. proposed two innovative chiral stent types with auxetic properties: an anti-tetrachiral stent with circular and elliptical nodes, and a hierarchical anti-tetrachiral stent with circular and elliptical nodes (Figure 6a). FEA was employed to study the effects of stent geometrical parameters on the tensile mechanical behaviour of the proposed stents. It was deduced that the proposed anti-tetrachiral stent can be tailored by adjusting the levels of hierarchical structures and unit cell design parameters. FEA was used to study the deformation mechanism during stenting. The proposed structures exhibited remarkable radial expanding abilities while maintaining axial stability, which is ideal for such applications [62].

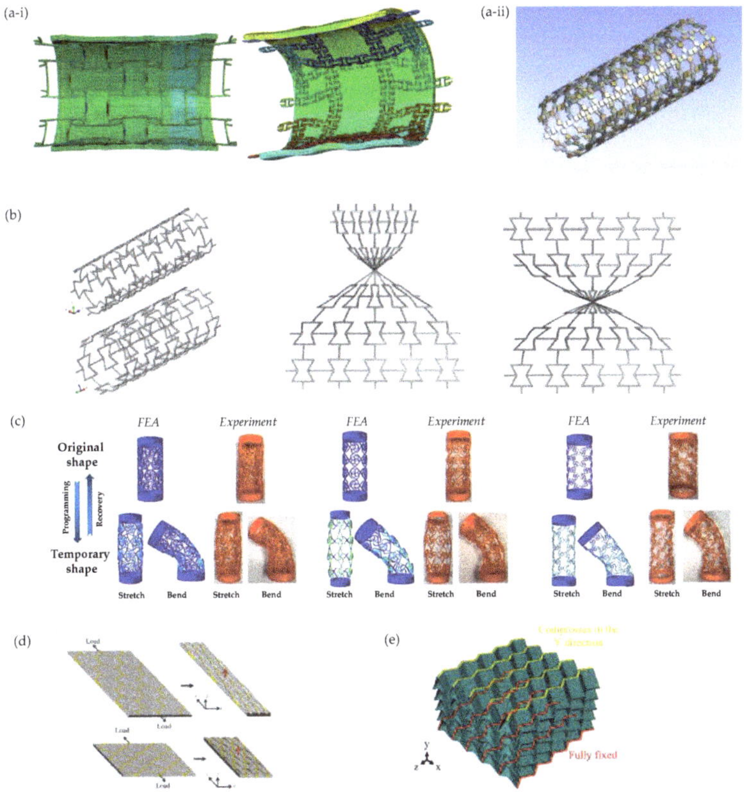

Figure 6. Other examples of 3D auxetics studied via FEA: (**a**) simulations by Wu et al. [62] of the mechanical properties of (**a-i**) regular and hierarchical anti-tetrachiral auxetic stents, the latter being based on the design concept by Gatt et al. [200] shown in (**a-ii**); (**b**) 2022 work by Changfang et al. [188] on surface auxetic structures; (**c**) 2022 work by Wan et al. [201] on programmable auxetic metamaterials with shape memory effects; (**d**) honeycomb composites with auxetic out-of-plane characteristics as proposed and modelled by Grima et al. [183]; and (**e**) 2023 work by Li et al. [187] on the auxetic and failure characteristics of composite stacked origami cellular materials under compression.

Other studies of this type include work on the surface auxetic structures (SAS) studied by Changfang et al. (2022) [188] (Figure 6b). These structures are obtained by morphing 2D auxetic surfaces into 3D shells. FEA was employed to study two types of SAS: RAS (reversed auxetic structure) and CAS (crimped auxetic structure). The focus was on simulating the compressive behaviour of both plane and surface auxetic structures to understand their mechanical behaviour and energy absorption characteristics. Through this study, it was deduced that RAS displayed the auxetic effect of compression shrinkage as well as the super-mechanical effect of compression twist. Such behaviour was found to appear only in the local positions of the beams, giving the structure great potential in engineering applications [188]. Recent work by Wan et al. delved into four-dimensional (4D) printed programmable auxetic metamaterials with shape memory effects, using both FEA (see Figure 6c) and experimental approaches [201]. The analysis revealed that the proposed cylindrical shells possess desired mechanical properties and configurations, indicating potential applications as biomedical scaffolds [201].

FEA has also proven particularly valuable when analysing auxetic composite structures. This is evident from the early contributions to the field by Evans et al. [94,97]. Grima and co-workers [183] used FEA simulations to demonstrate how an auxetic 3D composite consisting of metal honeycombs embedded in a soft rubbery matrix could exhibit a negative Poisson's ratio out-of-plane by forcing the softer matrix to move out of the pores whilst the honeycomb is stretched (Figure 6a). Through FEA, researchers could investigate the effects of changes in framework geometry in relation to changes in the Poisson's ratio both in-plane and out-of-plane. This enabled the optimisation of composite parameters for maximum auxetic behaviour [183]. Additionally, these simulations led to an analytical study outlining the requirements for auxetic behaviour.

Recent work on composites also includes that of Li et. al. (2023) [187] who used FEA to investigate the mechanical and auxetic characteristics of a composite fibre-reinforced stacked origami structure (Figure 6d). It was deduced that composite stacked origami structures have lower density and better energy absorption characteristics compared to those made from metal using additive manufacturing processing [187]. Another interesting work is that on cementitious composites [184], where FEA-based machine learning was used to generate accurate predictions of the auxetic behaviour in cementitious composites [184].

3. Simulation of Responses on Auxetics under Dynamic, Quasi-Static Loading, Impact and Indent Loading

The use of FEA as a tool to simulate the responses of auxetic materials and structures has been pivotal, primarily due to its efficiency and time/cost-effectiveness compared to actual physical testing as it reduces the need for production and testing of experimental prototypes [192,202]. FEA's efficiency lies in its ability to accurately acquire the desired mechanical responses and to analyse the behaviour of different sections of the material in a manner which is sometimes difficult to perform experimentally in a non-destructive manner. This is crucial when assessing the response of the material to applied loads and other desirable properties as required by the intended application.

The properties of auxetics are inherently linked to their geometry (including topology) and deformation mechanism (which may be controlled through a number of ways, such as materials contrast [203]). To achieve the desired properties, researchers often investigate the effect of the geometric parameters on the mechanical properties and behaviour of auxetics under specific conditions. Recently, there has been growing interest in the mechanical properties and energy absorption capacity of auxetic materials under dynamic and quasi-static loading. One of the key challenges in this area is to develop a deeper understanding of the factors that influence these properties under different loading conditions and the deformation modes of specific auxetic structures, challenges which may be successfully addressed using FEA.

In a 2022 study, Han et al. [204] used FEA to investigate the mechanical properties and deformation modes of gradient and uniform auxetic tubes subjected to axial and inclined loads (Figure 7a). Their findings revealed that the gradient auxetic tube had better energy absorption capacity and higher strength compared to the uniform auxetic tube [204]. Similarly, in the same year, Han, Ren et al. used FEA to investigate a design for ribbed metamaterials with high-quality energy-absorption capability speeds [205]. In a more recent study, Wang et al. [144] conducted numerical simulations to study the static and dynamic plateau stresses of windmill-like (WSH) honeycombs and their energy absorption capacity (Figure 7b). Comparing these with the STAR-4 auxetic honeycomb and the standard re-entrant honeycombs, they found that the windmill-like (WSH) honeycombs had excellent energy absorption performance under both static and dynamic loading conditions [144]. In the study by Chen et al., a set of auxetic lattices with enhanced stiffness were proposed. This was achieved by adding a strengthening rib into conventional auxetic unit cells in a direction perpendicular to the re-entrant direction. The effective mechanical properties of these variants were calculated using the fast Fourier transform-based homogenization method, which illustrated that their Young's modulus in 2D can be improved by approximately 200% along the strengthening direction without significantly sacrificing the auxetic characteristics. However, such an enhancement is weakened in 3D. The paper provides insight into the design of the new structures of unit cells with enhanced stiffness and a negative Poisson's ratio [206].

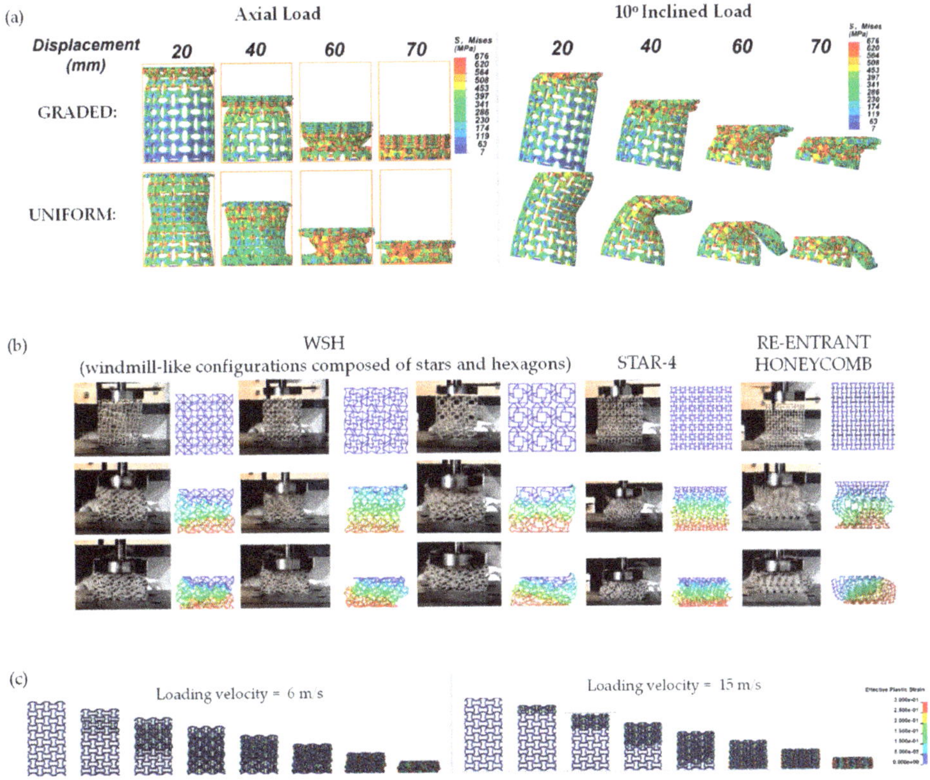

Figure 7. Examples of dynamic simulations: (**a**) deformation modes of a gradient auxetic tube and uniform auxetic tube under axial and inclined loads for 3D auxetics studied via FEA by Han et al.'s 2022 work [204]; (**b**) work by Wang et al. (2023) showing a comparison of experimental and FEA deformation diagrams under dynamic impact behaviour with an impact velocity of 1.7 m/s [144]; and (**c**) uniaxial compression at different velocities of the regular chiral as simulated by Novak et al. [207].

Novak et al. [18,207] explored the mechanical behaviour of chiral auxetic [207] (see Figure 7c for the results of the simulation of a regular chiral auxetic under dynamic uniaxial compression at different velocities) and graded/non-graded [18] cellular structures under different loading conditions. Their investigation included quasi-static low-velocity dynamic compression and high-strain rate loading and shearing scenarios. The study employed experimental measurements, infrared thermography, high-speed camera images and computational simulations to examine the deformation mechanism of chiral auxetic structures. Computational simulations were used to obtain a more detailed analysis of mechanical behaviour at different strain rates and estimations of plateau stress at arbitrary loading velocities. As a result of this analysis, Novak et al.'s work provides insights into the use of chiral auxetic structures in crashworthiness, ballistics and blast protection applications [18,207].

Other noteworthy studies that have made use of dynamic and quasi-static compression, as well as different impact velocities, include several modified re-entrant diamond structures which exhibited a superior specific energy absorption [208,209]. Similarly, investigations of modified re-entrant honeycombs exhibited potential for crashworthiness applications [210,211]. Additionally, studies examining the energy absorption capacity of star-circle honeycomb structures led to the development of different design strategies for the auxetic honeycomb [212].

FEA has also been implemented to expose such materials to dynamic and static crushing conditions. Li et al. conducted a study to analyse the in-plane uniaxial and biaxial crushing characteristics of three honeycombs through explicit dynamic FEA. This research aimed to compare the deformation mode, plateau stress, energy absorption and impact response [213]. In a 2019 study by Qi et al. [154], the in-plane crushing response of tetrachiral honeycombs was investigated under both quasi-static and dynamic loading conditions. This revealed the different modes of deformation in response to the different loading conditions [154] (Figure 8a). A subsequent study by the same authors [214] (Figure 8b) made use of the FEA-predicted deformation to reveal the underlying mechanisms by analysing patterns in the internal stresses for a re-entrant honeycomb with petal-shaped inclined ribs, which they termed as "re-entrant circular auxetic honeycombs". The article identifies three distinctive regions in their unit cell configuration map, each corresponding to a mesoscale interaction pattern and a macro-scale deformation mode [214].

Other studies analysing the crushing performance through FEA include a paper published in 2020 by Wei et al. [215] investigating the deformation upon the crushing of star-shaped honeycombs and a new type of auxetic honeycomb "star" structure (termed star-triangular honeycombs) (Figure 8c). Another study by Singh et al. [216] used FEA to investigate the deformation mechanisms observed during the static inclined compression of the re-entrant honeycomb auxetic structure (Figure 8d), a mode of loading which is likely to have a number of practical applications. This research introduced a novel method to extract micro deformation mechanisms under inclined loading, which were related to the macro deformation regime and the overall mechanical response of the re-entrant honeycomb structure. Through the identification of elements undergoing the plastic strain of more than 10%, micro modes were successfully identified [216]. Incline loading on re-entrant honeycomb systems has also been studied in work by Dhari et al. [217]. Furthermore, a 3D re-entrant structure was also analysed under dynamic crushing conditions to observe its behaviour under extreme deformation [218].

Auxetic materials are also associated with impact resistance and have been proposed in applications of protective gear and automotive bumpers. Over the recent years, FEA has seen extensive use in exploring how auxetic materials behave in impact scenarios, significantly aiding material testing across industries such as automotive and aerospace industries. FEA offers a repeatable, non-destructive and rapid test, in lieu or in conjunction with the more time-consuming experimental tests. FEA protocols commonly investigate how auxetic materials respond to impact by analysing the dispersal and redirection of the force, as well as the energy absorption properties. This has proven particularly useful in analysing the crushing patterns and the deformations. For instance, Liu et al. showcased the behaviour of re-entrant auxetic honeycombs under different loading speeds [219]. Meanwhile, Guo et al. examined the impact of velocity and indenter size on a double U honeycomb structure (Figure 2f), and found it to exhibit superior energy absorption capacity and stress distribution compared to the conventional counterpart [143]. Similar to the work above, FEA has been instrumental in assessing the impact velocity and crashworthiness of auxetic structures, including a number of honeycombs [220] and modified honeycombs [144], hierarchical honeycombs [211] and chiral auxetic structures [152].

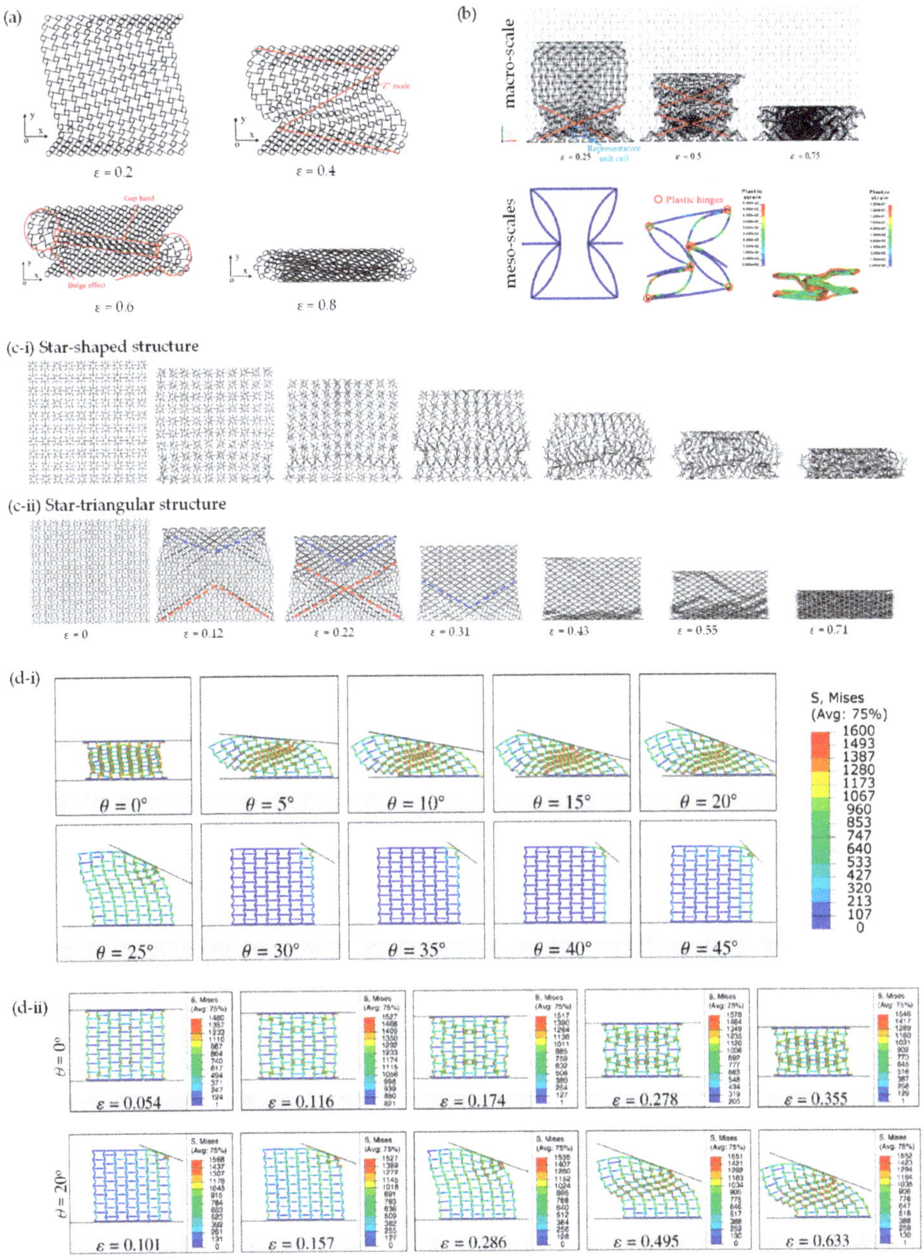

Figure 8. The crushing of auxetics as simulated by FEA: (**a**) quasi-static crushing at different extents of the global strains of tetrachiral honeycomb as summated by Qi et al. [154] at a velocity of 1 m/s; (**b**) predicted deformation processes of a re-entrant circular auxetic honeycombs specimen at typical states in the macro- and mesoscale as simulated by Qi et al. [214]; (**c**) the crushing of "star" honeycombs as simulated by Wei et al. [215]; (**d**) static inclined compression of re-entrant honeycomb auxetic structure as simulated by Singh et al. [216], where (**d-i**) shows the effect of angle of the inclination θ of how the load is applied and (**d-ii**) focuses on the system where θ = 0°, 20°, showing the deformation at different extents of applied strain.

A number of studies have used FEA to investigate the effect of indentation on the deformation mechanism [17]. In recent studies, FEA served to validate findings on the indentation resistance of the hexagonal honeycombs [221] and the indentation behaviour of 3D-printed auxetic reinforced composites [222]. Moreover, a recent study by Attard et al. looked at indentation from the perspective of the indenter. FEA was employed to examine the response of a finger-like indenter made from an inner hardcore (representing bone) surrounded by a softer outer layer (representing flesh) when pressing on a hard sample lined with an auxetic or conventional softer materials. This study revealed that although the auxetic material, with a highly negative Poisson's ratio, may feel harder when compared to a conventional material of the same Young's modulus, the auxetic does not have the "bottoming up" disadvantage and, thus, a thinner layer of auxetic can replace a much thicker conventional protective layer. Another emerging direction of studies [223] is the examination of the dynamic properties of auxetics. This is due to a multitude of potential applications that include acoustic absorbers [224–227], seismic insulators [228–231], energy harvesting devices [232,233] and artificial intelligence [234,235]. To assess the dynamic properties of the system through the FEM simulations, the first step typically corresponds to the determination of its phononic band structure. A common approach in this case is the use of a single unit cell with Floquet periodic boundary conditions. The resulting phononic band structure provides a plethora of information about the properties of the system ranging from the group and phase velocities of waves propagating through the system [236] to the phononic band gaps [237–239], as well as other more complex topological effects [240–242]. The possibility of finding the phononic band gaps is of practical importance since it allows the determination of ranges of frequencies at which waves are not transmitted through the system. The Bragg band gaps [243] can be observed once the effective wavelength of the wave propagating through the structure is approximately twice as large as the lattice constant of a metamaterial acting as a phononic crystal [244–246]. Over the years, a multitude of studies have been conducted on this topic to identify different classes of metamaterials that could act as wave insulators at various ranges of frequencies. These studies have been reported for devices at very different size scales [247] that corresponded to considerable band gaps at a broad range of frequencies including megahertz [248], kilohertz [249–252] as well as even lower frequencies related, for example, to seismic applications [228,253,254].

An important aspect that merges the worlds of static and dynamic properties of metamaterials is the possibility of controlling the phononic band structure through structural reconfiguration. In the case of a majority of metamaterials, to significantly modify the phononic band structure and the corresponding band gaps. it is necessary to change the system either by modifying the dimensions of its structural components [255] or by artificially changing the mass of the system, for example, by adding elements having a nonzero mass [256,257]. Despite the effectiveness of these approaches, these methods share the limitation that it is impossible to change the behaviour of the system without fabricating it again. There are also literature examples of metamaterials where the quasi-static mechanical reconfiguration influences the system to the point that its vibration modes change compared to the initial structure. For such scenarios, the entire band structure can indeed be considerably modified. It should be noted that this goal, to a moderate extent, can be achieved even for relatively simple geometries that are often studied from the perspective of their static properties [258–261]. Nonetheless, a greater change in the band structure can typically be observed for more complex or multistable systems that can be controlled even via external stimuli [262–264].

In conclusion, the studies reviewed here demonstrate the usefulness of FEA in analysing the mechanical properties and energy absorption capacity of these materials. FEA was pivotal in illustrating the great potential that auxetics have for applications requiring high-energy absorption capacity, such as in the automotive and aerospace industries.

4. Design of Products

FEA excels at replicating testing scenarios and analysing the behaviour of material within a specific scenario. Its application is not limited to replicating what is typically measured in conventional lab testing; rather, it can be extended to replicate the conditions in which the product is expected to perform, enabling the observation of its response.

Chow et al. proposed the use of 3D-printed thermoplastic polyurethane (TPU) with an auxetic architecture insert for pressure therapy to treat hypertrophic scars. The auxetic structure was designed to easily accommodate the contours of the human body during joint movements. The concept was initially tested in FEA examining the synclastic effect of out-of-plane bending. Subsequently, the formability, structural deformation and auxetic response of re-entrant and double arrowhead auxetic structures were numerically evaluated, followed by an experimental prototype. The study successfully demonstrated how the design was able to facilitate a stable level of pressure during body motion, promoting the recovery of hypertrophic scars [265].

In the construction industry, Menon et al. used FEA to observe the deflection behaviour of basic auxetic re-entrant beams and proposed an improvement to auxetic beam designs, comparing them with traditional beams when used in constructing a lightweight bridge (Figure 9a). The FEA analysis showed that the new auxetic beam designs exhibited better properties with minimal deflection, enhanced load-bearing capacity and a 64% reduction in material [266]. Another study [267] also used FEA to investigate auxetic honeycomb sandwich panels for structural applications. These panels offered reduced weight and displayed a remarkable reduction in the radiated sound power level due to the sandwich structure with an auxetic core [267]. It is important to note that FEA simulations of auxetics extend beyond mechanical scenarios such as uniaxial loading, shearing, pressing, crushing or indenting to encompass more complex scenarios essential for advanced applications. In particular, FEA was extensively used in acoustic and vibration frequency analysis of auxetics. For instance, in two studies by Li et al., the authors developed and applied a FEM to evaluate the propagation of acoustic and elastic waves through 3D phononic crystals. The method accurately computed band structures and identified band gaps and eigenmodes. The results showed that the finite element method was precise and efficient for computing band structures of complex phononic crystal structures with irregular unit shapes and could provide accurate results with commercial finite element code [268,269]. The research has been subsequently used by numerous studies [270,271] for numerical and experimental investigations of phononic crystal structures and the design of novel acoustic devices.

In a study by Li et al., fundamental frequencies were modelled for a structure comprising sandwich plates with a functionally graded (FG) auxetic 3D lattice core. Non-linear FEA revealed that the effects of FG configurations and strut incline angles are significant, with the FG-X specimen exhibiting the highest fundamental frequency. The study also investigated the large amplitude vibration characteristics of sandwich plates with an auxetic FG 3D lattice core in different thermal environments. Using full-scale non-linear FEA simulations, the effects of FG configurations on the natural frequencies, non-linear-to-linear frequency ratios of sandwich plates, and EPR amplitude curves were studied. Results indicate that the FG configurations distinctly affect the linear and non-linear vibration behaviour of sandwich plates, with EPR-amplitude curves stabilising when the vibration amplitude is sufficiently large. Overall, the study sheds light on the vibration behaviour of functionally graded auxetic 3D lattice metamaterials and sandwich plates with such core, offering insights for further investigations [272].

FEA has been applied in several studies to replicate the loading conditions of passenger vehicles to investigate different structures of auxetic non-pneumatic tyres [273–275] (Figure 9b). Additionally, it played a role in the design of anti-tetrachiral stents and hierarchical anti-tetrachiral stents with circular and elliptical nodes. Through FEA, the effects of stent geometrical parameters on the tensile mechanical behaviour of these stents were studied [62], as discussed previously.

Given the excellent ability of FEA to mimic high impacts, with adjustable impact velocities, indents and more, FEM has also been used to facilitate and explore the use of auxetics in military and sports equipment. FEA has already been used to predict material and product behaviour under certain conditions and to analyse design parameters in snowboard wrist protectors [276], helmets [41,277,278] and other sports equipment [279–281]. Mosleh et al. [278] used FEA to compare three scenarios involving helmets and head impact, namely an oblique head impact on foam at an angle of 45°, a linear impact of the helmeted head at an angle of 0° and a 45° oblique impact of the helmeted head (Figure 9c). In 2020, Airoldi et al. [282] studied foam-filled energy absorbers with auxetic behaviour for localized impacts and compared the results of FEA simulation to experimental results. A more recent study by Chen et al. (2023) employed FEA to investigate the effect of re-entrant arrowhead liners on helmet protection performance [283]. Through this study, it was illustrated that the auxetic lattice liners offer resistance to indentation, thereby enhancing the protection performance of the helmet [283].

When developing PPE, it is also crucially important to consider how materials behave when bent. Auxetic materials possess the remarkable ability to adopt a dome-shaped curvature under bending, making them particularly desirable for integration into PPE. Research conducted by Easey et al. employed FEA to explore the dome shape configurations exhibited by various cellular geometries, including re-entrant, arrowhead, tri-chiral and hexagonal patterns [284]. Their study highlighted that auxetic cellular domes demonstrate lower indentation resistance under compressive loading compared to solid counterparts, underscoring their potential suitability for PPE applications [284].

Another property sought after in the development of PPE is energy absorption and crashworthiness. Auxetic materials demonstrate enhanced energy absorption and crashworthiness due to their distinctive negative Poisson's ratio, resulting in compression-induced compaction. This property is also sought after in the automotive and aircraft industries. In a recent study by Tan et al., FEA was employed to study the energy absorption and crushing performance of hierarchical honeycombs [285]. It was illustrated that the electric vehicle crashworthiness is remarkably improved by the application of the auxetic hierarchical crash box [285].

Within this context, FEA has been employed to investigate the effects of unit-cell geometry and Poisson's ratio on mechanical properties in auxetic structures and plates. Additionally, FEA has been utilised to analyse the potential of auxetic constituents in composite materials, as demonstrated in the comparison between FEA and experimental analysis (Figure 9d) [286,287].

Several studies have investigated the ballistic impact behaviour of auxetic materials. In a study by Novak, chiral auxetic cellular structures were tested to analyse the effect of ballistic velocity and the deformation behaviour of composite sandwich panels. The experimental results validated the computational model of cover plates, which was further utilised to develop computational models of auxetic composite sandwich panels. The study shows that by using the auxetic sandwich panel, the ballistic performance is enhanced compared to monolithic cover plates [288,289]. In a recent study, the ballistic impact behaviour of auxetic sandwich composite human body armour was analysed using FEA. Numerical simulations showed improved indentation resistance and higher energy absorption in the auxetic armour compared to conventional monolithic armour [61]. Similar auxetic sandwich panels were also considered for blast protection in military vehicles, showcasing superior performance both in terms of being more lightweight and offering better protection compared to the solid plate [290]. Additionally, other studies also investigated the use of auxetics in body protection pads [291].

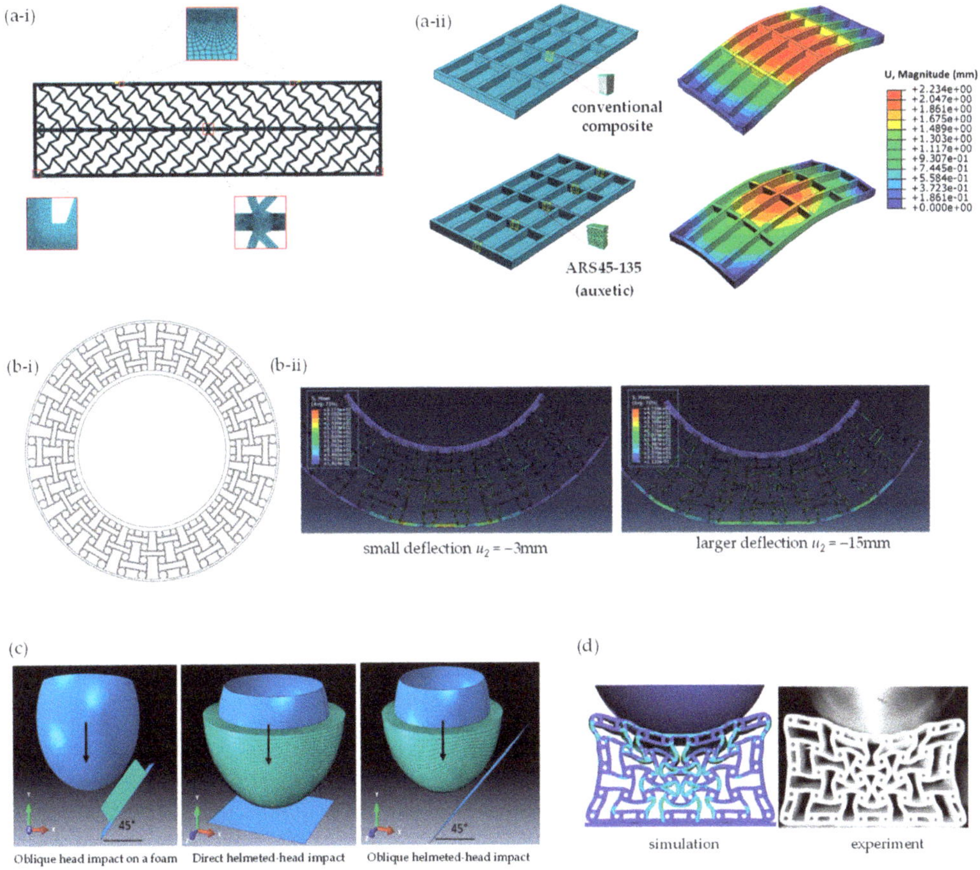

Figure 9. FEA of auxetics and product design: (**a**) auxetic beams for use in the construction industry by Menon et al. (2022) [266], where (**a-i**) shows the ARS45-135 auxetic beam and (**a-ii**) compares this beam to a conventional sandwich-panel beam in lightweight bridge construction; (**b**) non-pneumatic tyres investigated by Wu et al. (2022) [274], where (**b-i**) shows the profile in 2D and (**b-ii**) shows the results of FEA-simulated deformations; (**c**) helmet design as studied by Mosleh et al. (2018) [278] who examined direct and oblique head/helmet impacts; and (**d**) a comparison of FEA with simulated impact for application in sports equipment as reported by Shepherd et al., 2020.

5. Conclusions, Additional Considerations and Future Directions

This work demonstrated how FEA gradually became an essential tool in developing and understanding the mechanical properties of 2D and 3D auxetic cellular solids, such as honeycomb structures, as well as other auxetic materials, metamaterials and structures. This was only possible through advances in computational sciences. More specifically, FEA is particularly valuable in understanding auxetic behaviour, enabling the precise identification of stresses and strains within materials and metamaterials when loaded, which may not be easily demonstrated through experiments [292]. It is also especially beneficial to simulate characteristics which can only be tested using expensive equipment.

With time, FEA was used to simulate more real and complex scenarios required for end-product design. Some key studies attempted to optimise the geometry of auxetics to elicit superior properties. Thus, FEA was used to optimise the shape of novel auxetic structures and to determine their mechanical properties prior to prototyping and experimental testing. Since the early years of auxetics, FEA emerged as a fundamental research tool for quick and reliable analysis during the design stage, reducing the need for physical prototyping, thereby accelerating advances in the field. Further advances are still needed, particularly in niche scenarios such as the exploration of new material compositions, granular materials [293], or hybrid auxetic structures combining different mechanisms of auxeticity. Such investigations could further optimize the existing properties of auxetic materials and even uncover new potential applications. Moreover, while FEA provides valuable insights into the mechanical behaviour of auxetic materials, it would be beneficial to extend the analysis to more specific and realistic application scenarios. This includes exploring variable environmental conditions, such as temperature or pressure variations, as well as exposure to UV, moisture, chemicals, biological species (e.g., mould), etc. to assess the reliability and durability of auxetic materials under real operational conditions.

Like any other experimental or simulation method, FEA has its strengths and limitations, and properly appreciating these limitations ensures that any results obtained are not artefacts of the simulation. One of its limitations is its relative complexity. When considering model building and inputting, the accuracy of the simulation results depend on how the system is entered into the FEA package. Structures are modelled as a continuum and, therefore, it is not possible to accurately model molecular systems. Even for substructures which can be treated as a continuum (for example, in metamaterials and high resolution additively manufactured cellular structures), the structures can still be so small that the number of elements and processing capacities outstrip "normal" computing capabilities. Furthermore, to further validate the results of FEA simulations, it is recommended to integrate the work with a greater amount of experimental data. This could involve advanced mechanical testing to verify the response of auxetic materials under various loads and in different environmental conditions. Integrating experimental data could help to better calibrate FEA simulations and increase their accuracy. The validation of such FEM solutions requires good sets of experimental data, but in this area, modelling and simulation are ahead of the experimental base, and experiments might not always be possible/feasible. This is particularly the case where non-linear, non-elastic, time-dependent and other dynamic phenomena (for example, snap-through processes in folded structures) are being modelled. Some other examples of complex systems which may be trivial to represent within an FEA environment include highly parametrised models, which study the effect of geometric parameters on properties (as is normally the case with modelling of auxetics); disordered systems such as cellular foams (auxetic and non-auxetic), where a representative unit does not exist and assumptions/simplifications need to be made; molecular-level systems where, as discussed elsewhere [197], scalability issues become prominent as the molecular-level interactions cannot be adequately represented through a mechanical model; and realistic systems having some degree of imperfections due to the manufacturing process. To address some of these limitations, it is necessary to construct more complex models than what is currently in the literature, something which will increase the computational cost of the simulations.

Another limitation of FEA is that the quality of the simulations typically relies on the properties of the intrinsic materials. In some cases, these are not easy to define and the material properties are unmeasured (as is the case with the simulations of systems made from hypothetical auxetic materials, e.g., [13,93]) or are entered in an over simplified way, such as by assuming that the material is behaving linearly, thereby mitigating against high computational costs. Such assumptions may reduce the applicability and accuracy of the simulated characteristics, which calls for further validation.

Two critical aspects in FEA are the boundary conditions applied (which in the case of the modelling of auxetics, includes situations where one needs to apply full periodic boundary conditions) and the manner in which loads/deformations are applied (uniaxial stretching of auxetics is typically simulated by applying a fixed displacement on a boundary to mimic the application of uniaxial strain, or by applying a force to mimic the application of stress). Moreover, successful finite element simulation requires many key factors to be taken into consideration such as mesh size, mesh type and the constitutive relationship of the matrix material. To ensure that these key aspects of the methodology are correctly applied, it is essential to validate the protocol used, for example, by carrying out convergence testing to ensure that the meshing is sufficiently fine, particularly in regions close to points of impact, joints, regions of high-stress concentration, etc., yet not excessively fine to avoid increasing computational requirements unnecessarily.

Future directions for the modelling of auxetic materials are likely to focus on several key areas. There is a great need to better integrate atomic length-scale modelling and microstructural modelling to bridge the gaps between the length scales and time scales. While commercial FEA packages offer sophisticated pre- and post-processing capabilities, molecular-level modelling remains challenging. There is also significant potential for improvement in the modelling of non-linear, non-elastic, large strain, time-dependent and other dynamic phenomena. A major challenge in this area is the lack of sufficiently large sets of experimental data to validate such models. Establishing a free-to-use central database could greatly facilitate progress in this regard. Current FEM also lacks significant "predictive" capabilities, particularly in the development of radically new architectures. It is anticipated that the integration of AI techniques combined with FEM approaches could offer promising avenues for progress in this area. Finally, the field of metamaterials is enabling the development of a much broader range of "anomalous" properties such as a negative Poisson's ratio, negative thermal expansion coefficients, negative stiffness, negative mass, etc. These novel metamaterials have the potential to transcend traditional thermodynamic constraints. They can interact both mechanically and electromagnetically and have the ability to draw on external energy sources to drive their anomalous behaviour. This lack of constraint effectively bypasses normal physical constraints at the "material" level (although not at the "system" level). FEM and related methods will need considerable development to cope with the complexities of these new scenarios.

On a final note, it is crucial to consider the environmental impact and sustainability in the design and the application of auxetic materials. This could involve conducting a life cycle analysis of auxetic materials, from production, to use, to disposal. Such analysis can promote the development of materials that are not only technologically advanced but also environmentally sustainable.

Author Contributions: Conceptualization, J.N.G. and J.N.G.-C.; investigation, R.G.M., G.A.M., J.N.G.-C., M.A.C., D.A., K.K.D., P.-S.F., R.G. and J.N.G.; resources, J.N.G.; data curation, R.G.M., G.A.M., J.N.G.-C. and J.N.G.; writing—original draft preparation, R.G.M., G.A.M., J.N.G.-C., K.K.D. and J.N.G.; writing—review and editing, R.G.M., G.A.M., M.A.C., D.A., P.-S.F., R.G., K.E.E. and J.N.G.; supervision, D.A., P.-S.F., R.G. and J.N.G.; project administration, D.A. and J.N.G.; funding acquisition, J.N.G. and D.A. All authors have read and agreed to the published version of the manuscript.

Funding: This research was funded by the University of Malta and the Malta Council for Science & Technology (A-ROW, FUSION: The R&I Technology Development Programme 2018 project, grant number R&I-2017-033T). Part of this work was financed by the Malta Council for Science & Technology, for and on behalf of the Foundation for Science and Technology, through the Internationalisation Partnership Awards Scheme + (IPAS+) Grant Number IPAS-2023-051, the Tertiary Education Scholarship Scheme awarded to R.G.M. and G.A.M., and the Horizon 2020 Marie-Sklodowska Curie Individual Fellowship, AMPLIFI, Grant Agreement 101026382 awarded to M.A.C.

Institutional Review Board Statement: Not applicable.

Informed Consent Statement: Not applicable.

Data Availability Statement: Not applicable.

Conflicts of Interest: The authors declare no conflicts of interest. The funders had no role in the design of the study; in the collection, analyses or interpretation of data; in the writing of the manuscript; or in the decision to publish the results.

Appendix A

Table A1. Examples of works on auxetics classified according to the software used.

Program Used	2D/3D	References
ABAQUS	2D	[17,55,58,64,65,111,112,138–140,142,143,155,184,185,209,216,219,220,222,266,269]
	3D	[56,60,62,63,66,92,107,109,110,113,127–129,134,137,144–146,149,151,152,187,188,190–192,194,196,199,201,208,210,213,218,272,274,282,290,292]
ANSYS	2D	[94,95,97,135,150,157–159,161–166,172,174,182,212,221]
	3D	[57,61,96,105,108,131,136,141,160,167,171,177,183,193,267,273]
Other	2F	[59,130,170,173,176,186,211,214,215,270]
	3D	[18,132,133,153,154,189,202,207,265,268,275,276,288,289]

Table A2. Examples of works on auxetics classified according to the characteristic studied.

Property Studied	2D/3D	References
Auxeticity and stiffness	2D	[58,64,65,94,95,97,111,130,135,138–140,142,143,150,155,157–159,161–165,170,172–174,176,182,184,185,266]
	3D	[56,57,60,62,63,66,92,108,110,113,128,129,131,134,136,137,141,145,149,160, 171,177,183,187,189–194,196,199,201,274,275,292]
Energy absorption and indentation, impact and blast resistance	2D	[17,112,209,211,212,214,215,219–222]
	3D	[18,57,61,63,92,107,109,127,132,133,144,151–154,188,194,202,204,205,207,208,210,213,218,273,276,282,288–290]
Other	2D	[166,269,270]
	3D	[265,267,268,272]

References

1. Evans, K.E.; Nkansah, M.A.; Hutcherson, I.J.; Rogers, S.C. Molecular Network Design. *Nature* **1991**, *353*, 124. [CrossRef]
2. Wojciechowski, K.W. Two-Dimensional Isotropic System with a Negative Poisson Ratio. *Phys. Lett. A* **1989**, *137*, 60–64. [CrossRef]
3. Lakes, R. Foam Structures with a Negative Poisson's Ratio. *Science* **1987**, *235*, 1038–1040. [CrossRef] [PubMed]
4. Lempriere, B.M. Poisson's Ratio in Orthotropic Materials. *Am. Inst. Aeronaut. Astronaut. J.* **1968**, *6*, 2226–2227. [CrossRef]
5. Wojciechowski, K.W. Remarks on "Poisson Ratio beyond the Limits of the Elasticity Theory". *J. Phys. Soc. Jpn.* **2003**, *72*, 1819–1820. [CrossRef]
6. Keskar, N.R.; Chelikowsky, J.R. Negative Poisson Ratios in Crystalline SiO_2 from First-Principles Calculations. *Nature* **1992**, *358*, 222–224. [CrossRef]
7. Baughman, R.H.; Shacklette, J.M.; Zakhidov, A.A.; Stafströ, S. Negative Poisson's Ratios as a Common Feature of Cubic Metals. *Nature* **1998**, *392*, 362–365. [CrossRef]
8. Brańka, A.C.; Heyes, D.M.; Wojciechowski, K.W. Auxeticity of Cubic Materials. *Phys. Status Solidi* **2009**, *246*, 2063–2071. [CrossRef]
9. Brańka, A.C.; Heyes, D.M.; Wojciechowski, K.W. Auxeticity of Cubic Materials under Pressure. *Phys. Status Solidi* **2011**, *248*, 96–104. [CrossRef]
10. Goldstein, R.V.; Gorodtsov, V.A.; Lisovenko, D.S. Classification of Cubic Auxetics. *Phys. Status Solidi Basic Res.* **2013**, *250*, 2038–2043. [CrossRef]
11. Alderson, A. A Triumph of Lateral Thought. *Chem. Ind.* **1999**, *10*, 384–391.
12. Argatov, I.I.; Guinovart-Díaz, R.; Sabina, F.J. On Local Indentation and Impact Compliance of Isotropic Auxetic Materials from the Continuum Mechanics Viewpoint. *Int. J. Eng. Sci.* **2012**, *54*, 42–57. [CrossRef]
13. Attard, D.; Gatt, R.; Caruana-Gauci, R.; Grima-Cornish, J.N.; Cauchi, R.; Sillato, D.; Cerasola, D.; Ficarra, G.; Bezzina, D.S.; Formosa, C.; et al. What It could Feel like to Press on an Auxetic: Effect of Poisson's Ratio on the Indenter. *Smart Mater. Struct.* **2023**, *32*, 095038. [CrossRef]
14. Chan, N.; Evans, K.E. Indentation Resilience of Conventional and Auxetic Foams. *J. Cell. Plast.* **1998**, *34*, 231–260. [CrossRef]
15. Imbalzano, G.; Tran, P.; Ngo, T.D.; Lee, P.V.S. Three-Dimensional Modelling of Auxetic Sandwich Panels for Localised Impact Resistance. *J. Sandw. Struct. Mater.* **2015**, *19*, 291–316. [CrossRef]
16. Lakes, R.S.; Elms, K. Indentability of Conventional and Negative Poisson's Ratio Foams. *J. Compos. Mater.* **1993**, *27*, 1193–1202. [CrossRef]

17. Li, Z.; Wang, K.F.; Wang, B.L. Indentation Resistance of Brittle Auxetic Structures: Combining Discrete Representation and Continuum Model. *Eng. Fract. Mech.* **2021**, *252*, 107824. [CrossRef]
18. Novak, N.; Krstulović-Opara, L.; Ren, Z.; Vesenjak, M. Compression and Shear Behaviour of Graded Chiral Auxetic Structures. *Mech. Mater.* **2020**, *148*, 103524. [CrossRef]
19. Photiou, D.; Prastiti, N.; Sarris, E.; Constantinides, G. On the Conical Indentation Response of Elastic Auxetic Materials: Effects of Poisson's Ratio, Contact Friction and Cone Angle. *Int. J. Solids Struct.* **2016**, *81*, 33–42. [CrossRef]
20. Yang, S.; Chalivendra, V.B.; Kim, Y.K. Fracture and Impact Characterization of Novel Auxetic Kevlar®/Epoxy Laminated Composites. *Compos. Struct.* **2017**, *168*, 120–129. [CrossRef]
21. Alderson, K.L.; Simkins, V.R.; Coenen, V.L.; Davies, P.J.; Alderson, A.; Evans, K.E. How to Make Auxetic Fibre Reinforced Composites. *Phys. Status Solidi* **2005**, *242*, 509–518. [CrossRef]
22. Choi, J.B. Fracture Toughness of Re-Entrant Foam Materials with a Negative Poisson's Ratio: Experiment and Analysis. *Int. J. Fract.* **1996**, *80*, 73–83. [CrossRef]
23. Zouaoui, M.; Saifouni, O.; Gardan, J.; Makke, A.; Recho, N.; Kauffmann, J. Improvement of Fracture Toughness Based on Auxetic Patterns Fabricated by Metallic Extrusion in 3D Printing. *Procedia Struct. Integr.* **2022**, *42*, 680–686. [CrossRef]
24. Choi, J.B.; Lakes, R.S. Non-Linear Properties of Metallic Cellular Materials with a Negative Poisson's Ratio. *J. Mater. Sci.* **1992**, *27*, 5375–5381. [CrossRef]
25. Novak, N.; Duncan, O.; Allen, T.; Alderson, A.; Vesenjak, M.; Ren, Z. Shear Modulus of Conventional and Auxetic Open-Cell Foam. *Mech. Mater.* **2021**, *157*, 103818. [CrossRef]
26. Alomarah, A.; Yuan, Y.; Ruan, D. A Bio-Inspired Auxetic Metamaterial with Two Plateau Regimes: Compressive Properties and Energy Absorption. *Thin-Walled Struct.* **2023**, *192*, 111175. [CrossRef]
27. Zhang, X.G.; Jiang, W.; Zhang, Y.; Luo, C.; Zhang, X.Y.; Han, D.; Hao, J.; Teng, X.C.; Xie, Y.M.; Ren, X. Energy Absorption Properties of Composite Tubes with Hexagonal and Re-Entrant Honeycomb Fillers. *Constr. Build. Mater.* **2022**, *356*, 129298. [CrossRef]
28. Asad, M.; Dhanasekar, M.; Zahra, T.; Thambiratnam, D. Impact Mitigation of Masonry Walls with Carbon Fibre and Auxetic Fibre Composite Renders—A Numerical Study. *Structures* **2020**, *28*, 2733–2751. [CrossRef]
29. Tao, Z.; Ren, X.; Zhao, A.G.; Sun, L.; Zhang, Y.; Jiang, W.; Han, D.; Zhang, X.Y.; Xie, Y.M. A Novel Auxetic Acoustic Metamaterial Plate with Tunable Bandgap. *Int. J. Mech. Sci.* **2022**, *226*, 107414. [CrossRef]
30. Xu, Y.; Schlangen, E.; Luković, M.; Šavija, B. Tunable Mechanical Behavior of Auxetic Cementitious Cellular Composites (CCCs): Experiments and Simulations. *Constr. Build. Mater.* **2021**, *266*, 121388. [CrossRef]
31. Zhang, Q.; Sun, Y. A Series of Auxetic Metamaterials with Negative Thermal Expansion Based on L-Shaped Microstructures. *Thin-Walled Struct.* **2024**, *197*, 111596. [CrossRef]
32. Arjunan, A.; Zahid, S.; Baroutaji, A.; Robinson, J. 3D Printed Auxetic Nasopharyngeal Swabs for COVID-19 Sample Collection. *J. Mech. Behav. Biomed. Mater.* **2021**, *114*, 104175. [CrossRef] [PubMed]
33. Liu, J.; Yao, X.; Wang, Z.; Ye, J.; Luan, C.; He, Y.; Lin, H.; Fu, J. A Flexible Porous Chiral Auxetic Tracheal Stent with Ciliated Epithelium. *Acta Biomater.* **2021**, *124*, 153–165. [CrossRef] [PubMed]
34. Scarpa, F. Auxetic Materials for Bioprostheses. *IEEE Signal Process. Mag.* **2008**, *25*, 126–128. [CrossRef]
35. Song, L.; Ahmed, M.F.; Li, Y.; Zeng, C.; Li, Y. Vascular Differentiation from Pluripotent Stem Cells in 3-D Auxetic Scaffolds. *J. Tissue Eng. Regen. Med.* **2018**, *12*, 1679–1689. [CrossRef]
36. Yan, Y.; Li, Y.; Song, L.; Zeng, C.; Li, Y. Pluripotent Stem Cell Expansion and Neural Differentiation in 3-D Scaffolds of Tunable Poisson's Ratio. *Acta Biomater.* **2017**, *49*, 192–203. [CrossRef] [PubMed]
37. Yao, Y.; Yuan, H.; Huang, H.; Liu, J.; Wang, L.; Fan, Y. Biomechanical Design and Analysis of Auxetic Pedicle Screw to Resist Loosening. *Comput. Biol. Med.* **2021**, *133*, 104386. [CrossRef]
38. Allen, T.; Martinello, N.; Zampieri, D.; Hewage, T.; Senior, T.; Foster, L.; Alderson, A. Auxetic Foams for Sport Safety Applications. *Procedia Eng.* **2015**, *112*, 104–109. [CrossRef]
39. Cross, T.M.; Hoffer, K.W.; Jones, D.P.; Kirschner, P.B.; Langvin, E.; Meschter, J.C. Auxetic Structures and Footwear with Soles Having Auxetic Structures. U.S. Patent Application No. 9402439B2, 2 August 2016.
40. Duncan, O.; Bailly, N.; Allen, T.; Petit, Y.; Wagnac, E.; Alderson, A. Effect of Compressive Strain Rate on Auxetic Foam. *Appl. Sci.* **2021**, *11*, 1207. [CrossRef]
41. Foster, L.; Peketi, P.; Allen, T.; Senior, T.; Duncan, O.; Alderson, A. Application of Auxetic Foam in Sports Helmets. *Appl. Sci.* **2018**, *8*, 354. [CrossRef]
42. Hanna, B.; Adams, R.; Townsend, S.; Robinson, M.; Soe, S.; Stewart, M.; Burek, R.; Theobald, P. Auxetic Metamaterial Optimisation for Head Impact Mitigation in American Football. *Int. J. Impact Eng.* **2021**, *157*, 103991. [CrossRef]
43. Hassanin, H.; Abena, A.; Elsayed, M.A.; Essa, K. 4D Printing of NiTi Auxetic Structure with Improved Ballistic Performance. *Micromachines* **2020**, *11*, 745. [CrossRef] [PubMed]
44. Jacobs, M.J.N.; Van Dingenen, J.L.J. Ballistic Protection Mechanisms in Personal Armour. *J. Mater. Sci.* **2001**, *36*, 3137–3142. [CrossRef]
45. Steffens, F.; Oliveira, F.R.; Fangueiro, R. Energy Absorption from Composite Reinforced with High Performance Auxetic Textile Structure. *J. Compos. Mater.* **2021**, *55*, 1003–1013. [CrossRef]
46. Underhill, R. Defense Applications of Auxetic Materials. *Def. Syst. Inf. Anal. Cent. J.* **2014**, *1*, 7–13.

47. Minetola, P.; Giubilini, A. Multimaterial 3D Printing of Auxetic Jounce Bumpers for Automotive Suspensions. *Rapid Prototyp. J.* **2023**, *29*, 131–142. [CrossRef]
48. Wang, C.; Wang, W.; Zhao, W.; Wang, Y.; Zhou, G. Structure Design and Multi-Objective Optimization of a Novel NPR Bumper System. *Compos. Part B Eng.* **2018**, *153*, 78–96. [CrossRef]
49. Zhao, Y.; Zhang, Q.; Li, Y.; Wang, B.; Ma, F. Theoretical, Emulation and Experimental Analysis on Auxetic Re-Entrant Octagonal Honeycombs and Its Applications on Pedestrian Protection of Engine Hood. *Compos. Struct.* **2021**, *260*, 113534. [CrossRef]
50. Lantada, A.D.; de Blas Romero, A.; Schwentenwein, M.; Jellinek, C.; Homa, J. Lithography-Based Ceramic Manufacture (LCM) of Auxetic Structures: Present Capabilities and Challenges. *Smart Mater. Struct.* **2016**, *25*, 54015. [CrossRef]
51. Lira, C.; Scarpa, F.; Rajasekaran, R. A Gradient Cellular Core for Aeroengine Fan Blades Based on Auxetic Configurations. *J. Intell. Mater. Syst. Struct.* **2011**, *22*, 907–917. [CrossRef]
52. Wang, Z.; Zulifqar, A.; Hu, H. Auxetic Composites in Aerospace Engineering. In *Advanced Composite Materials for Aerospace Engineering*; Elsevier: Amsterdam, The Netherlands, 2016; pp. 213–240.
53. Kim, Y.; Son, K.; Lee, J. Auxetic Structures for Tissue Engineering Scaffolds and Biomedical Devices. *Materials* **2021**, *14*, 6821. [CrossRef] [PubMed]
54. Shukla, S.; Behera, B.K. Auxetic Fibrous Materials and Structures in Medical Engineering—A Review. *J. Text. Inst.* **2022**, *114*, 1078–1089. [CrossRef]
55. Abbaslou, M.; Hashemi, R.; Etemadi, E. Novel Hybrid 3D-Printed Auxetic Vascular Stent Based on Re-Entrant and Meta-Trichiral Unit Cells: Finite Element Simulation with Experimental Verifications. *Mater. Today Commun.* **2023**, *35*, 105742. [CrossRef]
56. Ali, M.N.; Busfield, J.J.C.C.; Rehman, I.U. Auxetic Oesophageal Stents: Structure and Mechanical Properties. *J. Mater. Sci. Mater. Med.* **2014**, *25*, 527–553. [CrossRef] [PubMed]
57. Alomarah, A.; Ruan, D.; Masood, S.; Gao, Z. Compressive Properties of a Novel Additively Manufactured 3D Auxetic Structure. *Smart Mater. Struct.* **2019**, *28*, 085019. [CrossRef]
58. Alomarah, A.; Xu, S.; Masood, S.H.; Ruan, D. Dynamic Performance of Auxetic Structures: Experiments and Simulation. *Smart Mater. Struct.* **2020**, *29*, 055031. [CrossRef]
59. Khan, S.Z.; Mustahsan, F.; Mahmoud, E.R.I.; Masood, S.H. A Novel Modified Re-Entrant Honeycomb Structure to Enhance the Auxetic Behavior: Analytical and Numerical Study by FEA. *Mater. Today Proc.* **2019**, *39*, 1041–1045. [CrossRef]
60. Li, C.; Shen, H.S.; Wang, H. Full-Scale Finite Element Modeling and Nonlinear Bending Analysis of Sandwich Plates with Functionally Graded Auxetic 3D Lattice Core. *J. Sandw. Struct. Mater.* **2021**, *23*, 3113–3138. [CrossRef]
61. Shah, I.A.; Khan, R.; Koloor, S.S.R.; Petrů, M.; Badshah, S.; Ahmad, S.; Amjad, M. Finite Element Analysis of the Ballistic Impact on Auxetic Sandwich Composite Human Body Armor. *Materials* **2022**, *15*, 2064. [CrossRef]
62. Wu, W.; Song, X.; Liang, J.; Xia, R.; Qian, G.; Fang, D. Mechanical Properties of Anti-Tetrachiral Auxetic Stents. *Compos. Struct.* **2018**, *185*, 381–392. [CrossRef]
63. Yang, W.; Huang, R.; Liu, J.J.; Liu, J.J.; Huang, W. Ballistic Impact Responses and Failure Mechanism of Composite Double-Arrow Auxetic Structure. *Thin-Walled Struct.* **2022**, *174*, 109087. [CrossRef]
64. Zhang, J.; Lu, G.; Ruan, D.; Wang, Z. Tensile Behavior of an Auxetic Structure: Analytical Modeling and Finite Element Analysis. *Int. J. Mech. Sci.* **2018**, *136*, 143–154. [CrossRef]
65. Zhang, J.; Lu, G.; Wang, Z.; Ruan, D.; Alomarah, A.; Durandet, Y. Large Deformation of an Auxetic Structure in Tension: Experiments and Finite Element Analysis. *Compos. Struct.* **2018**, *184*, 92–101. [CrossRef]
66. Crespo, J.; Duncan, O.; Alderson, A.; Montáns, F.J. Auxetic Orthotropic Materials: Numerical Determination of a Phenomenological Spline-Based Stored Density Energy and Its Implementation for Finite Element Analysis. *Comput. Methods Appl. Mech. Eng.* **2020**, *371*, 113300. [CrossRef]
67. Wojciechowski, K.W. Constant Thermodynamic Tension Monte Carlo Studies of Elastic Properties of a Two-Dimensional System of Hard Cyclic Hexamers. *Mol. Phys.* **1987**, *61*, 1247–1258. [CrossRef]
68. Strek, T.; Maruszewski, B.T.B.; Narojczyk, J.W.; Wojciechowski, K.W. Finite Element Analysis of Auxetic Plate Deformation. *J. Non-Cryst. Solids* **2008**, *354*, 4475–4480. [CrossRef]
69. Narojczyk, J.W.; Alderson, A.; Imre, A.R.; Scarpa, F.; Wojciechowski, K.W. Negative Poisson's Ratio Behavior in the Planar Model of Asymmetric Trimers at Zero Temperature. *J. Non-Cryst. Solids* **2008**, *354*, 4242–4248. [CrossRef]
70. Madenci, E.; Guven, I. *The Finite Element Method and Applications in Engineering Using ANSYS®*, 2nd ed.; Springer: New York, NY, USA, 2015; ISBN 9781489975492.
71. Huebner, K.H.; Dewhirst, D.L.; Smith, D.E.; Byrom, T.G. *The Finite Element Method for Engineers*, 4th ed.; Wiley-Interscience: New York, NY, USA, 2001; ISBN 978-0471370789.
72. Courant, R. Variational Methods for the Solution of Problems of Equilibrium and Vibrations. *Bull. Am. Math. Soc.* **1943**, *49*, 1–23. [CrossRef]
73. Hrennikoff, A. Solution of Problems of Elasticity by the Framework Method. *J. Appl. Mech.* **1941**, *8*, A169–A175. [CrossRef]
74. Trivedi, S. Finite Element Analysis: A Boon to Dentistry. *J. Oral Biol. Craniofacial Res.* **2014**, *4*, 200–203. [CrossRef]
75. Wood, I.; Jawad, Z.; Paisley, C.; Brunton, P. Non-Carious Cervical Tooth Surface Loss: A Literature Review. *J. Dent.* **2008**, *36*, 759–766. [CrossRef] [PubMed]
76. Aparna, J.; Maiti, S.; Jessy, P. Polyether Ether Ketone—As an Alternative Biomaterial for Metal Richmond Crown-3-Dimensional Finite Element Analysis. *J. Conserv. Dent.* **2021**, *24*, 553. [CrossRef]

77. Srirekha, A.; Bashetty, K. Infinite to Finite: An Overview of Finite Element Analysis. *Indian J. Dent. Res.* **2010**, *21*, 425. [CrossRef] [PubMed]
78. Tatarciuc, M.; Maftei, G.A.; Vitalariu, A.; Luchian, I.; Martu, I.; Diaconu-Popa, D. Inlay-Retained Dental Bridges—A Finite Element Analysis. *Appl. Sci.* **2021**, *11*, 3770. [CrossRef]
79. Thresher, R.W.; Saito, G.E. The Stress Analysis of Human Teeth. *J. Biomech.* **1973**, *6*, 443–449. [CrossRef] [PubMed]
80. Valera-Jiménez, J.F.; Burgueño-Barris, G.; Gómez-González, S.; López-López, J.; Valmaseda-Castellón, E.; Fernández-Aguado, E. Finite Element Analysis of Narrow Dental Implants. *Dent. Mater.* **2020**, *36*, 927–935. [CrossRef] [PubMed]
81. Galbusera, F.; Casaroli, G.; Chande, R.; Lindsey, D.; Villa, T.; Yerby, S.; Mesiwala, A.; Panico, M.; Gallazzi, E.; Brayda-Bruno, M. Biomechanics of Sacropelvic Fixation: A Comprehensive Finite Element Comparison of Three Techniques. *Eur. Spine J.* **2020**, *29*, 295–305. [CrossRef] [PubMed]
82. Huiskes, R.; Chao, E.Y.S. A Survey of Finite Element Analysis in Orthopedic Biomechanics: The First Decade. *J. Biomech.* **1983**, *16*, 385–409. [CrossRef]
83. Mengoni, M. Biomechanical Modelling of the Facet Joints: A Review of Methods and Validation Processes in Finite Element Analysis. *Biomech. Model. Mechanobiol.* **2020**, *20*, 389–401. [CrossRef]
84. Phellan, R.; Hachem, B.; Clin, J.; Mac-Thiong, J.M.; Duong, L. Real-Time Biomechanics Using the Finite Element Method and Machine Learning: Review and Perspective. *Med. Phys.* **2021**, *48*, 7–18. [CrossRef]
85. Cremonesi, M.; Franci, A.; Idelsohn, S.; Oñate, E. A State of the Art Review of the Particle Finite Element Method (PFEM). *Arch. Comput. Methods Eng.* **2020**, *27*, 1709–1735. [CrossRef]
86. Fadiji, T.; Coetzee, C.J.; Berry, T.M.; Ambaw, A.; Opara, U.L. The Efficacy of Finite Element Analysis (FEA) as a Design Tool for Food Packaging: A Review. *Biosyst. Eng.* **2018**, *174*, 20–40. [CrossRef]
87. Liang, X.; Ali, M.Z.; Zhang, H. Induction Motors Fault Diagnosis Using Finite Element Method: A Review. *IEEE Trans. Ind. Appl.* **2020**, *56*, 1205–1217. [CrossRef]
88. Müzel, S.D.; Bonhin, E.P.; Guimarães, N.M.; Guidi, E.S. Application of the Finite Element Method in the Analysis of Composite Materials: A Review. *Polymers* **2020**, *12*, 818. [CrossRef] [PubMed]
89. Gao, J.; Xu, W.; Ding, Z. 3D Finite Element Mesh Generation of Complicated Tooth Model Based on CT Slices. *Comput. Methods Programs Biomed.* **2006**, *82*, 97–105. [CrossRef] [PubMed]
90. Moens, D.; Vandepitte, D. Recent Advances in Non-Probabilistic Approaches for Non-Deterministic Dynamic Finite Element Analysis. *Arch. Comput. Methods Eng.* **2006**, *13*, 389–464. [CrossRef]
91. Viceconti, M.; Zannoni, C.; Testi, D.; Petrone, M.; Perticoni, S.; Quadrani, P.; Taddei, F.; Imboden, S.; Clapworthy, G. The Multimod Application Framework: A Rapid Application Development Tool for Computer Aided Medicine. *Comput. Methods Programs Biomed.* **2007**, *85*, 138–151. [CrossRef] [PubMed]
92. Wang, L.; Ulliac, G.; Wang, B.; Iglesias Martínez, J.A.; Dudek, K.K.; Laude, V.; Kadic, M. 3D Auxetic Metamaterials with Elastically-Stable Continuous Phase Transition. *Adv. Sci.* **2022**, *9*, 2204721. [CrossRef]
93. Attard, D.; Caruana-Gauci, R.; Cerasola, D.; Grima-Cornish, J.N.; Bezzina, D.S.; Ficarra, G.; Grima, J.N. On the Effect of the Poisson's Ratio on Samples Subjected to Shearing. *Eng. Res. Express* **2023**, *5*, 035044. [CrossRef]
94. Evans, K.E.; Nkansah, M.A.; Hutchinson, I.J. Modelling Negative Poisson Ratio Effects in Network-Embedded Composites. *Acta Metall. Mater.* **1992**, *40*, 2462–2469. [CrossRef]
95. Nkansah, M.A.; Evans, K.E.; Hutchinson, I.J. Modelling the Mechanical Properties of an Auxetic Molecular Network. *Model. Simul. Mater. Sci. Eng.* **1994**, *2*, 337–352. [CrossRef]
96. Evans, K.E.; Nkansah, M.A.; Hutchinson, I.J. Auxetic Foams: Modelling Negative Poisson's Ratios. *Acta Metall. Mater. mater.* **1994**, *42*, 1289–1294. [CrossRef]
97. Nkansah, M.A.; Evans, K.E.; Hutchinson, I.J. Modelling the Effects of Negative Poisson's Ratios in Continuous-Fibre Composites. *J. Mater. Sci.* **1993**, *28*, 2687–2692. [CrossRef]
98. Alderson, A.; Evans, K.E. Microstructural Modelling of Auxetic Microporous Polymers. *J. Mater. Sci.* **1995**, *30*, 3319–3332. [CrossRef]
99. Sigmund, O. *Design of Material Structures Using Topology Optimization*; Technical University of Denmark: Kongens Lyngby, Denmark, 1994.
100. Larsen, U.D.; Sigmund, O.; Bouwstra, S. Design and Fabrication of Compliant Micromechanisms and Structures with Negative Poisson's Ratio. *J. Microelectromech. Syst.* **1997**, *6*, 99–106. [CrossRef]
101. Sigmund, O.; Torquato, S. Design of Smart Composite Materials Using Topology Optimization. *Smart Mater. Struct.* **1999**, *8*, 365–379. [CrossRef]
102. Grima, J.N.; Gatt, R.; Farrugia, P.-S. On the Properties of Auxetic Meta-Tetrachiral Structures. *Phys. Status Solidi* **2008**, *245*, 511–520. [CrossRef]
103. Gatt, R.; Attard, D.; Farrugia, P.-S.; Azzopardi, K.M.; Mizzi, L.; Brincat, J.-P.; Grima, J.N. A Realistic Generic Model for Anti-Tetrachiral Systems. *Phys. Status Solidi* **2013**, *250*, 2012–2019. [CrossRef]
104. Alderson, A.; Alderson, K.L.; Attard, D.; Evans, K.E.; Gatt, R.; Grima, J.N.; Miller, W.; Ravirala, N.; Smith, C.W.; Zied, K. Elastic Constants of 3-, 4- and 6-Connected Chiral and Anti-Chiral Honeycombs Subject to Uniaxial in-Plane Loading. *Compos. Sci. Technol.* **2010**, *70*, 1042–1048. [CrossRef]

105. Afshar, A.; Rezvanpour, H. Computational Study of Non-Porous Auxetic Plates with Diamond Shape Inclusions. *J. Compos. Sci.* **2022**, *6*, 192. [CrossRef]
106. Chen, K.; Gao, Q.; Fang, S.; Zou, D.; Yang, Z.; Liao, W.H. An Auxetic Nonlinear Piezoelectric Energy Harvester for Enhancing Efficiency and Bandwidth. *Appl. Energy* **2021**, *298*, 117274. [CrossRef]
107. Gunaydin, K.; Turkmen, H.S. In-Plane Quasi-Static Crushing Finite Element Analysis of Auxetic Lattices. In Proceedings of the 2019 9th International Conference on Recent Advances in Space Technologies (RAST), Istanbul, Turkey, 11–14 June 2019; pp. 645–648. [CrossRef]
108. Kavakli, H.S.; Ali, D. Enhancing the Mechanical Properties of Auxetic Metamaterials by Incorporating Nonrectangular Cross Sections into Their Component Rods: A Finite Element Analysis. *Phys. Status Solidi* **2023**, *260*, 2200194. [CrossRef]
109. Logakannan, K.P.; Ramachandran, V.; Rengaswamy, J.; Ruan, D. Stiffened Star-Shaped Auxetic Structure with Tri-Directional Symmetry. *Compos. Struct.* **2022**, *279*, 114773. [CrossRef]
110. Rad, M.S.; Hatami, H.; Ahmad, Z.; Yasuri, A.K. Analytical Solution and Finite Element Approach to the Dense Re-Entrant Unit Cells of Auxetic Structures. *Acta Mech.* **2019**, *230*, 2171–2185. [CrossRef]
111. Wei, Y.L.; Yang, Q.S.; Tao, R. SMP-Based Chiral Auxetic Mechanical Metamaterial with Tunable Bandgap Function. *Int. J. Mech. Sci.* **2021**, *195*, 106267. [CrossRef]
112. Xu, W.; Dong, Z.; Ma, P. Finite Element Analyses of Auxetic Warp-Knitted Fabric Deformation Behaviors under Low-Velocity Impact Loading. *J. Text. Inst.* **2020**, *111*, 1578–1586. [CrossRef]
113. Zhang, X.Y.; Ren, X.; Wang, X.Y.; Zhang, Y.; Xie, Y.M. A Novel Combined Auxetic Tubular Structure with Enhanced Tunable Stiffness. *Compos. Part B Eng.* **2021**, *226*, 109303. [CrossRef]
114. Javanbakht, Z.; Öchsner, A. *Advanced Finite Element Simulation with MSC Marc*; Springer International Publishing: Cham, Switzerland, 2017; ISBN 9783319476674.
115. Borovinsek, M.; Novak, N.; Vesenjak, M.; Ren, Z.; Ulbin, M. Designing 2D Auxetic Structures Using Multi-Objective Topology Optimization. *Mater. Sci. Eng. A* **2020**, *795*, 139914. [CrossRef]
116. Almgren, R.F. An Isotropic Three-Dimensional Structure with Poisson's Ratio = −1. *J. Elast.* **1985**, *15*, 427–430. [CrossRef]
117. Gibson, L.J.J.; Ashby, M.F.F.; Schajer, G.S.S.; Robertson, C.I.I. The Mechanics of Two-Dimensional Cellular Materials. *Proc. R. Soc. A Math. Phys. Eng. Sci.* **1982**, *382*, 25–42. [CrossRef]
118. Grima, J.N.; Attard, D.; Ellul, B.; Gatt, R. An Improved Analytical Model for the Elastic Constants of Auxetic and Conventional Hexagonal Honeycombs. *Cell. Polym.* **2011**, *30*, 287–310. [CrossRef]
119. Sigmund, O. Tailoring Materials with Prescribed Elastic Properties. *Mech. Mater.* **1995**, *20*, 351–368. [CrossRef]
120. Sigmund, O. Materials with Prescribed Constitutive Parameters: An Inverse Homogenization Problem. *Int. J. Solids Struct.* **1994**, *31*, 2313–2329. [CrossRef]
121. Zhu, B.; Zhang, X.; Zhang, H.; Liang, J.; Zang, H.; Li, H.; Wang, R. Design of Compliant Mechanisms Using Continuum Topology Optimization: A Review. *Mech. Mach. Theory* **2020**, *143*, 103622. [CrossRef]
122. Zheng, Y.; Wang, Y.; Lu, X.; Liao, Z.; Qu, J. Evolutionary Topology Optimization for Mechanical Metamaterials with Auxetic Property. *Int. J. Mech. Sci.* **2020**, *179*, 105638. [CrossRef]
123. Schwerdtfeger, J.; Wein, F.; Leugering, G.; Singer, R.F.; Körner, C.; Stingl, M.; Schury, F. Design of Auxetic Structures via Mathematical Optimization. *Adv. Mater.* **2011**, *23*, 2650–2654. [CrossRef] [PubMed]
124. Clausen, A.; Wang, F.; Jensen, J.S.; Sigmund, O.; Lewis, J.A. Topology Optimized Architectures with Programmable Poisson's Ratio over Large Deformations. *Adv. Mater.* **2015**, *27*, 5523–5527. [CrossRef] [PubMed]
125. Bruggi, M.; Zega, V.; Corigliano, A. Synthesis of Auxetic Structures Using Optimization of Compliant Mechanisms and a Micropolar Material Model. *Struct. Multidiscip. Optim.* **2017**, *55*, 1–12. [CrossRef]
126. Wang, Z.-P.; Poh, L.H.; Dirrenberger, J.; Zhu, Y.; Forest, S. Isogeometric Shape Optimization of Smoothed Petal Auxetic Structures via Computational Periodic Homogenization. *Comput. Methods Appl. Mech. Engrg.* **2017**, *323*, 250–271. [CrossRef]
127. Gohar, S.; Hussain, G.; Ilyas, M.; Ali, A. Performance of 3D Printed Topologically Optimized Novel Auxetic Structures under Compressive Loading: Experimental and FE Analyses. *J. Mater. Res. Technol.* **2021**, *15*, 394–408. [CrossRef]
128. Harkati, E.; Daoudi, N.; Bezazi, A.; Haddad, A.; Scarpa, F. In-Plane Elasticity of a Multi Re-Entrant Auxetic Honeycomb. *Compos. Struct.* **2017**, *180*, 130–139. [CrossRef]
129. Li, Z.Y.; Wang, X.T.; Ma, L.; Wu, L.Z. Study on the Mechanical Properties of CFRP Composite Auxetic Structures Consist of Corrugated Sheets and Tubes. *Compos. Struct.* **2022**, *292*, 115655. [CrossRef]
130. Lee, J.H.; Choi, J.B.; Choi, K. Application of Homogenization FEM Analysis to Regular and Re-Entrant Honeycomb Structures. *J. Mater. Sci.* **1996**, *31*, 4105–4110. [CrossRef]
131. Zied, K.; Osman, M.; Elmahdy, T. Enhancement of the In-Plane Stiffness of the Hexagonal Re-Entrant Auxetic Honeycomb Cores. *Phys. Status Solidi* **2015**, *252*, 2685–2692. [CrossRef]
132. Usta, F.; Türkmen, H.S.; Scarpa, F. Low-Velocity Impact Resistance of Composite Sandwich Panels with Various Types of Auxetic and Non-Auxetic Core Structures. *Thin-Walled Struct.* **2021**, *163*, 107738. [CrossRef]
133. Usta, F.; Türkmen, H.S.; Scarpa, F. High-Velocity Impact Resistance of Doubly Curved Sandwich Panels with Re-Entrant Honeycomb and Foam Core. *Int. J. Impact Eng.* **2022**, *165*, 104230. [CrossRef]
134. Yang, H.; Wang, B.; Ma, L. Mechanical Properties of 3D Double-U Auxetic Structures. *Int. J. Solids Struct.* **2019**, *180–181*, 13–29. [CrossRef]

135. Bezazi, A.; Scarpa, F.; Remillat, C. A Novel Centresymmetric Honeycomb Composite Structure. *Compos. Struct.* **2005**, *71*, 356–364. [CrossRef]
136. Chen, Y.; Fu, M.; Hu, H.; Xiong, J. Curved Inserts in Auxetic Honeycomb for Property Enhancement and Design Flexibility. *Compos. Struct.* **2022**, *280*, 114892. [CrossRef]
137. Etemadi, E.; Gholikord, M.; Zeeshan, M.; Hu, H. Improved Mechanical Characteristics of New Auxetic Structures Based on Stretch-Dominated-Mechanism Deformation under Compressive and Tensile Loadings. *Thin-Walled Struct.* **2023**, *184*, 110491. [CrossRef]
138. Xu, N.; Liu, H.; An, M.; Wang, L. Novel 2D Star-Shaped Honeycombs with Enhanced Effective Young's Modulus and Negative Poisson's Ratio. *Extrem. Mech. Lett.* **2021**, *43*, 101164. [CrossRef]
139. Zhu, Y.; Luo, Y.; Gao, D.; Yu, C.; Ren, X.; Zhang, C. In-Plane Elastic Properties of a Novel Re-Entrant Auxetic Honeycomb with Zigzag Inclined Ligaments. *Eng. Struct.* **2022**, *268*, 114788. [CrossRef]
140. Lu, Z.; Li, X.; Yang, Z.; Xie, F. Novel Structure with Negative Poisson's Ratio and Enhanced Young's Modulus. *Compos. Struct.* **2016**, *138*, 243–252. [CrossRef]
141. Huang, J.; Zhang, Q.; Scarpa, F.; Liu, Y.; Leng, J. In-Plane Elasticity of a Novel Auxetic Honeycomb Design. *Compos. Part B-Eng.* **2017**, *110*, 72–82. [CrossRef]
142. Mustahsan, F.; Khan, S.Z.; Zaidi, A.A.; Alahmadi, Y.H.; Mahmoud, E.R.I.; Almohamadi, H. Re-Entrant Honeycomb Auxetic Structure with Enhanced Directional Properties. *Materials* **2022**, *15*, 8022. [CrossRef] [PubMed]
143. Guo, M.F.; Yang, H.; Ma, L. Design and Analysis of 2D Double-U Auxetic Honeycombs. *Thin-Walled Struct.* **2020**, *155*, 106915. [CrossRef]
144. Wang, W.; Zhang, W.; Guo, M.; Yang, J.; Ma, L. Energy Absorption Characteristics of a Lightweight Auxetic Honeycomb under Low-Velocity Impact Loading. *Thin-Walled Struct.* **2023**, *185*, 110577. [CrossRef]
145. Zhang, W.; Zhao, S.; Scarpa, F.; Wang, J.; Sun, R. In-Plane Mechanical Behavior of Novel Auxetic Hybrid Metamaterials. *Thin-Walled Struct.* **2021**, *159*, 107191. [CrossRef]
146. Zhan, C.; Li, M.; Mccoy, R.; Zhao, L.; Lu, W. 3D Printed Hierarchical Re-Entrant Honeycombs: Enhanced Mechanical Properties and the Underlying Deformation Mechanisms. *Compos. Struct.* **2022**, *290*, 115550. [CrossRef]
147. Lakes, R.S. Deformation Mechanisms in Negative Poisson's Ratio Materials: Structural Aspects. *J. Mater. Sci.* **1991**, *26*, 2287–2292. [CrossRef]
148. Prall, D.; Lakes, R.S. Properties of a Chiral Honeycomb with a Poisson's Ratio of −1. *Int. J. Mech. Sci.* **1997**, *39*, 305–314. [CrossRef]
149. Alomarah, A.; Ruan, D.; Masood, S. Tensile Properties of an Auxetic Structure with Re-Entrant and Chiral Features—A Finite Element Study. *Int. J. Adv. Manuf. Technol.* **2018**, *99*, 2425–2440. [CrossRef]
150. Attard, D.; Farrugia, P.S.; Gatt, R.; Grima, J.N. Starchirals—A Novel Class of Auxetic Hierarchal Structures. *Int. J. Mech. Sci.* **2020**, *179*, 105631. [CrossRef]
151. Gang, X.; Ren, X.; Jiang, W.; Yu, X.; Luo, C.; Zhang, Y.; Min, Y. A Novel Auxetic Chiral Lattice Composite: Experimental and Numerical Study. *Compos. Struct.* **2022**, *282*, 115043. [CrossRef]
152. Gao, D.; Wang, S.; Zhang, M.; Zhang, C. Experimental and Numerical Investigation on In-Plane Impact Behaviour of Chiral Auxetic Structure. *Compos. Struct.* **2021**, *267*, 113922. [CrossRef]
153. Nečemer, B.; Glodež, S.; Novak, N.; Kramberger, J. Numerical Modelling of a Chiral Auxetic Cellular Structure under Multiaxial Loading Conditions. *Theor. Appl. Fract. Mech.* **2020**, *107*, 102514. [CrossRef]
154. Qi, C.; Jiang, F.; Yu, C.; Yang, S. In-Plane Crushing Response of Tetra-Chiral Honeycombs. *Int. J. Impact Eng.* **2019**, *130*, 247–265. [CrossRef]
155. Shim, J.; Shan, S.; Košmrlj, A.; Kang, S.H.; Chen, E.R.; Weaver, J.C.; Bertoldi, K. Harnessing Instabilities for Design of Soft Reconfigurable Auxetic/Chiral Materials. *Soft Matter* **2013**, *9*, 8198–8202. [CrossRef]
156. Zhang, K.; Zhao, P.; Zhao, C.; Hong, F.; Deng, Z. Study on the Mechanism of Band Gap and Directional Wave Propagation of the Auxetic Chiral Lattices. *Compos. Struct.* **2020**, *238*, 111952. [CrossRef]
157. Gatt, R.; Brincat, J.P.; Azzopardi, K.M.; Mizzi, L.; Grima, J.N. On the Effect of the Mode of Connection between the Node and the Ligaments in Anti-Tetrachiral Systems. *Adv. Eng. Mater.* **2015**, *17*, 189–198. [CrossRef]
158. Mizzi, L.; Attard, D.; Gatt, R.; Pozniak, A.A.; Wojciechowski, K.W.; Grima, J.N. Influence of Translational Disorder on the Mechanical Properties of Hexachiral Honeycomb Systems. *Compos. Part B Eng.* **2015**, *80*, 84–91. [CrossRef]
159. Mizzi, L.; Attard, D.; Gatt, R.; Farrugia, P.S.; Grima, J.N. An Analytical and Finite Element Study on the Mechanical Properties of Irregular Hexachiral Honeycombs. *Smart Mater. Struct.* **2018**, *27*, 105016. [CrossRef]
160. Mizzi, L.; Attard, D.; Gatt, R.; Dudek, K.K.; Ellul, B.; Grima, J.N. Implementation of Periodic Boundary Conditions for Loading of Mechanical Metamaterials and Other Complex Geometric Microstructures Using Finite Element Analysis. *Eng. Comput.* **2021**, *37*, 1765–1779. [CrossRef]
161. Grima, J.N.; Gatt, R. Perforated Sheets Exhibiting Negative Poisson's Ratios. *Adv. Eng. Mater.* **2010**, *12*, 460–464. [CrossRef]
162. Grima, J.; Mizzi, L.; Azzopardi, K.M.; Gatt, R. Auxetic Perforated Mechanical Metamaterials with Randomly Oriented Cuts. *Adv. Mater.* **2016**, *28*, 385–389. [CrossRef] [PubMed]
163. Mizzi, L.; Azzopardi, K.M.; Attard, D.; Grima, J.N.; Gatt, R. Auxetic Metamaterials Exhibiting Giant Negative Poisson's Ratios. *Phys. Status Solidi—Rapid Res. Lett.* **2015**, *9*, 425–430. [CrossRef]

164. Grima, J.N.; Gatt, R.; Ellul, B.; Chetcuti, E. Auxetic Behaviour in Non-Crystalline Materials Having Star or Triangular Shaped Perforations. *J. Non-Cryst. Solids* **2010**, *356*, 1980–1987. [CrossRef]
165. Mizzi, L.; Attard, D.; Evans, K.E.; Gatt, R.; Grima, J.N. Auxetic Mechanical Metamaterials with Diamond and Elliptically Shaped Perforations. *Acta Mech.* **2021**, *232*, 779–791. [CrossRef]
166. Billon, K.; Zampetakis, I.; Scarpa, F.; Ouisse, M.; Sadoulet-Reboul, E.; Collet, M.; Perriman, A.; Hetherington, A. Mechanics and Band Gaps in Hierarchical Auxetic Rectangular Perforated Composite Metamaterials. *Compos. Struct.* **2017**, *160*, 1042–1050. [CrossRef]
167. Yao, J.; Sun, R.; Scarpa, F.; Remillat, C.; Gao, Y.; Su, Y. Two-Dimensional Graded Metamaterials with Auxetic Rectangular Perforations. *Compos. Struct.* **2021**, *261*, 113313. [CrossRef]
168. Grima, J.N.; Evans, K.E. Auxetic Behavior from Rotating Squares. *J. Mater. Sci. Lett.* **2000**, *19*, 1563–1565. [CrossRef]
169. Grima, J.N.; Evans, K.E. Auxetic Behavior from Rotating Triangles. *J. Mater. Sci.* **2006**, *41*, 3193–3196. [CrossRef]
170. Wang, H.; Xiao, S.; Wang, J. Disordered Auxetic Metamaterials Architected by Random Peanut-Shaped Perturbations. *Mater. Des.* **2021**, *212*, 110291. [CrossRef]
171. Atilla Yolcu, D.; Okutan Baba, B. Measurement of Poisson's Ratio of the Auxetic Structure. *Meas. J. Int. Meas. Confed.* **2022**, *204*, 112040. [CrossRef]
172. Acuna, D.; Gutiérrez, F.; Silva, R.; Palza, H.; Nunez, A.S.; Düring, G. A Three Step Recipe for Designing Auxetic Materials on Demand. *Commun. Phys.* **2022**, *5*, 1–9. [CrossRef]
173. Mrozek, A.; Strek, T. Numerical Analysis of Dynamic Properties of an Auxetic Structure with Rotating Squares with Holes. *Materials* **2022**, *15*, 8712. [CrossRef] [PubMed]
174. Chetcuti, E.; Ellul, B.; Manicaro, E.; Brincat, J.P.J.-P.; Attard, D.; Gatt, R.; Grima, J.N. Modeling Auxetic Foams through Semi-Rigid Rotating Triangles. *Phys. Status Solidi* **2014**, *251*, 297–306. [CrossRef]
175. Grima-Cornish, J.N.; Grima, J.N.; Attard, D. Negative Mechanical Materials and Metamaterials: Giant Out-of-Plane Auxeticity from Multi-Dimensional Wine-Rack-like Motifs. *MRS Adv.* **2020**, *5*, 717–725. [CrossRef]
176. Hur, J.M.; Seo, D.S.; Kim, K.; Lee, J.K.; Lee, K.J.; Kim, Y.Y.; Kim, D.N. Harnessing Distinct Deformation Modes of Auxetic Patterns for Stiffness Design of Tubular Structures. *Mater. Des.* **2021**, *198*, 109376. [CrossRef]
177. Mizzi, L.; Gatt, R.; Grima, J.N. Non-Porous Grooved Single-Material Auxetics. *Phys. Status Solidi* **2015**, *252*, 1559–1564. [CrossRef]
178. Grima, J.N.; Jackson, R.; Alderson, A.; Evans, K.E. Do Zeolites Have Negative Poisson's Ratios? *Adv. Mater.* **2000**, *12*, 1912–1918. [CrossRef]
179. Alderson, A.; Alderson, K.L.; Evans, K.E.; Grima, J.N.; Williams, M.R.; Davies, P.J. Modelling the Deformation Mechanisms, Structure-Property Relationships and Applications of Auxetic Nanomaterials. *Phys. Status Solidi Basic Res.* **2005**, *242*, 499–508. [CrossRef]
180. Grima-Cornish, J.N.; Vella-Żarb, L.; Grima, J.N.; Evans, K.E. Boron Arsenate and Its Pressure-Dependent Auxetic Properties. *APL Mater.* **2022**, *10*, 091109. [CrossRef]
181. Hoberman, C. Reversibly Expandable Doubly-Curved Truss Structure. U.S. Patent 4,942,700, 24 July 1990.
182. Alderson, A.; Alderson, K.L.; Chirima, G.; Ravirala, N.; Zied, K.M. The In-Plane Linear Elastic Constants and out-of-Plane Bending of 3-Coordinated Ligament and Cylinder-Ligament Honeycombs. *Compos. Sci. Technol.* **2010**, *70*, 1034–1041. [CrossRef]
183. Grima, J.N.; Cauchi, R.; Gatt, R.; Attard, D. Honeycomb Composites with Auxetic Out-of-Plane Characteristics. *Compos. Struct.* **2013**, *106*, 150–159. [CrossRef]
184. Lyngdoh, G.A.; Kelter, N.K.; Doner, S.; Krishnan, N.M.A.; Das, S. Elucidating the Auxetic Behavior of Cementitious Cellular Composites Using Finite Element Analysis and Interpretable Machine Learning. *Mater. Des.* **2022**, *213*, 110341. [CrossRef]
185. Poźniak, A.A.A.; Wojciechowski, K.W.; Grima, J.N.; Mizzi, L. Planar Auxeticity from Elliptic Inclusions. *Compos. Part B Eng.* **2016**, *94*, 379–388. [CrossRef]
186. Roche, J.; Von Lockette, P.; Lofland, S. Study of Hard-and Soft-Magnetorheological Elastomers (MRE's) Actuation Capabilities. In *Proceedings of the 2011 COMSOL Conference in Boston*; COMSOL, Inc.: Burlington, MA, USA, 2011.
187. Li, Z.-Y.Y.; Wang, X.-T.T.; Ma, L.; Wu, L.-Z.Z.; Wang, L. Auxetic and Failure Characteristics of Composite Stacked Origami Cellular Materials under Compression. *Thin-Walled Struct.* **2023**, *184*, 110453. [CrossRef]
188. Changfang, Z.; Changlin, Z.; Jianlin, Z.; Hongwei, Z.; Kebin, Z.; Yangzuo, L. Compressive Mechanical Behavior for Surface Auxetic Structures. *J. Alloys Compd.* **2022**, *894*, 162427. [CrossRef]
189. Yang, L.; Harrysson, O.; West, H.; Cormier, D. Mechanical Properties of 3D Re-Entrant Honeycomb Auxetic Structures Realized via Additive Manufacturing. *Int. J. Solids Struct.* **2015**, *69–70*, 475–490. [CrossRef]
190. Wang, X.T.; Li, X.W.; Ma, L. Interlocking Assembled 3D Auxetic Cellular Structures. *Mater. Des.* **2016**, *99*, 467–476. [CrossRef]
191. Wang, X.-T.; Wang, B.; Li, X.; Ma, L. Mechanical Properties of 3D Re-Entrant Auxetic Cellular Structures. *Int. J. Mech. Sci.* **2017**, *131–132*, 396–407. [CrossRef]
192. Nasim, M.S.; Etemadi, E. Three Dimensional Modeling of Warp and Woof Periodic Auxetic Cellular Structure. *Int. J. Mech. Sci.* **2018**, *136*, 475–481. [CrossRef]
193. Farrugia, P.S.; Gatt, R.; Grima, J.N. A Novel Three-Dimensional Anti-Tetrachiral Honeycomb. *Phys. Status Solidi Basic Res.* **2019**, *256*, 1800473. [CrossRef]
194. Wang, Q.; Yang, Z.; Lu, Z.; Li, X. Mechanical Responses of 3D Cross-Chiral Auxetic Materials under Uniaxial Compression. *Mater. Des.* **2020**, *186*, 108226. [CrossRef]

195. Grima, J.N.; Zammit, V.; Gatt, R.; Alderson, A.; Evans, K.E. Auxetic Behaviour from Rotating Semi-Rigid Units. *Phys. Status Solidi* **2007**, *244*, 866–882. [CrossRef]
196. Photiou, D.; Avraam, S.; Sillani, F.; Verga, F.; Jay, O.; Papadakis, L. Experimental and Numerical Analysis of 3D Printed Polymer Tetra-Petal Auxetic Structures under Compression. *Appl. Sci.* **2021**, *11*, 10362. [CrossRef]
197. Grima-Cornish, J.N.; Attard, D.; Vella-Żarb, L.; Grima, J.N.; Evans, K.E. Boron Arsenate Scaled-Up: An Enhanced Nano-Mimicking Mechanical Metamaterial. *Phys. Status Solidi Basic Res.* **2022**, *259*, 1–12. [CrossRef]
198. Galea, R.; Farrugia, P.-S.; Dudek, K.K.; Attard, D.; Grima, J.N.; Gatt, R. A Novel Design Method to Produce 3D Auxetic Metamaterials with Continuous Pores Exemplified through 3D Rotating Auxetic Systems. *Mater. Des.* **2023**, *226*, 111596. [CrossRef]
199. Su, Y.; Xu, X.; Shi, J.; Huang, G. A 3D Mechanism-Driven Hexagonal Metamaterial: Evaluation of Auxetic Behavior. *Int. J. Mech. Sci.* **2021**, *209*, 106699. [CrossRef]
200. Gatt, R.; Mizzi, L.; Azzopardi, J.I.; Azzopardi, K.M.; Attard, D.; Casha, A.; Briffa, J.; Grima, J.N. Hierarchical Auxetic Mechanical Metamaterials. *Sci. Rep.* **2015**, *5*, 8395. [CrossRef] [PubMed]
201. Wan, M.; Yu, K.; Sun, H. 4D Printed Programmable Auxetic Metamaterials with Shape Memory Effects. *Compos. Struct.* **2022**, *279*, 114791. [CrossRef]
202. Gao, Q.; Zhao, X.; Wang, C.; Wang, L.; Ma, Z. Multi-Objective Crashworthiness Optimization for an Auxetic Cylindrical Structure under Axial Impact Loading. *Mater. Des.* **2018**, *143*, 120–130. [CrossRef]
203. Peng, X.L.; Soyarslan, C.; Bargmann, S. Phase Contrast Mediated Switch of Auxetic Mechanism in Composites of Infilled Re-Entrant Honeycomb Microstructures. *Extrem. Mech. Lett.* **2020**, *35*, 100641. [CrossRef]
204. Han, D.; Zhang, Y.; Yu, X.; Min, Y.; Ren, X. Mechanical Characterization of a Novel Thickness Gradient Auxetic Tubular Structure under Inclined Load. *Eng. Struct.* **2022**, *273*, 115079. [CrossRef]
205. Han, D.; Ren, X.; Zhang, Y.; Yu Zhang, X.; Gang Zhang, X.; Luo, C.; Min Xie, Y. Lightweight Auxetic Metamaterials: Design and Characteristic Study. *Compos. Struct.* **2022**, *293*, 115706. [CrossRef]
206. Chen, Z.; Wu, X.; Min, Y.; Wang, Z.; Zhou, S. Re-Entrant Auxetic Lattices with Enhanced Stiffness: A Numerical Study. *Int. J. Mech. Sci.* **2020**, *178*, 105619. [CrossRef]
207. Novak, N.; Vesenjak, M.; Tanaka, S.; Hokamoto, K.; Ren, Z. Compressive Behaviour of Chiral Auxetic Cellular Structures at Diferent Strain Rates. *Int. J. Impact Eng.* **2020**, *141*, 103566. [CrossRef]
208. Liu, J.; Liu, H. Energy Absorption Characteristics and Stability of Novel Bionic Negative Poisson's Ratio Honeycomb under Oblique Compression. *Eng. Struct.* **2022**, *267*, 114682. [CrossRef]
209. Logakannan, K.P.; Ramachandran, V.; Rengaswamy, J.; Gao, Z.; Ruan, D. Quasi-Static and Dynamic Compression Behaviors of a Novel Auxetic Structure. *Compos. Struct.* **2020**, *254*, 112853. [CrossRef]
210. Jiang, H.; Ren, Y.; Jin, Q.; Zhu, G.; Hu, Y.; Cheng, F. Crashworthiness of Novel Concentric Auxetic Reentrant Honeycomb with Negative Poisson's Ratio Biologically Inspired by Coconut Palm. *Thin-Walled Struct.* **2020**, *154*, 106911. [CrossRef]
211. Tan, H.L.; He, Z.C.; Li, K.X.; Li, E.; Cheng, A.G.; Xu, B. In-Plane Crashworthiness of Re-Entrant Hierarchical Honeycombs with Negative Poisson's Ratio. *Compos. Struct.* **2019**, *229*, 111415. [CrossRef]
212. Lu, H.; Wang, X.; Chen, T. In-Plane Dynamics Crushing of a Combined Auxetic Honeycomb with Negative Poisson's Ratio and Enhanced Energy Absorption. *Thin-Walled Struct.* **2021**, *160*, 107366. [CrossRef]
213. Li, Z.; Gao, Q.; Yang, S.; Wang, L.; Tang, J. Comparative Study of the In-Plane Uniaxial and Biaxial Crushing of Hexagonal, Re-Entrant, and Mixed Honeycombs. *J. Sandw. Struct. Mater.* **2019**, *21*, 1991–2013. [CrossRef]
214. Qi, C.; Jiang, F.; Yang, S.; Remennikov, A. Multi-Scale Characterization of Novel Re-Entrant Circular Auxetic Honeycombs under Quasi-Static Crushing. *Thin-Walled Struct.* **2021**, *169*, 108314. [CrossRef]
215. Wei, L.; Zhao, X.; Yu, Q.; Zhu, G. A Novel Star Auxetic Honeycomb with Enhanced In-Plane Crushing Strength. *Thin-Walled Struct.* **2020**, *149*, 106623. [CrossRef]
216. Singh, R.; Javanbakht, Z.; Hall, W. On the Inclined Static Loading of Honeycomb Re-Entrant Auxetics. *Compos. Struct.* **2021**, *273*, 114289. [CrossRef]
217. Dhari, R.S.; Javanbakht, Z.; Hall, W. On the Deformation Mechanism of Re-Entrant Honeycomb Auxetics under Inclined Static Loads. *Mater. Lett.* **2021**, *286*, 129214. [CrossRef]
218. Wang, T.; Li, Z.; Wang, L.; Ma, Z.; Hulbert, G.M. Dynamic Crushing Analysis of a Three-Dimensional Re-Entrant Auxetic Cellular Structure. *Materials* **2019**, *12*, 460. [CrossRef]
219. Liu, W.; Wang, N.; Luo, T.; Lin, Z. In-Plane Dynamic Crushing of Re-Entrant Auxetic Cellular Structure. *Mater. Des.* **2016**, *100*, 84–91. [CrossRef]
220. Ou, Y.; Yan, S.; Wen, P. In-Plane Impact Dynamics Analysis of Re-Entrant Honeycomb with Variable Cross-Section. *Comput. Model. Eng. Sci.* **2021**, *127*, 209–222. [CrossRef]
221. Hu, L.L.; Zhou, M.Z.; Deng, H. Dynamic Indentation of Auxetic and Non-Auxetic Honeycombs under Large Deformation. *Compos. Struct.* **2019**, *207*, 323–330. [CrossRef]
222. Li, T.; Liu, F.; Wang, L. Enhancing Indentation and Impact Resistance in Auxetic Composite Materials. *Compos. Part B Eng.* **2020**, *198*, 108229. [CrossRef]

223. Krushynska, A.O.; Torrent, D.; Aragón, A.M.; Ardito, R.; Bilal, O.R.; Bonello, B.; Bosia, F.; Chen, Y.; Christensen, J.; Colombi, A.; et al. Emerging Topics in Nanophononics and Elastic, Acoustic, and Mechanical Metamaterials: An Overview. *Nanophotonics* **2023**, *12*, 659–686. [CrossRef]
224. Mei, J.; Ma, G.; Yang, M.; Yang, Z.; Wen, W.; Sheng, P. Dark Acoustic Metamaterials as Super Absorbers for Low-Frequency Sound. *Nat. Commun.* **2012**, *3*, 756. [CrossRef]
225. Tang, Y.; Ren, S.; Meng, H.; Xin, F.; Huang, L.; Chen, T.; Zhang, C.; Lu, T.J. Hybrid Acoustic Metamaterial as Super Absorber for Broadband Low-Frequency Sound. *Sci. Rep.* **2017**, *7*, 43340. [CrossRef] [PubMed]
226. Cummer, S.A.; Christensen, J.; Alù, A. Controlling Sound with Acoustic Metamaterials. *Nat. Rev. Mater.* **2016**, *1*, 16001. [CrossRef]
227. Christensen, J.; de Abajo, F.J.G. Anisotropic Metamaterials for Full Control of Acoustic Waves. *Phys. Rev. Lett.* **2012**, *108*, 124301. [CrossRef]
228. Achaoui, Y.; Ungureanu, B.; Enoch, S.; Brûlé, S.; Guenneau, S. Seismic Waves Damping with Arrays of Inertial Resonators. *Extrem. Mech. Lett.* **2016**, *8*, 30–37. [CrossRef]
229. Brûlé, S.; Javelaud, E.H.; Enoch, S.; Guenneau, S. Experiments on Seismic Metamaterials: Molding Surface Waves. *Phys. Rev. Lett.* **2014**, *112*, 133901. [CrossRef] [PubMed]
230. Colombi, A.; Colquitt, D.; Roux, P.; Guenneau, S.; Craster, R. V A Seismic Metamaterial: The Resonant Metawedge. *Sci. Rep.* **2016**, *6*, 27717. [CrossRef] [PubMed]
231. Krödel, S.; Thomé, N.; Daraio, C. Wide Band-Gap Seismic Metastructures. *Extrem. Mech. Lett.* **2015**, *4*, 111–117. [CrossRef]
232. De Ponti, J.M.; Colombi, A.; Ardito, R.; Braghin, F.; Corigliano, A.; Craster, R. V Graded Elastic Metasurface for Enhanced Energy Harvesting. *New J. Phys.* **2020**, *22*, 13013. [CrossRef]
233. Jensen, J.S. Phononic Band Gaps and Vibrations in One- and Two-Dimensional Mass–Spring Structures. *J. Sound Vib.* **2003**, *266*, 1053–1078. [CrossRef]
234. Li, L.; Ruan, H.; Liu, C.; Li, Y.; Shuang, Y.; Alù, A.; Qiu, C.-W.; Cui, T.J. Machine-Learning Reprogrammable Metasurface Imager. *Nat. Commun.* **2019**, *10*, 1082. [CrossRef]
235. Coulombe, J.C.; York, M.C.A.; Sylvestre, J. Computing with Networks of Nonlinear Mechanical Oscillators. *PLoS ONE* **2017**, *12*, e0178663. [CrossRef]
236. Milton, G.W. *The Theory of Composites*; Cambridge University Press: Cambridge, UK, 2002.
237. Hussein, M.I.; Leamy, M.J.; Ruzzene, M. Dynamics of Phononic Materials and Structures: Historical Origins, Recent Progress, and Future Outlook. *Appl. Mech. Rev.* **2014**, *66*, 040802. [CrossRef]
238. Khelif, A.; Aoubiza, B.; Mohammadi, S.; Adibi, A.; Laude, V. Complete Band Gaps in Two-Dimensional Phononic Crystal Slabs. *Phys. Rev. E* **2006**, *74*, 46610. [CrossRef]
239. Baravelli, E.; Ruzzene, M. Internally Resonating Lattices for Bandgap Generation and Low-Frequency Vibration Control. *J. Sound Vib.* **2013**, *332*, 6562–6579. [CrossRef]
240. Süsstrunk, R.; Huber, S.D. Classification of Topological Phonons in Linear Mechanical Metamaterials. *Proc. Natl. Acad. Sci. USA* **2016**, *113*, E4767–E4775. [CrossRef]
241. Chaplain, G.J.; De Ponti, J.M.; Aguzzi, G.; Colombi, A.; Craster, R. V Topological Rainbow Trapping for Elastic Energy Harvesting in Graded Su-Schrieffer-Heeger Systems. *Phys. Rev. Appl.* **2020**, *14*, 54035. [CrossRef]
242. Khanikaev, A.B.; Fleury, R.; Mousavi, S.H.; Alù, A. Topologically Robust Sound Propagation in an Angular-Momentum-Biased Graphene-like Resonator Lattice. *Nat. Commun.* **2015**, *6*, 8260. [CrossRef] [PubMed]
243. Krushynska, A.O.; Miniaci, M.; Bosia, F.; Pugno, N.M. Coupling Local Resonance with Bragg Band Gaps in Single-Phase Mechanical Metamaterials. *Extrem. Mech. Lett.* **2017**, *12*, 30–36. [CrossRef]
244. Lu, M.-H.; Feng, L.; Chen, Y.-F. Phononic Crystals and Acoustic Metamaterials. *Mater. Today* **2009**, *12*, 34–42. [CrossRef]
245. Robillard, J.F.; Matar, O.B.; Vasseur, J.O.; Deymier, P.A.; Stippinger, M.; Hladky-Hennion, A.C.; Pennec, Y.; Djafari-Rouhani, B. Tunable Magnetoelastic Phononic Crystals. *Appl. Phys. Lett.* **2009**, *95*, 124104. [CrossRef]
246. Chen, Y.; Kadic, M.; Wegener, M. Roton-like Acoustical Dispersion Relations in 3D Metamaterials. *Nat. Commun.* **2021**, *12*, 3278. [CrossRef] [PubMed]
247. Zheludev, N.I.; Kivshar, Y.S. From Metamaterials to Metadevices. *Nat. Mater.* **2012**, *11*, 917–924. [CrossRef]
248. Iglesias Martínez, J.A.; Moughames, J.; Ulliac, G.; Kadic, M.; Laude, V. Three-Dimensional Phononic Crystal with Ultra-Wide Bandgap at Megahertz Frequencies. *Appl. Phys. Lett.* **2021**, *118*, 063507. [CrossRef]
249. Craster, R.V.; Guenneau, S. *Acoustic Metamaterials*; Craster, R.V., Guenneau, S., Eds.; SSMaterial; Springer: Dordrecht, The Netherlands, 2012; ISBN 978-94-007-9467-2.
250. D'Alessandro, L.; Belloni, E.; Ardito, R.; Corigliano, A.; Braghin, F. Modeling and Experimental Verification of an Ultra-Wide Bandgap in 3D Phononic Crystal. *Appl. Phys. Lett.* **2016**, *109*, 2–6. [CrossRef]
251. Warmuth, F.; Wormser, M.; Körner, C. Single Phase 3D Phononic Band Gap Material. *Sci. Rep.* **2017**, *7*, 3843. [CrossRef] [PubMed]
252. Lucklum, F.; Vellekoop, M.J. Bandgap Engineering of Three-Dimensional Phononic Crystals in a Simple Cubic Lattice. *Appl. Phys. Lett.* **2018**, *113*, 201902. [CrossRef]
253. Liu, Z.; Zhang, X.; Mao, Y.; Zhu, Y.Y.; Yang, Z.; Chan, C.T.; Sheng, P. Locally Resonant Sonic Materials. *Science* **2000**, *289*, 1734–1736. [CrossRef]
254. Mu, D.; Shu, H.; Zhao, L.; An, S. A Review of Research on Seismic Metamaterials. *Adv. Eng. Mater.* **2020**, *22*, 1901148. [CrossRef]

255. Oudich, M.; Gerard, N.J.R.K.; Deng, Y.; Jing, Y. Tailoring Structure-Borne Sound through Bandgap Engineering in Phononic Crystals and Metamaterials: A Comprehensive Review. *Adv. Funct. Mater.* **2023**, *33*, 35–39. [CrossRef]
256. Kunin, V.; Yang, S.; Cho, Y.; Deymier, P.; Srolovitz, D.J. Static and Dynamic Elastic Properties of Fractal-Cut Materials. *Extrem. Mech. Lett.* **2016**, *6*, 103–114. [CrossRef]
257. Huang, Y.; Li, J.; Chen, W.; Bao, R. Tunable Bandgaps in Soft Phononic Plates with Spring-Mass-like Resonators. *Int. J. Mech. Sci.* **2019**, *151*, 300–313. [CrossRef]
258. Bertoldi, K.; Boyce, M.C. Wave Propagation and Instabilities in Monolithic and Periodically Structured Elastomeric Materials Undergoing Large Deformations. *Phys. Rev. B* **2008**, *78*, 184107. [CrossRef]
259. Bertoldi, K.; Boyce, M.C. Mechanically Triggered Transformations of Phononic Band Gaps in Periodic Elastomeric Structures. *Phys. Rev. B* **2008**, *77*, 52105. [CrossRef]
260. Dudek, K.K.; Iglesias Martínez, J.A.; Ulliac, G.; Hirsinger, L.; Wang, L.; Laude, V.; Kadic, M. Micro-Scale Mechanical Metamaterial with a Controllable Transition in the Poisson's Ratio and Band Gap Formation. *Adv. Mater.* **2023**, *35*, 65–69. [CrossRef]
261. Zhang, H.; Cheng, X.; Yan, D.; Zhang, Y.; Fang, D. A Nonlinear Mechanics Model of Soft Network Metamaterials with Unusual Swelling Behavior and Tunable Phononic Band Gaps. *Compos. Sci. Technol.* **2019**, *183*, 107822. [CrossRef]
262. Hu, W.; Ren, Z.; Wan, Z.; Qi, D.; Cao, X.; Li, Z.; Wu, W.; Tao, R.; Li, Y. Deformation Behavior and Band Gap Switching Function of 4D Printed Multi-Stable Metamaterials. *Mater. Des.* **2021**, *200*, 109481. [CrossRef]
263. Nimmagadda, C.; Matlack, K.H. Thermally Tunable Band Gaps in Architected Metamaterial Structures. *J. Sound Vib.* **2019**, *439*, 29–42. [CrossRef]
264. Chen, Y.; Li, T.; Scarpa, F.; Wang, L. Lattice Metamaterials with Mechanically Tunable Poisson's Ratio for Vibration Control. *Phys. Rev. Appl.* **2017**, *7*, 24012. [CrossRef]
265. Chow, L.; Yick, K.L.; Wong, K.H.; Leung, M.S.; Sun, Y.; Kwan, M.; Ning, K.; Yu, A.; Yip, J.; Chan, Y.; et al. 3D Printing Auxetic Architectures for Hypertrophic Scar Therapy. *Macromol. Mater. Eng.* **2022**, *307*, 2100866. [CrossRef]
266. Menon, H.G.; Dutta, S.; Krishnan, A.; Hariprasad, M.P. Proposed Auxetic Cluster Designs for Lightweight Structural Beams with Improved Load Bearing Capacity. *Eng. Struct.* **2022**, *260*, 114241. [CrossRef]
267. Sadegh, M.; Ranjbar, M.; Boldrin, L.; Scarpa, F.; Patsias, S.; Ozada, N. Vibroacoustics of 2D Gradient Auxetic Hexagonal Honeycomb Sandwich Panels. *Compos. Struct.* **2018**, *187*, 593–603. [CrossRef]
268. Jianbao, L.; Wang, Y.S.; Zhang, C. Finite Element Method for Analysis of Band Structures of Three Dimensonal Phononic Crystals. In Proceedings of the 2008 IEEE Ultrasonics Symposium, Beijing, China, 2–5 November 2008; pp. 1468–1471. [CrossRef]
269. Li, J.; Wang, Y.S.; Zhang, C. Finite Element Method for Analysis of Band Structures of Phononic Crystal Slabs with Archimedean-like Tilings. In Proceedings of the 2009 IEEE International Ultrasonics Symposium, Rome, Italy, 20–23 September 2009; pp. 1548–1551. [CrossRef]
270. Koutsianitis, P.I.; Tairidis, G.K.; Drosopoulos, G.A.; Stavroulakis, G.E. Conventional and Star-Shaped Auxetic Materials for the Creation of Band Gaps. *Arch. Appl. Mech.* **2019**, *89*, 2545–2562. [CrossRef]
271. Koutsianitis, P.I.; Tairidis, G.K.; Stavroulakis, G.E. Shunted Piezoelectric Patches on Auxetic Microstructures for the Enhancement of Band Gaps. *Arch. Appl. Mech.* **2021**, *91*, 739–751. [CrossRef]
272. Li, C.; Shen, H.S.; Wang, H.; Yu, Z. Large Amplitude Vibration of Sandwich Plates with Functionally Graded Auxetic 3D Lattice Core. *Int. J. Mech. Sci.* **2020**, *174*, 105472. [CrossRef]
273. Andriya, N.; Dutta, V.; Vani, V.V. Study on 3D Printed Auxetic Structure-Based Non-Pneumatic Tyres (NPT'S). *Mater. Manuf. Process.* **2022**, *37*, 1280–1297. [CrossRef]
274. Wu, T.; Li, M.; Zhu, X.; Lu, X. Research on Non-Pneumatic Tire with Gradient Anti-Tetrachiral Structures. *Mech. Adv. Mater. Struct.* **2021**, *28*, 2351–2359. [CrossRef]
275. Zang, L.; Wang, X.; Yan, P.; Zhao, Z. Structural Design and Characteristics of a Non-Pneumatic Tire with Honeycomb Structure. *Mech. Adv. Mater. Struct.* **2021**, *29*, 4066–4073. [CrossRef]
276. Newton-Mann, C.; Winwood, K.; Driscoll, H.; Hamilton, N.; Allen, T. Finite Element Model of an Impact on a Palmar Pad from a Snowboard Wrist Protector. *Proceedings* **2018**, *2*, 314. [CrossRef]
277. Hernández, S.; Brebbia, C.A.; El-Sayed, M.E.M. Helmet Optimisation Based on Head-Helmet Modelling. *WIT Trans. Built Environ.* **2003**, *67*, 339.
278. Mosleh, Y.; Cajka, M.; Depreitere, B.; Van der Sloten, J.; Ivens, J. Designing Safer Composite Helmets to Reduce Rotational Accelerations during Oblique Impacts. *J. Eng. Med.* **2018**, *232*, 479–491. [CrossRef] [PubMed]
279. Allen, T.; Haake, S.; Goodwill, S. Comparison of a Finite Element Model of a Tennis Racket to Experimental Data. *Sports Eng.* **2009**, *12*, 87–98. [CrossRef]
280. Kays, B.T.; Smith, L.V. Effect of Ice Hockey Stick Stiffness on Performance. *Sports Eng.* **2017**, *20*, 245–254. [CrossRef]
281. Valentini, P.P.; Pennestrì, E.; Quattrociocchi, L. Biomechanical Model for Simulating Impacts against Protective Padding of Sport Facility. *Sports Eng.* **2016**, *19*, 47–57. [CrossRef]
282. Airoldi, A.; Novak, N.; Sgobba, F.; Gilardelli, A.; Borovinšek, M. Foam-Filled Energy Absorbers with Auxetic Behaviour for Localized Impacts. *Mater. Sci. Eng. A* **2020**, *788*, 139500. [CrossRef]
283. Chen, Z.; Li, J.; Wu, B.; Chen, X.; Ren, X.; Xie, Y. A Novel Bio-Inspired Helmet with Auxetic Lattice Liners for Mitigating Traumatic Brain Injury. *Smart Mater. Struct.* **2023**, *32*, 105020. [CrossRef]

284. Easey, N.; Chuprynyuk, D.; Musa, W.M.S.W.; Bangs, A.; Dobah, Y.; Shterenlikht, A.; Scarpa, F. Dome-Shape Auxetic Cellular Metamaterials: Manufacturing, Modeling, and Testing. *Front. Mater.* **2019**, *6*, 86. [CrossRef]
285. Tan, H.; He, Z.; Li, E.; Cheng, A.; Chen, T.; Tan, X.; Li, Q.; Xu, B. Crashworthiness Design and Multi-Objective Optimization of a Novel Auxetic Hierarchical Honeycomb Crash Box. *Struct. Multidiscip. Optim.* **2021**, *64*, 2009–2024. [CrossRef]
286. Nallavan, G. Impact of Recent Developments in Fabrication of Auxetic Materials on Safety and Protection in Sport. *AIP Conf. Proc.* **2020**, *2271*, 030006. [CrossRef]
287. Shepherd, T.; Winwood, K.; Venkatraman, P.; Alderson, A.; Allen, T. Validation of a Finite Element Modeling Process for Auxetic Structures under Impact. *Phys. Status Solidi* **2020**, *257*, 1900197. [CrossRef]
288. Novak, N.; Vesenjak, M.; Kennedy, G.; Thadhani, N.; Ren, Z. Response of Chiral Auxetic Composite Sandwich Panel to Fragment Simulating Projectile Impact. *Phys. Status Solidi* **2019**, *1*, 1900099. [CrossRef]
289. Novak, N.; Starčevič, L.; Vesenjak, M.; Ren, Z. Blast Response Study of the Sandwich Composite Panels with 3D Chiral Auxetic Core. *Compos. Struct.* **2019**, *210*, 167–178. [CrossRef]
290. Wang, Y.; Zhao, W.; Zhou, G.; Wang, C. International Journal of Mechanical Sciences Analysis and Parametric Optimization of a Novel Sandwich Panel with Double-V Auxetic Structure Core under Air Blast Loading. *Int. J. Mech. Sci.* **2018**, *142–143*, 245–254. [CrossRef]
291. Yang, C.; Chang, Y.B.; Lee, D. Nonlinearity of Enhanced Cell Structures Having Auxetic Material Properties. In Proceedings of the ASME International Mechanical Engineering Congress and Exposition, Salt Lake City, UT, USA, 11–14 November 2019; Volume 12. [CrossRef]
292. Chang, Y.; Liu, Y.; Hu, H. A Finite Element Analysis of Auxetic Composite Fabric with Rotating Square Structure. *J. Ind. Text.* **2023**, *53*, 152808372311731. [CrossRef]
293. Lupo, M.; Sofia, D.; Barletta, D.; Poletto, M. Calibration of DEM Simulation of Cohesive Particles. *Chem. Eng. Transect.* **2019**, *74*, 379–384. [CrossRef]

materials

MDPI

Article

Steel Automotive Wheel Rims—Data Fusion for the Precise Identification of the Technical Condition and Indication of the Approaching End of Service Life

Michal Borecki [1,*], Arkadiusz Rychlik [2], Li Zan [3,†] and Michael L. Korwin-Pawlowski [4]

1 Institute of Microelectronics and Optoelectronics, Warsaw University of Technology, 00-662 Warszawa, Poland
2 Faculty of Technical Sciences, University of Warmia and Mazury, 10-719 Olsztyn, Poland
3 Institute of Telecommunications, Warsaw University of Technology, 00-661 Warszawa, Poland
4 Département d'informatique et d'ingénierie, Université du Québec en Outaouais, Gatineau, QC J8X 3X7, Canada
* Correspondence: michal.borecki@pw.edu.pl; Tel.: +48-22-234-7749
† Current address: China Telecom (EU), 00-838 Warszawa, Poland.

Abstract: Steel automotive wheel rims are subject to wear and tear, down to the end of their service life. Manufacturers use standard destructive tests to determine the probable lifetime of the car wheel rim. With this approach, to predict the remaining use time, it is necessary to know the initial parameters of the wheel rim, actual mileage, and its use characteristics, which is difficult information to obtain in the real world. Moreover, this work shows that a vehicle's technical condition can affect the rim's remaining service time. This work describes a new method of precise binary identification of the technical condition of steel car wheel rims using the dispersion of damping factors which result from experimental modal analysis. This work also proposes a new method of indicating the approaching end of wheel rim service life with limited parameters: run-out, average of damping factors, and dispersion of damping factors. The proposed procedure requires two sequential examinations of the rim in standard periods related to the average annual mileage of the vehicle. On this basis, it is possible to indicate the approaching end of the life of the steel rims about 10,000 km in advance.

Keywords: steel; wheel rim; technical condition; data fusion; dispersion of damping factors

1. Introduction

The wheel is an essential part of the vehicle. It can be made as a monolithic structure, such as a train wheel [1], or a hybrid construction, such as an automotive wheel, which consists of a tire and a rim [2]. The automotive wheel rim allows the tire to be fixed to the rim and the wheel to the hub. Automotive wheel rims can be made in different configurations [3]. A simple classification of automotive wheel rims includes single-piece and multi-piece types. The single-piece constructions can be realized as a cast from a mold or machined from solid material. The standard design of the multi-piece wheel rim includes the use of different materials, such as various types of steels, aluminum and magnesium alloys, and glass or carbon fiber components used for metal reinforcement [4]. As the density of steel is almost 2.5 times that of aluminum, automotive wheel rims made of these materials behave differently in use.

A view of a steel automotive wheel rim with a glossary is presented in Figure 1. The wheel rim is not an ideal axis-symmetric structure. Axial symmetry is disturbed by the ventilation holes, bolt holes, and valve seat. The number of bolt holes and their shape can differ depending on the automotive manufacturer. The bolt hole can be conically or cylindrically shaped.

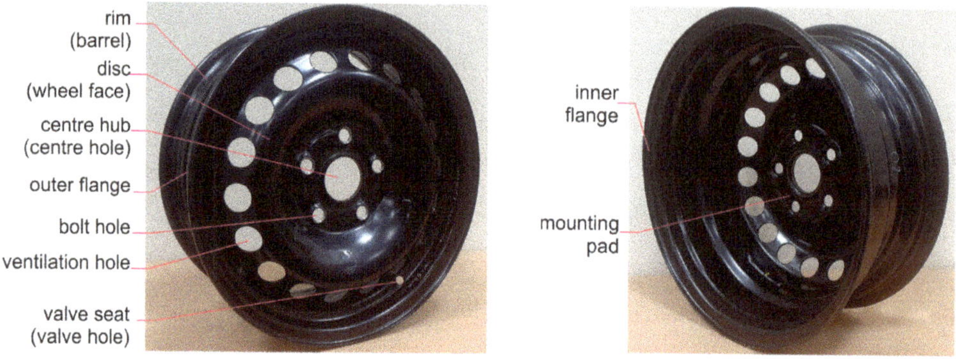

(a) (b)

Figure 1. View of a steel automotive wheel rim: (**a**) Outer view; (**b**) Inner view.

Due to the complex geometry of the structure, the investigation and optimization of automotive wheel rims' construction are currently conducted by computer-aided simulations using finite element method (FEM). There are two basic types of FEM used in wheel rim simulations that result in a determination of responses for forced vibrations and loads.

The vibration analysis results obtained from FEM are displacement, velocity, and acceleration of nodes in a time function. Vibration data can be analyzed in the frequency or the time domain. In contrast, modal analysis is a process of obtaining modal parameters such as natural frequencies, damping loss factors, and modal constants from the vibration data. Natural frequencies are mainly related to object shape, dimensions, and material [5]. The modes of complex geometrical objects are numbered according to the increasing resonant frequency of vibrations [6].

The damping factor of a mechanical object depends on three main effects: the internal friction of the material, the acoustic radiation of the system, and the energy loss caused by boundary connections of the system [7]. Generally, the component of internal friction of material dominates over other factors [8]. In the macroscopic scale, the damping loss factor increases when Young's modulus decreases. However, when the object is not symmetrical, the damping factor is characterized by a set of values [9]. Thus, in a precise analysis of complex structures, the damping factors are not always characterized by a monotonic course [10]. The 30% damping factor dispersion and some dispersion of the natural frequency can be measured in specified short-term situations of mechanical objects. However, in standard cases, the damping loss factor increases with the accumulated fatigue of the system and knowing the start and end values, the trajectory of the damping factor can be used to predict the time to failure of mechanical objects [5].

The mechanically forced fatigue of an object can be followed by self-healing when the object is not operated. The short-time healing state of the mechanical properties of the object may be associated with the presence of local maxima of the damping factor that are followed by an increase in the resonant frequency [11]. The long-term result of the healing of mechanical objects is commonly related to a decrease in damping factor and stabilization of natural frequency. While the precise interpretation of the experimental results of the modal analysis is not trivial, a system's damping factors depict the mechanical object's structural integrity, and modal analysis can be essential to characterize the wheel rim's actual technical condition.

The loading analysis of automotive wheel rims performed with FEM enables the calculation of forced total deformation, equivalent stress, and equivalent strain [12]. Obtained mechanical properties are fundamental for correctly assessing the lifespan of the wheel rim for the loaded car in a stable position and in motion [13]. After simulations, automotive wheel rim investigation includes performing some destructive tests [14]. For example, experimental axial or radial axle load of the rim is performed using dedicated equipment that consists of a controlled actuator and a deformation measuring system [15]. The standard investigation of the wheel rim involves a rolling test, cornering test, and radial fatigue test [16]. Deformations obtained from the tests can be recorded via an optical 3D measuring system [17].

In all constructions, the technical condition of a component relates to its initial state and wear state [18]. Producers may examine the initial state of the steel automotive wheel rim using measurement equipment for the dimensional control of rim and disc, unbalanced measurement, air tightness test oriented to search for cracks, and the rolling test for pointing resistance for radial fatigue [19]. During regular operation, the wheel rim is exposed to standard wear related to its operation conditions, periodic tire changes, wheel balancing, accidental direct impacts, and impacts transmitted by the tire.

The degradation of the automotive wheel rim in natural conditions is correlated with the load and the vehicle's speed and type of route, whether a highway or an urban artery. Monitoring the vehicle's course in the city is not trivial [20]. Camera observation allows for the estimation of vehicle acceleration and lane-changing dynamics only at the place of observation [21]. Approximate results indicate that cars move on motorways mainly on almost straight sections, and the change in direction is associated with lane changes and lasts about a few seconds, while the average number of lane changes is up to a dozen or so per hour [22,23]. The ratio of time of vehicle motion on a curve to linear motion can be estimated by analyzing a typical one-hour drive, where a quick lane change takes about 4 s. On the motorways, the driver swiftly changes lanes about ten times per hour. Thus, the ratio can be estimated for motorways as 0.01. The number of quick lane changes in the city is around 100 times per hour. Therefore, the ratio should be increased to 0.1. Therefore, the average curvature to linear movement ratio can be estimated as 0.055. According to the physics of circular motion, higher forces act on automotive rims in a quick lane change than in a slow change. Hence, simulators of accelerated rim wear use forced wheel movement along a circular trajectory or the radial fatigue test to simulate the dynamics of the wheel during cornering [24]. The expected results of these tests are the internal stress of the rim [25].

Internal stress can lead to the first visual sign of wheel rim wearing: finish layer peeling [26]. On the other hand, when the finish layer of the wheel rim is cracked, the metal construction begins to erode in some road conditions with a speed of 0.1 mm/year [27]. Loss of material due to corrosion is an issue in the area where the wheel rim itself mates to the hub and under the nuts securing the wheel to the hub [28,29]. Standard procedures of wheel care, including regular tire change and wheel balancing, require wheel removal from the hub and wheel mounting on the hub [30]. This process requires tightening wheel locking nuts with a defined torque, which can result in degrading the areas under the nuts and sides of corresponding holes [31]. The result can be the plastic deformation of bolts, nuts, and the area under them [32], and further missing nuts, which can have severe implications [33].

Accidental situations of wheel rim degradation most often occur due to hitting straight-on potholes, curbs, or other obstacles on the road, or side collisions with the curb [34]. Possible results are visible and internal cracks in the metal, bent inner or outer flanges, and deep marks on the surface [35]. In most cases, the leading cause of severe damage is driver negligence and underinflated tires [36]. However, when the wheel rim is worn and overloaded, the probability of a critical break due to an impact increases [37].

The last stage of rim degradation is its complete uselessness or destruction. Destruction of the wheel rim can be defined as visible signs of deformation and damage to the material structure [38]. Thus, the binary classification of rim technical condition as fit to use or damaged was proposed. The technical condition of the automotive rim can be examined using non-destructive vibration analysis performed in laboratory conditions [39]. Suggestions to improve the analysis include modifying the measuring system using a wheel balancer to a rim condition measurement system that supports run-out and vibration modes at four rim points and recording them in a dedicated database. These data are subject to further processing with an artificial neural network (ANN) [40]. In the indicated paper, the ANN was trained with a set of data obtained for new and used rims. The raw vibration data were obtained for vibration excitation at four points on each rim. The raw data were reduced to pattern form for the ANN training process and implementation task. Different types of data patterns were investigated. The analysis shows that automotive wheel rim fit for use can be characterized by a set of actual and reference parameters of experimental modal analysis, including the dispersion of excited natural frequencies and their amplitudes. However, a significant disadvantage of this system was the requirement of absolute values of amplitudes of vibration measurements at an exact time. Therefore, a precise mechanism of vibration excitation, its calibration, and synchronization with the data acquisition system were important.

The field tests of automotive rims are possible only for complete wheels and require constructing an attached device whose main component is a dedicated vibration-sensing device. The examination results show that the frequency spectrum of wheel fit for use is below 5 Hz. At the same time, precise results indicate that the vibration signal is expected between 0 and 1.65 Hz with a main peak around 0.1 Hz [41]. A similar set-up based on an integrated acceleration measurement system and ANN signal processing may be used to detect the bolts' loosening securing the rim to the hub [42].

The long-term wear of rims is today mainly analyzed for wheels of high-speed trains [43]. The analysis shows that the dominant fatigue process of steel train wheel rims is micro-damage evolution under off-axis fatigue [44]. In the study of the prediction of railway wheel service life, difficulties related to the availability of samples with a known type of degradation and wear state were indicated, as well as the necessity of the precise definition of their wear examination intervals [45]. It has also been shown that the prediction of the remaining useful life of the railway wheel, based on small samples, leads to significant discrepancies in theoretical and real results [46]. Based on the above premises, it has been proven that generating examples of railway wheels that comply with the standard degradation distribution is an essential step in predicting the lifetime of any railway wheel [47]. So far, deterministic procedures for predicting a railway or automotive wheel's remaining life are unknown.

In most cases, car owners and vehicle inspection station employees determine an automotive wheel's technical condition based on its visual inspection and the soapy water test oriented to the detection of air leaks [48]. The check should also concern the inside section of the rim and the sides that are in contact with a dual wheel case [49]. Sometimes, visual inspection is followed by examining the wheel rim geometry, a standard approach to assessing the technical condition of an automotive wheel rim. Axial and radial run-out values of the rim on the inner and outer sides of the rim flange relative to the centering hole must be lower than {2.0 mm; 1.2 mm} according to the ETRTO standard [50]. Other standards exist as country regulations, such as Poland's PN-93S standard {1.25 mm; 1.25 mm} [51]. Excessive outer rim flange deformations with run-out values greater than 2.2 mm can be felt as vehicle user's steering wheel vibrations [13]. The wear examination intervals relate to seasonal tire replacements or annual vehicle technical inspections. It can be concluded that even though the current recommendations vary, and the test methods are not consistent, wheel owners need non-destructive testing to indicate the technical condition and remaining life of the automotive wheel rim or at least to point to the moment in time when the rim must be changed [52].

2. Methodology of Examinations and Following the Structure of the Publication

The classic methodology of wheel rim examination used by manufacturers consists of destructive tests, e.g., ISO 3006 [53], SAE J267 [54], or SAE J328 standards [19], to determine the lifetime of the produced series. Thus, the maximum expected distance to drive in standard conditions is known. However, the situation on the roads is not really known and the actual technical condition of the rim is not a linear function of distance traveled.

To predict the remaining distance of use of the rim, the distance that was traveled and the characteristics of previous use of the wheel rim must be known. The characteristics of prior use are often unknown. Thus, much information in automobile blogs points to the lifetime of car wheel rims from 4 years up towards almost infinity [55]. Since steel is the most common material used in automotive rims, the proposed study is limited to rims made of this material. A rim made of aluminum–magnesium alloy has a longer lifetime than a steel rim [56].

In this work, a comparison of the results of destructive tests with non-destructive measurements of steel car wheel rims is provided. The investigation considered cyclic destructive fatigue tests separated by non-destructive classical visual inspection, rim run-out measurements, and the application of experimental modal analysis. The presented examinations were limited to the widely used sizes of steel rims in passenger cars and trolleys of two diameters: 14 and 15 inches. The automobile industry does not define matching rules between wheel hub construction and car types [57]. Therefore, 15-inch steel automotive rims of type 6Jx15H2 used in popular cars and 14-inch rims of type 51/2Jx14H2 often mounted in small vehicles and trailers were investigated. The rim type 6Jx15H2 can be described as a 15-inch rim with a 6-inch-wide barrel, see Figure 1.

The selection of rims for examination was limited to the so-called "second assembly" type. This was due to the regulations that mandatory tests apply only to wheel rims marked as original that are mounted and sold with a new vehicle [58]. New automobile wheel rims commercially available at the original equipment manufacturer, despite showing the same markings, often differ from those that are mounted on vehicles at the factory. Rims under investigation were second assembly type and were made by an original equipment manufacturer (OEM), an acceptable replacement manufacturer (ARM), or a low-cost manufacturer (LCM). The examinations of steel automotive rims of new or used conditions with the characterized initial conditions according to standards are presented in Table 1.

Table 1. Rim types under examination with initial run-out characterization.

Rim Manufacturer	Rim Type	Rim Category	ETRTO	PN-93S	Run-Out on Flange < 2 mm	Examination in Section-Type of Study
OEM	6Jx15H2	new	met	not met	met	6.1—sequential
ARM	6Jx15H2	new	not met	not met	met	6.2—sequential and then continuous
ARM	6Jx15H2	new	met	met	met	6.3—rusted hub
LCM *	6Jx15H2	new	met	met	met	6.4—sequential
OEM	6Jx15H2	used	met	not met	met	6.5—sequential
OEM	6Jx15H2	new	met	met	met	6.6—artificial fatigue cracks
OEM	51/2Jx14H2	used	met	not met	met	7.2—field testing

* The second rim of this manufacturer, despite being new, was not in acceptable technical condition to be tested as the run-out on the outer flange was over 2.2 mm and because of other issues.

The run-out examinations were performed using eddy current sensors type CW10 coupled to NI card type USB 6343, which was part of the measurement system for experimental modal analysis. The initial run-out examination showed that the rims of the second assembly met the European ETRTO standards when examined in factory conditions. However, parameters of rims from ARM sometimes did not meet the standards but the run-out values on any flanges were below 2 mm. In contrast, the rim of the LCM's initial parameters could surprise both positively and negatively.

These rims were examined with experimental modal analysis (EMA) and run-out methods, while fatigue tests of the rims were conducted in laboratory or as on-field tests. The methodology of examination corresponded to the sections of this publication as follows:

- The initial numerical modal analysis of the rims necessary for defining the acquisition system parameters used for EMA is presented in Section 3;
- The construction of a measurement system for EMA based on an exciter with repeatable parameters and assumed precision of acquisition data is presented in Section 4;

- The construction of a fatigue station aimed at performing an adapted cornering fatigue test is depicted in Section 5;
- A series of examinations of rims with the use of a fatigue station and EMA set-up based on step-by-step sequential fatigue procedure with initial and sequential rim state characterization with EMA and run-out methods are depicted in Section 6;
- The discussion of laboratory test results and on-field testing are presented in Section 7;
- The conclusions are given in Section 8.

3. Initial Numerical Modal Analysis—Materials and Constructions

The modal analysis was selected to determine the initial investigations. However, the use of a primary function that describes normal modes of automotive rim components, coupled with linear vibration theory and linear superposition, seems to not be recommended [59]. Therefore, the SOLIDWORKS software (Simulation Premium 5.0 or SP5.0) has been used to provide a comprehensive set of structural and finite element analyses and to analyze the natural frequency of the object, which was modeled using a solid mesh. Five eigenfrequencies were searched using the FFEPlus solver (Intel Sparse, https://help.solidworks.com/2019/english/SolidWorks/cworks/c_Analysis_Solvers.htm, (accessed on 14 November 2023).

Numerical mode analysis was conducted for rim type 6Jx15H2; see Figure 1. A solid mesh generated for mixed curvature was used to model the rim geometry. The maximum mesh element size of 1.76098 mm and the minimum element size of 0.352197 mm were indicated. The mesh generator created a mesh of 1,882,950 elements and 3,082,794 nodes. The maximum aspect ratio of an element was 9.6918. Carbon steel (SS) type 1023 sheet was assumed as the rim model material. The structure type of the model was defined as linear elastic and isotropic. The material parameters were given as: specific mass: 7.858 kg/m^3, yield strength: 2.82685×10^8 N/m^2, tensile strength: 4.25×10^8 N/m^2, longitudinal elasticity coefficient: 2.05×10^{11} N/m^2, Poisson's ratio: 0.29. In the simulation model, the stationary area was the hub contact area.

The results of the simulation of the ideal-shape rim are presented in Figure 2. The simulation analysis of the excited modes indicates that the fundamental modes occur in pairs having slightly different resonance frequencies. The mentioned difference for modes 1 and 2 is 0.65 Hz, and for modes 3 and 4, it is 0.21 Hz. Notably, greater amplitudes of vibration characterize modes 3 and 4 more than modes 1 and 2. Therefore, modes 3 and 4 are more likely to be excited than 1 and 2. The lowest amplitude of vibration characterizes mode 5. Similar modes can produce differences and sum frequency components in use. It is essential because the results of frequency difference are always at the low-frequency band.

The intentionally degraded rim was simulated to show the low bandwidth of the EMA. The degradation was a 15 mm diameter hole in the rim barrel positioned in the barrel's inner, center, and outer sections. Selected results of the simulation of modes 3 and 4 are presented in Figure 3. The obtained results show that the natural frequency of modes 3 and 4 is similar in all cases, but when comparing Figures 2 and 3, the values of natural frequencies difference between modes are more remarkable in the rim-degraded case than in the ideal case.

A severely damaged rim can also be described as a rim with a broken-out part of the inner flange. This fact can be simulated with the use of two holes with a diameter equal to 15 mm, as assumed in the previous analysis. Such damage is easily overlooked during inspections. In addition, due to its size, such damage leads to air escaping from the tire. The simulation results of a rim with such damage are shown in Figure 4.

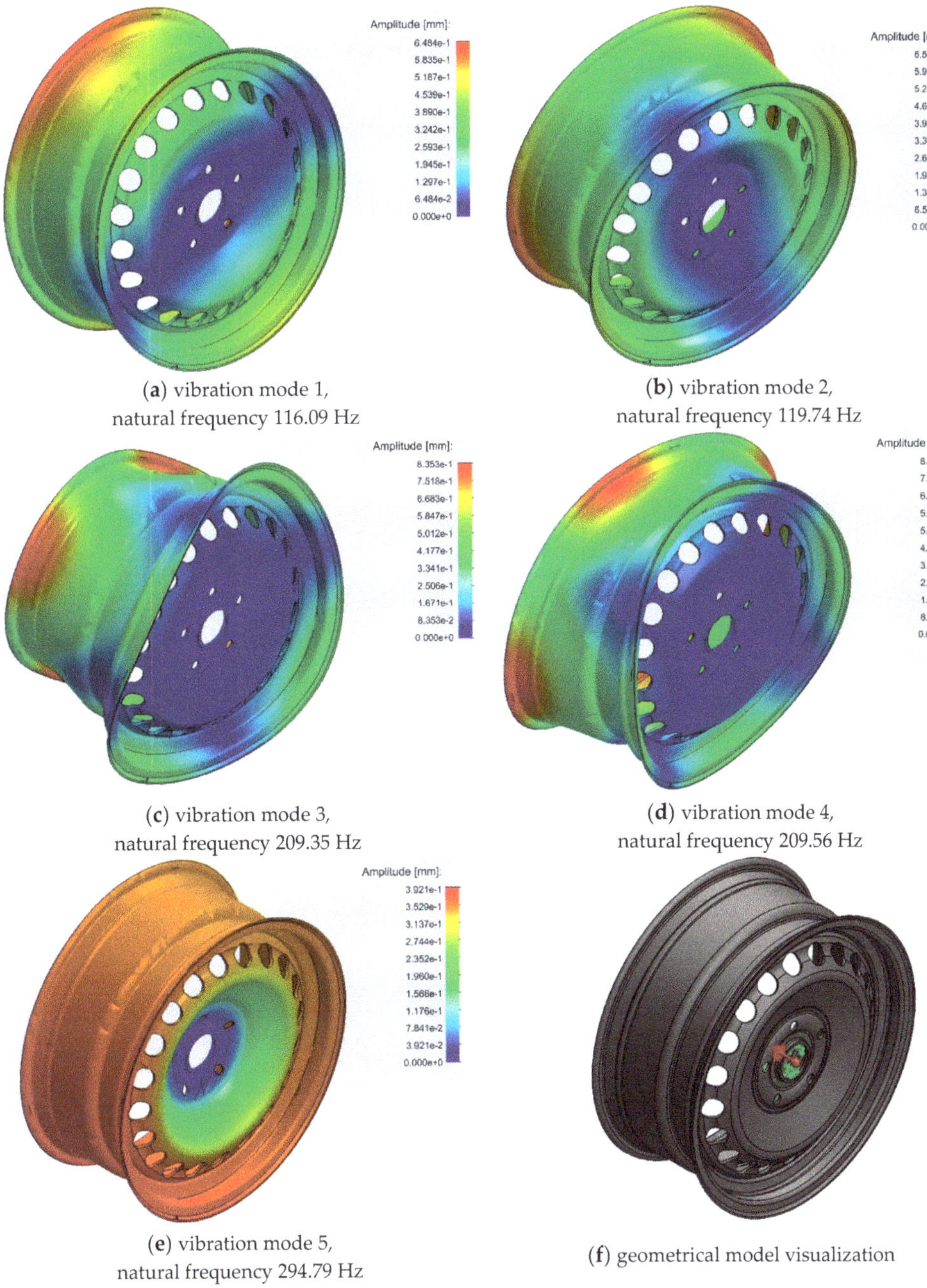

(**a**) vibration mode 1, natural frequency 116.09 Hz

(**b**) vibration mode 2, natural frequency 119.74 Hz

(**c**) vibration mode 3, natural frequency 209.35 Hz

(**d**) vibration mode 4, natural frequency 209.56 Hz

(**e**) vibration mode 5, natural frequency 294.79 Hz

(**f**) geometrical model visualization

Figure 2. Set of simulation results in natural frequencies of the rim in the ideal shape: (**a**) Vibration mode 1; (**b**) Vibration mode 2; (**c**) Vibration mode 3; (**d**) Vibration mode 4; (**e**) Vibration mode 5; (**f**) Geometrical model visualization.

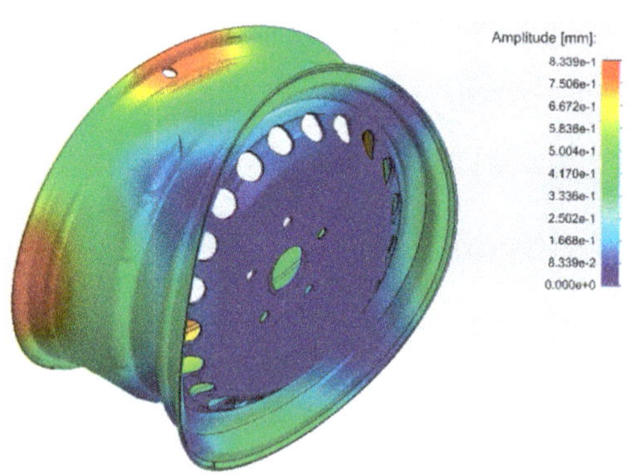

(**a**) inner hole, vibration mode 3, natural frequency 209.98 Hz

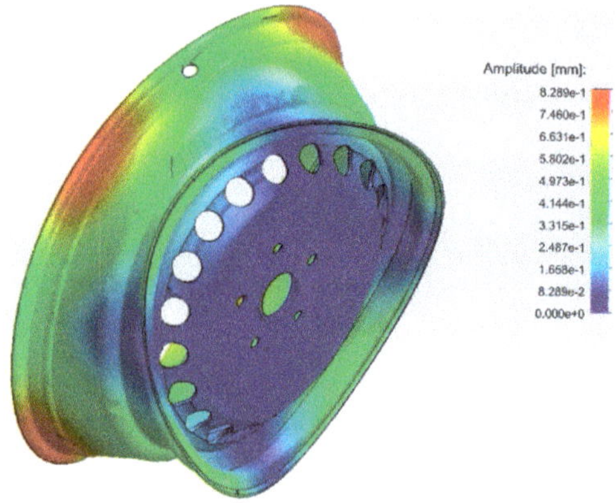

(**b**) inner hole, vibration mode 4, natural frequency 210,58 Hz

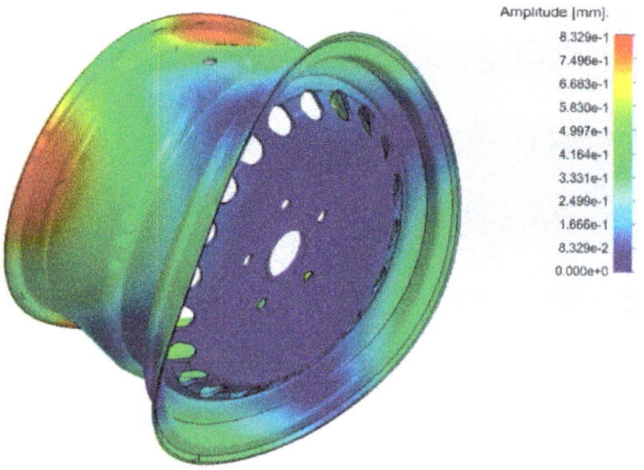

(**c**) center hole, vibration mode 3, natural frequency 209.29 Hz

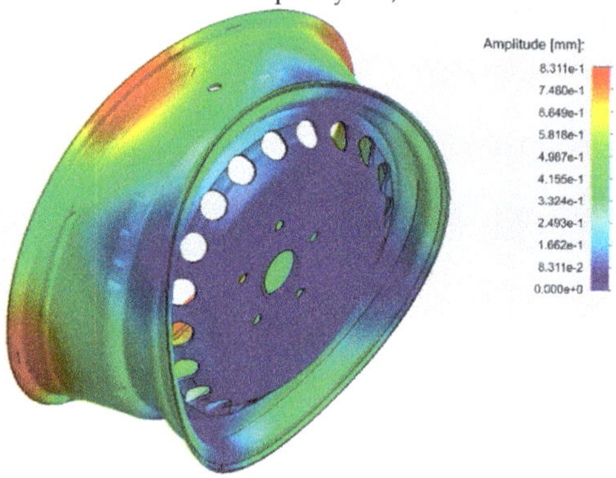

(**d**) center hole, vibration mode 4, natural frequency 209.80 Hz

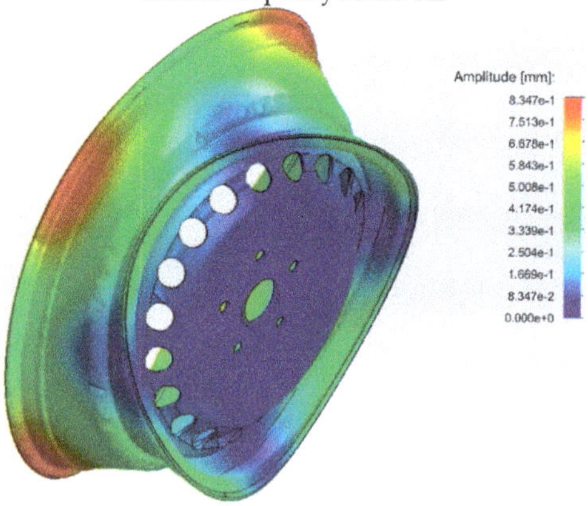

(**e**) outer hole, vibration mode 3, natural frequency 208.96 Hz

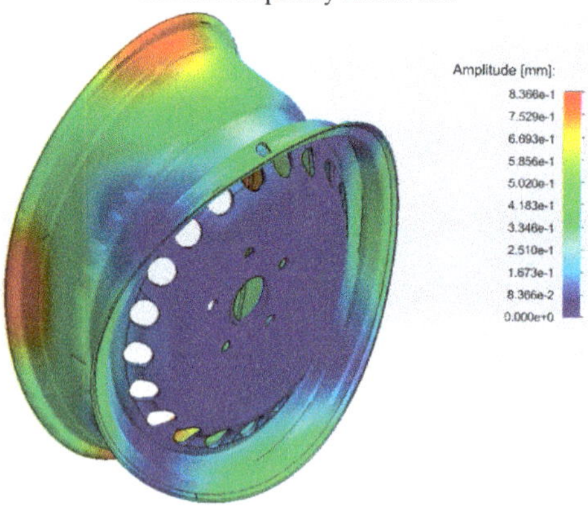

(**f**) outer hole, vibration mode 4, natural frequency 209.48 Hz

Figure 3. Simulation results of modes 3 and 4 of the rim-degraded situation with the hole positioned in the barrels: (**a**) Hole in inner section, mode 3; (**b**) Hole in inner section, mode 4; (**c**) Hole in center section, mode 3; (**d**) Hole in center section, mode 4; (**e**) Hole in outer section, mode 3; (**f**) Hole in outer section, mode 4.

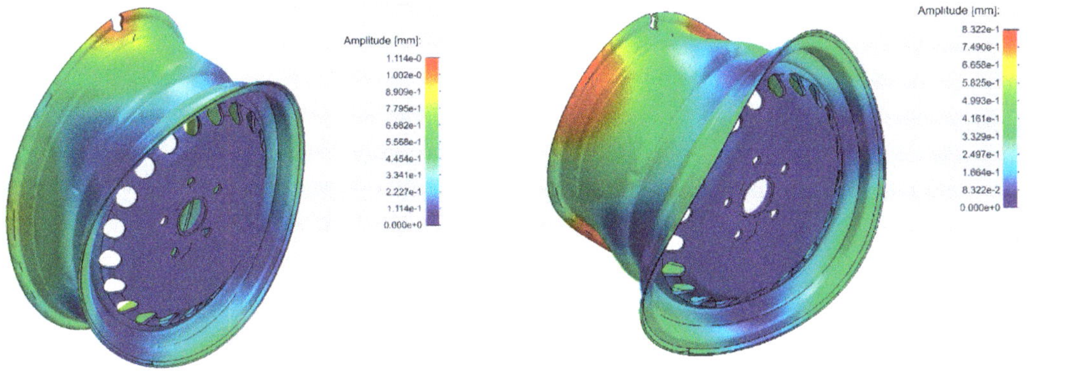

(**a**) Hole in inner flange, vibration mode 3, natural frequency 197.19 Hz

(**b**) Hole in the inner flange, vibration mode 4, natural frequency 209.18 Hz

Figure 4. Simulation results of a severely damaged wheel rim: (**a**) Hole in inner flange, vibration mode 3; (**b**) Hole in the inner flange, vibration mode 4.

The summary of the obtained results is presented in Table 2. As can be seen, the direct analysis of the amplitudes and frequencies of the excited modes for individual cases does not show regular dependency damage to amplitude tendency. On the other hand, studying the variability of the maximum amplitudes and frequencies of the excited modes indicates the possibility of detecting rim damage. As the frequency difference signal is cast in a low band, the presence of a signal in the low band can be used as an indicator of rim damage. Thus, the experimental modal analysis set-up should be characterized by the possibility of measurement in the low band of frequency signals in the range of from 0 to 12 Hz with at least 0.1 Hz resolution of frequency examination. According to the simulation results (see Table 2), measuring modal parameters as the variability of maximum amplitudes is much more complex than the variability of natural frequencies of excited modes.

Table 2. Summary of results presented in Figures 2–4.

Case	Mode Number	Maximum Amplitude of Vibration [mm]	Variability of Maximum Amplitudes [mm]	Natural Frequency [Hz]	Variability of Natural Frequencies [Hz]
Ideal rim	3	0.8353	0.0005	209.35	0.21
	4	0.8358		209.56	
Inner hole in the barrel	3	0.8339	0.0050	209.98	0.40
	4	0.8289		210,58	
Center hole in the barrel	3	0.8329	0.0018	209.29	0.51
	4	0.8311		209.80	
Outer hole in the barrel	3	0.8347	0.0019	208.96	0.48
	4	0.8366		209.48	
Broken part of the flange	3	1.1140	0.2818	197.19	11.99
	4	0.8322		209.18	

4. Measurement System for Experimental Modal Analysis (EMA) and Run-Out of Rim

The measurement of experimental modal analysis based on monitored vibration excitation of the rim is presented in Figure 5. The view of the mechanical part of the measurement system for EMA is presented in Figure 6.

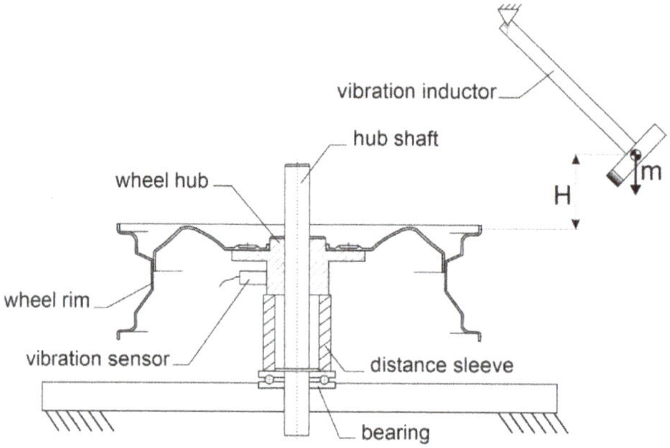

Figure 5. Scheme of the mechanical components of the measurement system used for experimental modal analysis.

Figure 6. View of the mechanical part of the measurement system for EMA with vibration excitation points (EP) marked.

The steel automotive wheel rim is positioned on the hub shaft and fixed by gravity on the hub. The vibration inductor is a roofing hammer of a mass of 425 g that is lifted before the excitation of vibrations to the height of H = 26 cm. In such conditions, the impact energy is 1.08 J.

The vibration sensor consists of the ICP accelerometer head model 625B01 attached to the profiled magnetic bed with a mounting screw. The head is characterized by a ±50 g measurement range, 100 mV/g sensitivity, and linear (±3 dB) frequency measurement range from 0.2 to 10,500 Hz. The interface between the head and computer is the NI USB 6343 acquisition card from National Instruments. Due to the limitation of the linear bandwidth for low frequencies, a calibration curve for the sensor was developed for operation in the range of 0.02–10 Hz. Acquired signals of amplitude versus time are used to determine logarithmic decrements of damped oscillations as well as to calculate frequency responses with the use of Fast Fourier Transformation (FFT). The logarithmic decrements are used further to calculate the damping factors. These calculations are performed on a personal computer in the LabView software (Version 2021).

Shaft rotation was measured with an encoder with a resolution of 1.40°. Vibrations were induced sequentially for four excitation points, defined by angles of wheel shaft rotation of 0, 90, 180, and 270 degrees. The 0-degree position was determined by the valve hole located on the excitation axis.

Example results of spectra investigation with the resolution of 0.02 Hz, using the proposed set-up and the string method for an ideal rim, are presented in Figure 7. The string method was called here the measurement of vibrations of a hanging rim, which was positioned on the hub and shaft to which the string was attached. The mentioned string was attached to the ceiling on the other side. The expected mathematical model maximum value does not characterize the constant component (at frequency 0 Hz) due to head coupling with the use of a capacitor to the acquisition card. In the analyzed case, the dominant frequency peak measured in the EMA set-up is 192.60 Hz, and for the string method is 193.3 Hz. The difference is most likely due to the influence of the bearing, which was not present on the shaft in the string method. Additionally, both values are lower than the initial simulations. The differences in natural frequencies of simulation and measurement come from the measurement of the wheel rim coupled with mounting accessories and simulation of only the wheel rim.

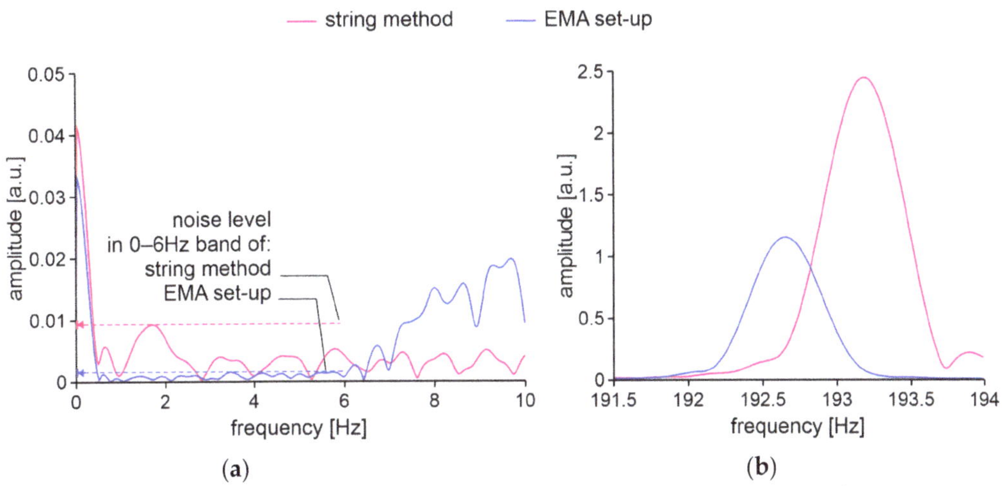

Figure 7. Ideal OEM rim examination with EMA set-up and using the string method: (**a**) Low band; (**b**) Natural frequency band.

In the 0–6 Hz band, the noise level for the string method is 4.5 times greater than for the one obtained from the EMA set-up. Thus, the EMA set-up was used for examinations of typical degraded automotive wheel rims. Moreover, based on measurement characteristics obtained with the acquisition system, for the mentioned vibration excitation points, the damping ratios on natural frequencies are calculated using logarithmic decrement of amplitudes [60].

The results obtained in ten series of measurements of the natural frequency and damping factor sets for four excitation points of the wheel rim from OEM, which is in factory condition, are similar but not identical. The excited frequencies differ maximally by 0.09 Hz. Thus, the low band result of such frequencies mixing can be masked in the measurement set-up by the actual constant component of Fourier transformation. Relating this difference to the resonant frequency gives a relative dispersion of 0.05%. In contrast, the dispersion of the damping factor is 11.4%.

Therefore, in the proposed EMA set-up, besides the average values of natural frequencies (*aNF*) and the average of damping factors (*aDF*) for the set of excitation points, the difference in natural frequencies (ΔFN) and dispersion of damping factors (*dDF*), and presence of a signal in the low band are results of measurements. The difference in natural frequencies is calculated according to the equation:

$$\Delta FN = \max(FN) - \min(FN), \quad (1)$$

where *FN* is the set of natural frequencies obtained for the set of excitation points (EP), while max(x) and min(x) are standard mathematical functions. The dispersion of damping factors is calculated with the equation:

$$dDF = \frac{\max(DF) - \min(DF)}{\text{avg}(DF)}, \quad (2)$$

where *DF* is the set of damping factors obtained for the set of excitation points (EP), and avg(x) is the standard mathematical function for calculating the average value of the set.

5. Fatigue Station

The dynamic wheel cornering fatigue test is one of the most significant tests to determine the usability expectation for newly manufactured automotive wheel rims. This test can be implemented with a rotating table configuration or as the rotating bending test [61]. The rotating bending test allows for the evaluation of the resistance of the automotive wheel rim to a bending moment achieved via an arm equipped with a rotating unbalanced mass. This mass is called eccentric or dead mass and is mounted on the load cell [62]. A strain developed at the rim due to the moment applied at the mounting pad may differ from case to case as 13%, while the moment applied to the mounting pad and disc results in a strain difference of 2% [63]. Thus, for precise examinations, the moment should be applied at the mounting pad and disc [64].

In our examination, we do not intend to follow standards of test implementation directly. The aim of the fatigue station we have built is to perform the accelerated fatigue of steel automotive wheel rims in conditions like the natural exploitation of standard cars. As a standard car, we assume 1500 kg of mass on four wheels based on 15-inch rims. We take that standard turns are made on a circle with a radius of 20 m, with a linear speed of 20 km/h, and the wheels do not skid. For such assumptions and basic estimations, the bending torque acting on the rim is 214 Nm. We also assume that the fatigue station should provide 60 percent of the designated load and run continuously. The proposed value is between what the UN ECE Regulation R124 recommends for replacement wheels, Annex 6, where two independent examinations at 50% and 75% of maximum side force were defined. The fatigue station simplified scheme and its view are presented in Figure 8. During the experiments, the fatigue station was attached to a massive base that weighed 5 tons. The fatigue station of the wheel rim is characterized by the following parameters: moment arm length of 110 cm, eccentric mass of 3 kg, eccentricity of 25 cm, and rotation speed of 450 revolutions per minute. Thus, the bending torque is 129 Nm, which is 60% of the assumed standard cornering of a typical car.

Based on the speed of the standard vehicle movement, a relationship between the duration of the fatigue test and the expected mileage of the vehicle can be estimated. The expected mileage for the test of the assumed vehicle, for a 15-inch wheel rim equipped with a standard tire, 0.055 the average ratio of curvature to linear movement, 450 revolutions per minute, is 1200 km per hour of test. Most used rim testing can be completed twice a year, during tire changes for heavily used vehicles and at the annual examination when all-season tires are in use. Heavy car use during the year in Germany equates to approximately 19,000 km, while the average annual mileage of cars in the European Union can be estimated as 11,000 km [65], where all-season tires are in use. This timestamp of rim examination can be related to 9500 km up to 11,000 km. In our case, we assumed that 8 h of duration of the fatigue test corresponded to 9600 km of expected mileage in mixed traffic conditions.

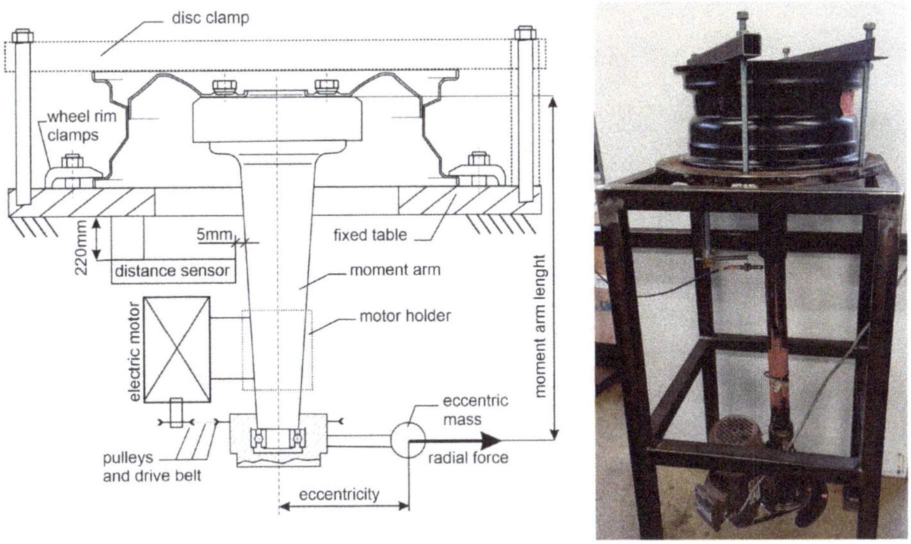

(**a**) (**b**)

Figure 8. Mechanical functions of the fatigue station: (**a**) Schematic drawing; (**b**) View.

The fatigue station was equipped with an eddy current distance sensor model CW10 cooperating via the OP200 interface with the input of an acquisition card NI USB 6343 used by the EMA measurement system. A distance sensor was used to monitor the deflection of the torque arm with 2% accuracy. This distance during the rotation of the eccentric mass is correlated with the deflection of the rim disc. In turn, this deflection is related to the material parameters of the wheel. According to the ISO 3006 standard, the rim disc deflection must not increase by more than 20% during the 50,000 eccentric mass rotations test. Exceeding this condition with properly tightened bolts means exceeding the permissible fatigue of the material and is the signal for fatigue procedure termination. Thus, in the fatigue test, it is also essential to maintain the pressure of the bolts. The bolts must be pre-tightened to a torque of 120 Nm and the torque of the screw during the experiment should be between 96 Nm and 120 Nm. The range was tested after 10,000 rotations of eccentric mass, 22 min after the start of the first step of the fatigue procedure. For this purpose, two Jonnesway type T27340N torque wrenches calibrated in accordance with the PN-ISO 6789 standard [66] were used. The first torque wrench was set to 120 Nm, while the second to 96 Nm. The second wrench was used to test the lowest value of the moment. In all the recorded experiments, that value was not crossed. Crossing this value automatically ended the fatigue experiment. However, sometimes the bolts became acceptably loose and had to be tightened with the first wrench with a torque of 120 Nm.

6. Experimental Laboratory Results

This section presents detailed data of laboratory examinations based on a single fatigue cycle shown in Figure 9. The fatigue planned unit duration is related to the expected mileage in mixed traffic conditions. The complete cycle of laboratory experimental investigations includes a series of tests consisting of EMA, run-out measurements, and a single fatigue cycle. The process of laboratory investigations continues until one of the end conditions of the test is met.

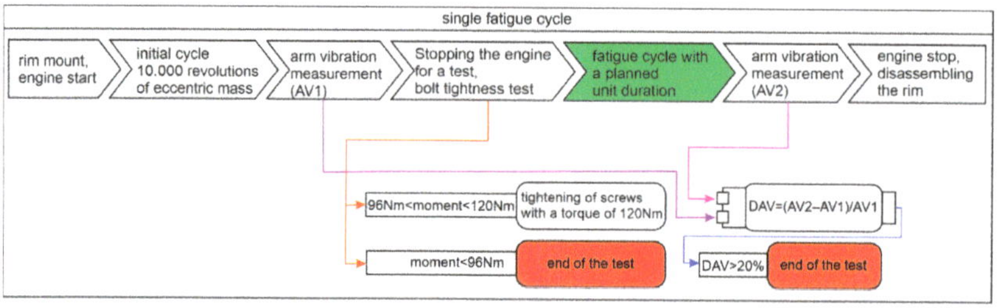

Figure 9. The scheme of a single fatigue cycle.

6.1. New Rim Type 6Jx15H2 from OEM Examinations with the Sequential Use of the Fatigue Station

The rim in the OEM new factory condition was examined using the fatigue station and the EMA and run-out measurement system. The duration of a single fatigue cycle was 8 h. The axial and radial run-out values of the rim on the inner and outer sides of the flange are presented in Figure 10. The condition for the end of the test was met in 40 h, as the increase in moment arm oscillations increased above 20%. The obtained results of run-out values show that after 32 h of testing, the rim lost tolerance of dimensions.

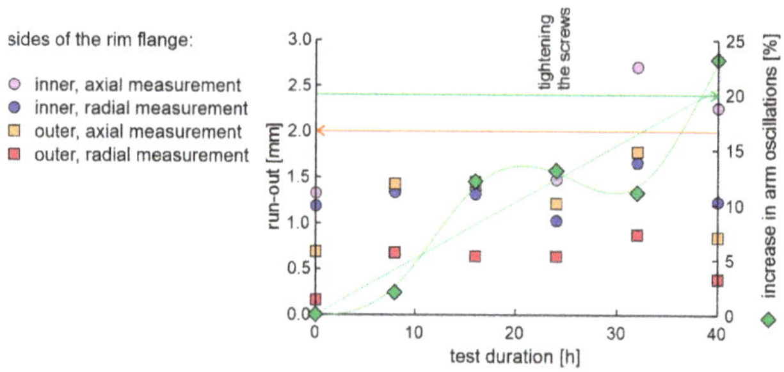

Figure 10. The axial and radial run-out values and increase in moment arm oscillations of the new rim from OEM in a fatigue test.

The average values and differences in natural frequencies are presented in Figure 11.

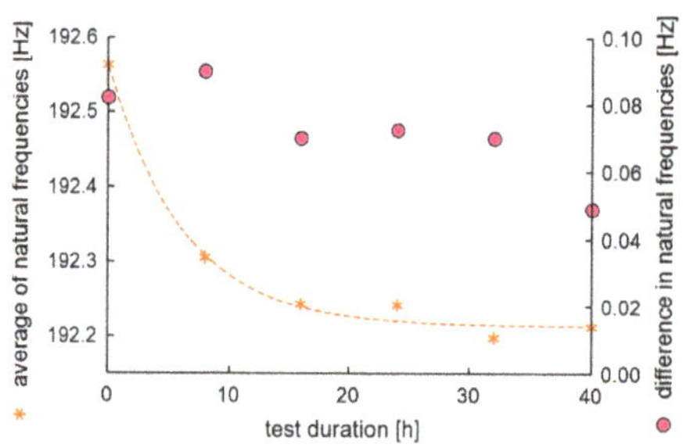

Figure 11. Average values and differences in natural frequencies of the rim of OEM in a standard fatigue test.

In the experiment, the average value of natural frequencies decreased, while the difference in natural frequencies can be characterized by decreased tendency. It can be postulated that the fatigue station performs a peculiar mechanical process related to rim running-in. The average values and dispersion of damping factors are presented in Figure 12.

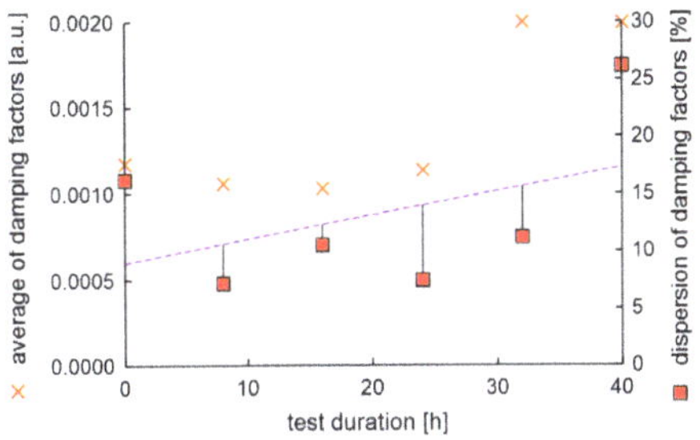

Figure 12. Average values and dispersion of damping factors of the rim from OEM in a standard fatigue test.

Results show that the average damping factor increases when the rim loses tolerance of dimensions. In this case, the dispersion of damping factors increases, but with a delay. It can also be seen that the local minimum dispersion of damping factors coincides with the cycle in which the rim mounting bolts were tightened. However, it should be noted that a significant increase in the dispersion of damping factors coincides with an increase in moment arm oscillations above 20%.

6.2. New Rim Type 6Jx15H2 from ARM—Sequential Tests Followed by Continuous Tests until the End of Fatigue Test Condition Is Met

The rim from ARM was examined in a factory-new condition. The axial and radial run-out values of the rim on the inner and outer sides of the rim flange and the increase in moment arm oscillations are presented in Figure 13. The same situation of tightening the screws at the third fatigue cycle occurs for the new rims from OEM and ARM.

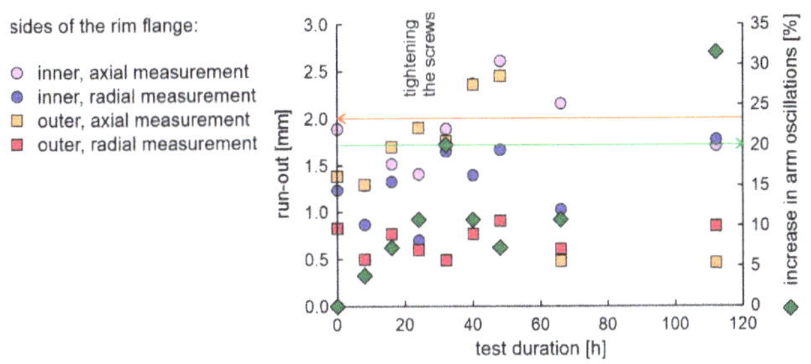

Figure 13. The axial and radial run-out values and increase in moment arm oscillations of the new rim from ARM.

Since the increase in moment arm oscillation after 32 h of fatigue is 20% and is not greater than the acceptable moment arm vibration, successive fatigue cycles were performed. The rim run-out that exceeds 2.5 mm occurs at 40 h of fatigue, while the highest values of shape distortion occur for 48 h of test. The time of these distortions is between the local maxima in moment arm oscillations. After 48 h, the following examination was at 64 h; then, the fatigue test was run until the oscillation increase of the momentum arm exceeded 20%, which occurred at 112 h. As can be seen, such an experiment does not result in signal values that could be considered as directly preceding crossing the limits of the permissible material fatigue. Interestingly, the run-out decreases with an increase in oscillations of the moment arm up to 32%, much greater than the allowable 20%. A similar situation occurred in a previous examination where the increase in momentum arm oscillations above 20% occurred together with a local decrease of run-out values.

The average values and differences in natural frequencies are presented in Figure 14. In this case, the average values and differences in natural frequencies initially decrease; this can be again related to mechanical running-in of the rim in initially new factory conditions. Higher values of differences in natural frequencies than 0.075 Hz are linked with a rim that does not meet run-out standards, see Figure 13.

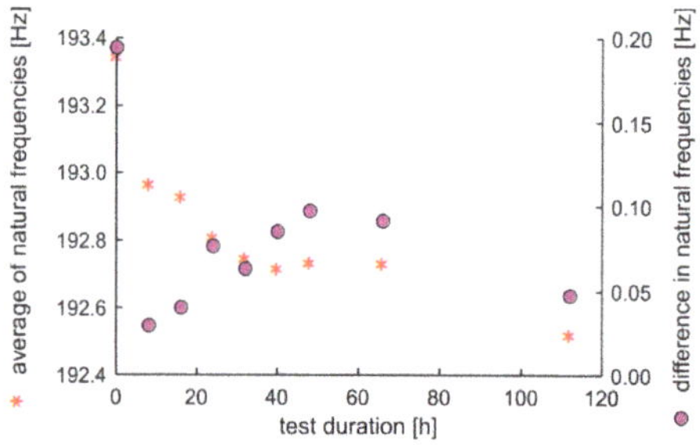

Figure 14. Average values and differences in natural frequencies of the new rim from ARM in an extended fatigue test.

The average values and dispersion of damping factors are presented in Figure 15. The results show that the average damping factor increases with the time of fatigue test duration. The dispersion of damping factor initially decreases when the test duration increases up to 16 h, then stabilizes and increases. The dispersion of damping factors versus test duration is not a monotonic function. The same as in the previous study, the maximum value of the dispersion of damping factors is related to the maximum increase of oscillations of the moment arm.

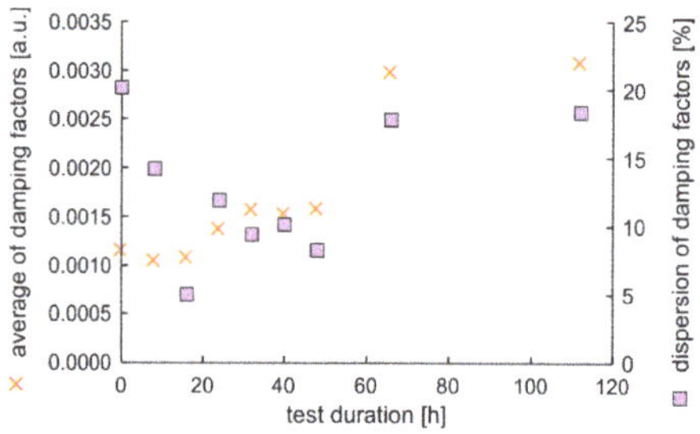

Figure 15. Average values and dispersion of damping factors of the new rim from ARM in an extended fatigue test.

6.3. New Rim Type 6Jx15H2 of ARM, with a Rusty Hub—Examinations with the Use of the Fatigue Station

The fatigue station for the dynamic wheel cornering fatigue test can also be used to simulate different rim–hub contacts. The current trial used a slightly rusty hub to mount the rim. The rim of the same ARM is examined as in Section 6.2. The axial and radial run-out values of the rim on the inner and outer sides of the rim flange are presented in Figure 16. Values indicate that the run-out norm was exceeded much earlier than in Sections 6.1 and 6.2.

The increase in oscillation of the moment arm occurs at a time during the fatigue test similar to the test described in Section 6.2. For the present test, it occurs after 48 h, whereas previously, it occurred after 40 h and after 32 h. The average values and differences in natural frequencies are presented in Figure 17.

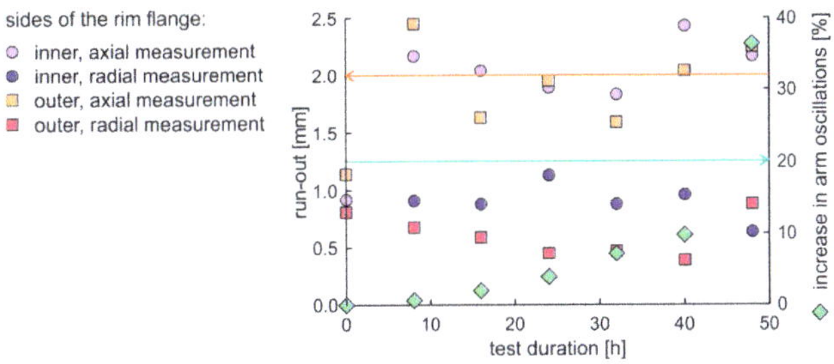

Figure 16. The axial and radial run-out and increase in moment arm oscillation values of the new rim from ARM in the fatigue test with a rusty hub.

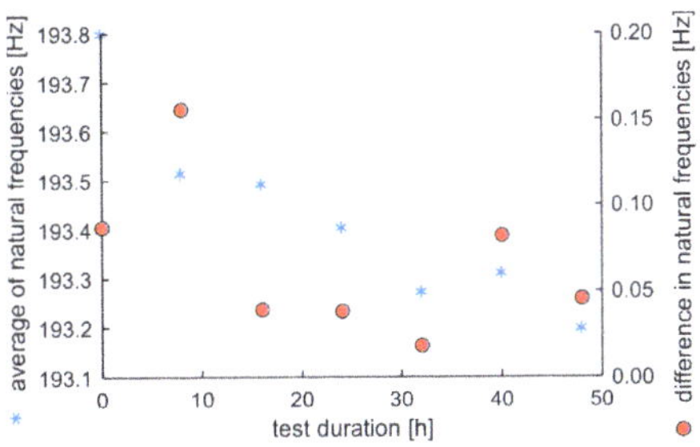

Figure 17. Average values and differences in natural frequencies of the new rim from ARM in the fatigue test with a rusty hub.

The course of the average natural frequency shows, as before (see Figure 10), a characteristic decrease during the fatigue test. However, the course of difference in natural frequencies is now characterized by greater values than presented in Figure 13, but it cannot be related to mechanical running-in of the rim in the initially new factory condition. The difference in natural frequencies is greater than 0.16 Hz at 8 h. This fact corresponds with the substantial run-out in the outer flange of the rim, whose axial deviation is 2.5 mm.

The average values and dispersion of damping factors are presented in Figure 18.

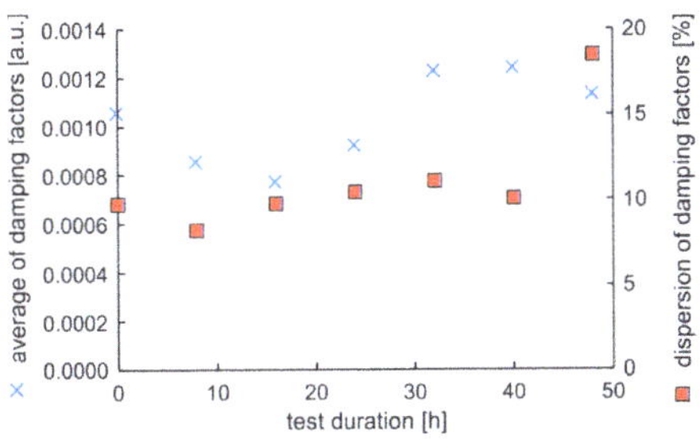

Figure 18. Average values and dispersion of damping factors of the new rim from ARM in the fatigue test with a rusty hub.

Again, the results show that the average of the damping factor initially decreases and then increases. Also, in this case, the dispersion of damping factors is highest when the moment arm's oscillations increase is maximal. Also, the maximum dispersion of the damping factor is preceded by an increase in the average value of damping factors.

From the 40 h of test time, signs of degradation of the rim bolt holes and acoustic clicking were observed. The degradation of the rim bolt holes after 40 h of test duration is presented in Figure 19.

Figure 19. Degradation of bolt holes after 40 h of the fatigue test with rusty hub use for the new rim from ARM.

6.4. New Rims Type 6Jx15H2 of LCM

New rims of LCM sometimes seem to be a good selection from an economic point of view. Thus, two new rims were used in examinations. The axial and radial run-out values of the rim on the inner and outer sides of the flange are presented in Figure 20. During the 40 h of the fatigue test, the PN standard is completely met, while the ETRTO standard is exceeded only for measurements after 8 and 24 h. These results are better than for the new rim of ARM.

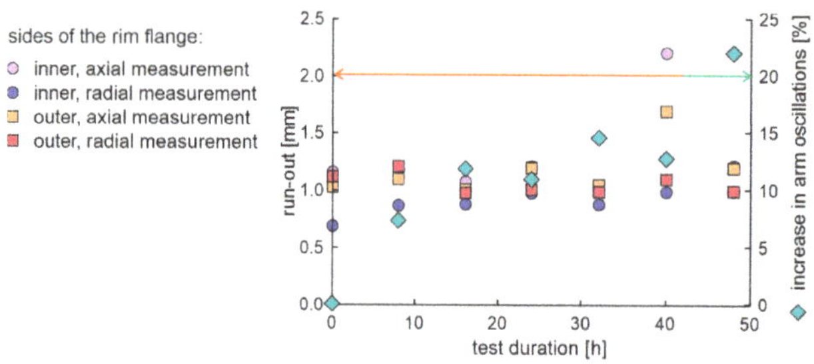

Figure 20. The axial and radial run-out values of the first new rim of LCM in a fatigue test.

The average values and differences in natural frequencies of the first rim of LCM are presented in Figure 21. The average of natural frequencies and the difference in natural frequencies' tendency to decrease with test duration is evident.

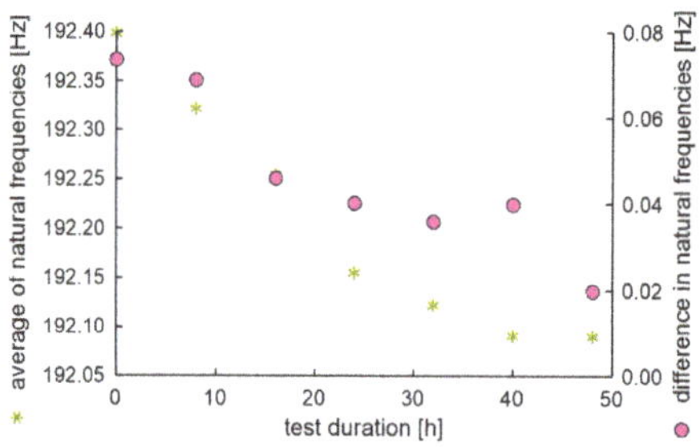

Figure 21. Average values and differences in natural frequencies of the first new rim of LCM.

The average values and dispersion of damping factors of the first rim of LCM are presented in Figure 22. The dispersion of damping factors' maximum is again at the end of the test and corresponds with the time of maximum increase of moment arm oscillations. The only issue that can be found in the first new rim of LCM is the high initial value of the dispersion of damping factors. However, this value unexpectedly decreases during the fatigue test.

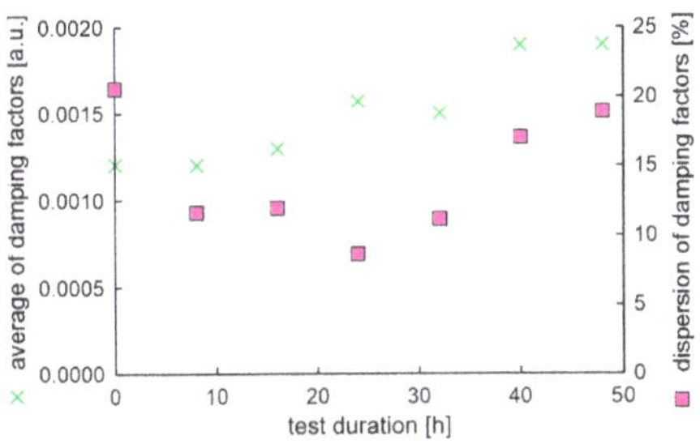

Figure 22. Average values and dispersion of damping factors of the first new rim of LCM in a fatigue test.

The second rim of LCM was not so fine. Its run-out values do not fit with the ETRTO and PN standards; its construction imperfections are visible in Figure 23. For reasons of safety of the testing personnel, this wheel was not subjected to fatigue tests.

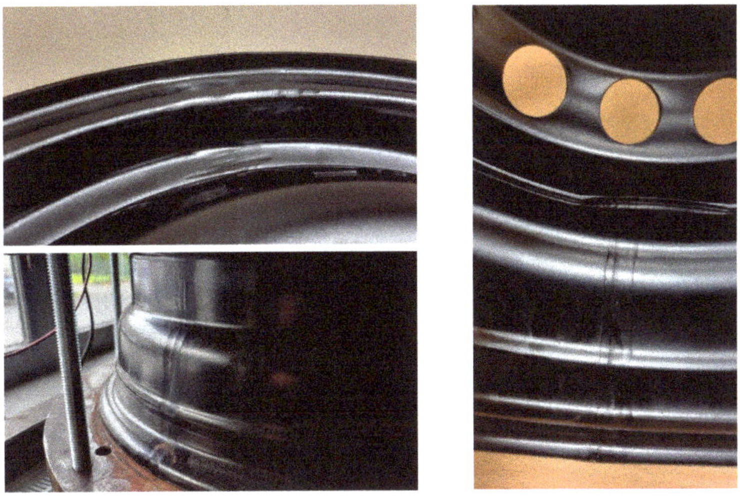

Figure 23. Views of imperfections of the second rim from LCM.

6.5. Used Rim from OEM That Meets the ETRTO Standard but Does Not Meet the PN-93S Standard

The used rim from OEM with an initial condition meeting the ETRTO standard but not meeting the PN-93S standard was examined. Based on visual inspection and run-out measurements, it can be intuitively concluded that the tested rim is minimally worn. However, as the exact wear characteristics of the rim before the tests were not known, the fatigue cycle interval was reduced from 8 to 3 h. The increase of moment arm oscillations and axial and radial run-out values of the rim on the inner and outer sides of the flange are presented in Figure 24.

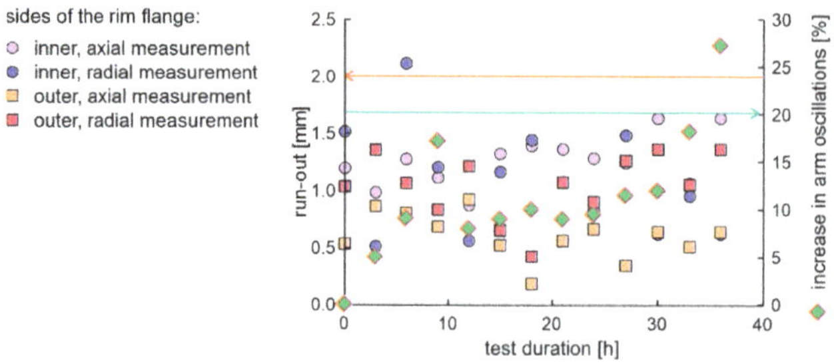

Figure 24. The increase of moment arm oscillations and run-out values of the used rim of OEM in a fatigue test.

The obtained results of the run-out indicate that after 6 h of the fatigue test, the rim lost its dimensional tolerance, but its shape somehow regenerated. It can be noticed that the used rim lost its dimensional tolerance much earlier than a new rim from the same OEM (Figure 10 Section 6.1). The average values and differences in natural frequencies are presented in Figure 25.

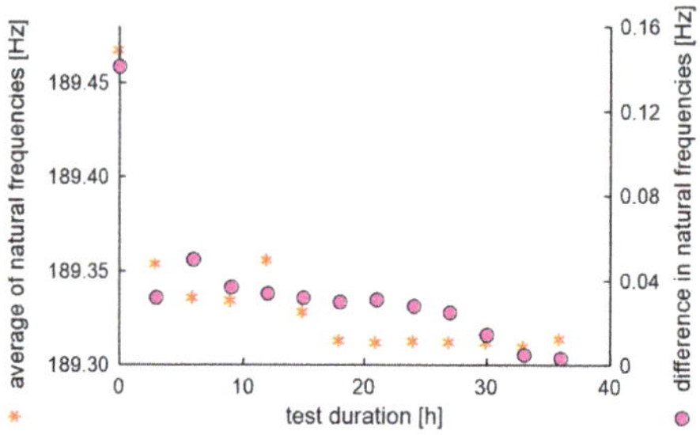

Figure 25. Average values and differences in natural frequencies of the used rim of OEM in a standard fatigue test with reduced time between EMA tests.

The EMA initial test results show that the difference in natural frequencies is more significant than that obtained for the new rim of the same producer, which can be expected, as the rim was used. Interestingly, the local maximum difference in natural frequencies at 6 h correlates with the maximum run-out. The average values and dispersion of damping factors are presented in Figure 26.

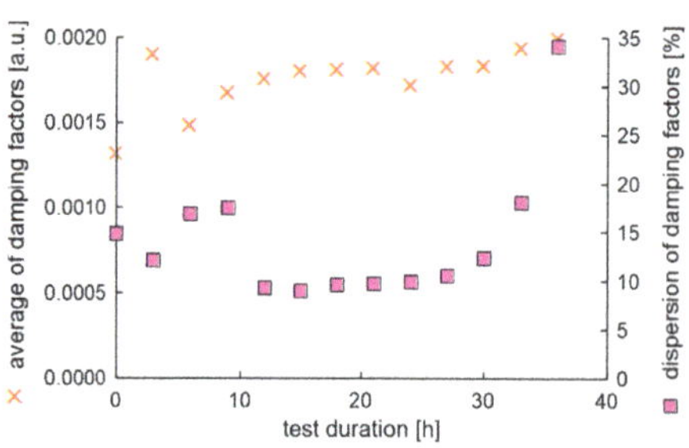

Figure 26. Average values and dispersion of damping factors of the used rim of OEM in the fatigue test.

As in the previous examination, the average of damping factor values increased in a non-monotonic way, while the dispersion of damping factors exceeded 17% sometime after the dimension degradation point. Also, the maximum dispersion of damping factors is correlated with the maximum increase of moment arm oscillations.

6.6. New Rims Type 6Jx15H2 of OEM—Examination of the Rim with Artificially Generated Cracks

One of the typical places of rim cracks is located between the vent holes. Waiting for such rim degradation to be produced by the fatigue station is problematic from the experiment's security point of view. However, artificially made cracks of the rim enable a controlled investigation of parameters of the degraded rim. The artificial cracks were realized by mechanical cutting along the red lines marked in Figure 27a. The effects are shown in Figure 27b.

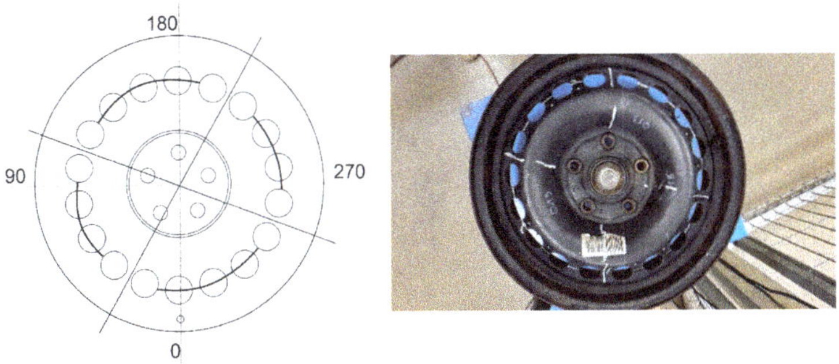

(**a**) Schematic drawing (**b**) View

Figure 27. Artificially formed cracks on the rim disc: (**a**) Schematic drawing; (**b**) View.

The cuttings resulted in a 4 Hz decrease in natural frequency while the damping factor remained constant. Due to rim design modifications, the fatigue cycle interval was reduced from 8 to 3 h. Axial and radial run-out values of the cut rim presented in Figure 28 are not subject to serious rim quality suspicions.

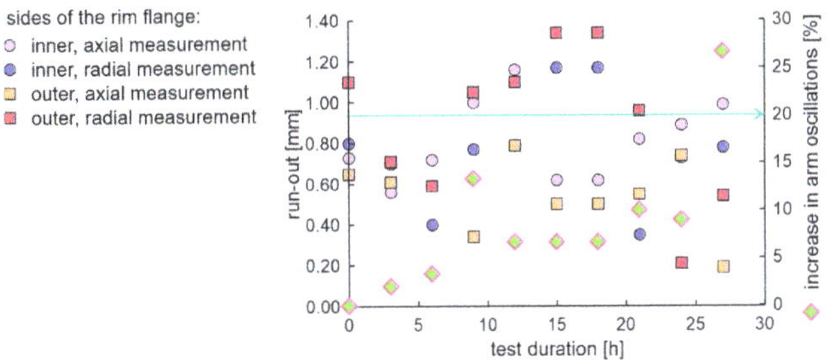

Figure 28. The increase in moment arm oscillations and run-out values of the new cut rim from OEM in a fatigue test.

The end condition of the test occurred in 27 h, which is faster than for any other new, undamaged rim. However, the run-out values, measured during the fatigue test of the rim with artificially generated cracks, are much smaller than for unmodified rims. The characteristics of natural frequencies during the fatigue test presented in Figure 29 appear normal.

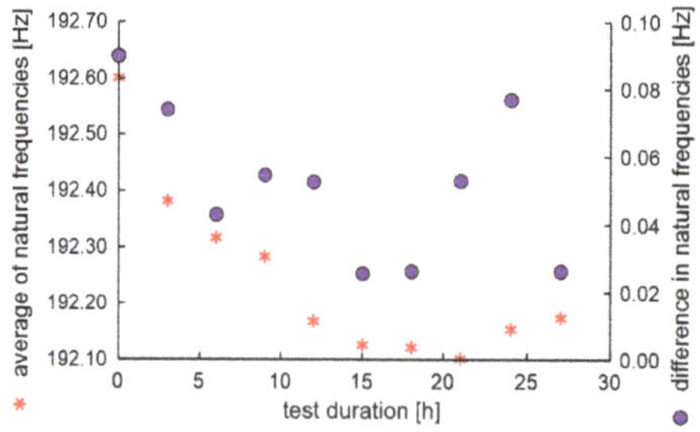

Figure 29. Average values and differences in natural frequencies of the cut rim of OEM in a standard fatigue test.

Measured damping factors during fatigue experiments presented in Figure 30 are also similar to the results obtained in previous sections.

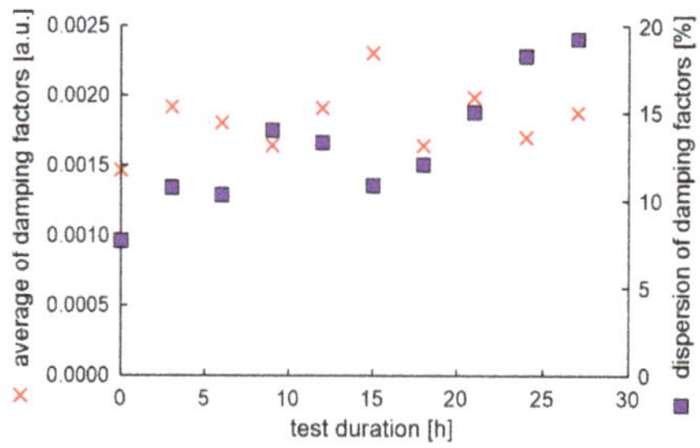

Figure 30. Average values and dispersion of damping factors of the cut rim of OEM in a fatigue test.

Again, most importantly, the maximum dispersion of the damping factor happens at the same time as the maximum increase of moment arm oscillations.

7. Discussion of Results of Section 6

The discussion considers cyclic destructive fatigue tests separated by non-destructive classical visual inspection, rim run-out tests, and the application of experimental modal tests presented in Section 6, as well as the verification results included in this section. The first result of the comparative studies presented in Sections 6.1 and 6.2 is that the cyclic use of the fatigue station and the coupled wheel rim tests give much more information than the continuous use of the fatigue station until the fatigue test end condition is reached.

The results of the cyclic investigations indicate several repetitive relationships. The EMA shows that a downward trend characterizes the course of the natural frequency versus fatigue test duration. However, regardless of which rim is tested, the first period of the fatigue test causes the greatest decrease in frequency. Thus, the initial reduction in natural frequencies can be interpreted as the fatigue station performing a peculiar mechanical process on the tested rim. Moreover, the natural frequencies of rims with cracks can differ significantly, for example, by 4 Hz (in Section 6.6–192.64 Hz). However, the direct comparison of the real initial natural frequency of rims can be deceptive; for example, a homogeneous rim from OEM natural frequency was 192.56 Hz (Section 6.1), while a cracked rim from OEM was 192.64 Hz. Therefore, an analysis of the technical condition of the wheel based on natural frequencies is only possible if its exact historical data is known.

Moreover, looking at the results of the measurement of natural frequency differences determined by Formula (1) in Section 4 and relating them to the examination in the low band of the new rim (Figure 7), one can point out that, in the proposed experimental set-up, the result of natural frequencies mixing is masked by the constant component of experimental FFT. The FFT width is about 0.4 Hz, while the difference in natural frequencies is lower than 0.2 Hz. However, based on the simulation and the results described in Section 7.2, geometric damage to the rims greater than 2.5 mm should produce an unmasked signal in a low band of mixing resonant modes. Therefore, the precise analysis of the technical condition of the wheel based on the low band requires a more advanced measurement system than the one presented in this publication.

Separate analysis of the rim run-out measurements obtained during the fatigue test can lead to the conclusion that determining the technical condition of the wheel rim according to run-out standards does not always make sense. The run-out values of the rim during the experiment are usually variable. Rim run-out values typically initially meet the standard, then may not meet it, or may be reduced and meet the standard again. It can be seen in almost all experiments, for example, in Section 6.5 in Figure 24 and Section 6.4 in Figure 20. In contrast, the averages of damping factors and dispersion of damping factors versus fatigue test duration are characterized by an upward trend.

7.1. Precise Technical Conditions Determination of Wheel Rims with the Use of Data Fusion

The real task is to determine the technical condition of a wheel of unknown history and any type using non-destructive testing. For this purpose, the data fusion of different methods can be helpful. The first and most important relation to be observed in the presented experiments is that the maximum increase in moment arm oscillations is related to the maximum dispersion of damping factors. Such data fusion is shown in Table 3.

Table 3. Data fusion of maximum increase in moment arm oscillations and maximum of dispersion of damping factors.

Examination Section	Time of Test [h]	Maximum of Increase in Moment Arm Oscillations [%]	Maximum of Dispersion of Damping Factors [%]
6.1	40	23.2	26.2
6.2	112	31.5	18.4
6.3	48	36.3	18.5
6.4	48	22.0	18.9
6.5	36	27.3	34.1

The first result of data fusion is that the minimum dispersion of damping factors, corresponding to an unacceptable increase in moment arm oscillations, is 18.4%. To threshold detection, the maximum value of the dispersion of damping factors that corresponds to an acceptable increase in moment arm oscillations is required. This data fusion of the cycle before the maximum increase in moment arm oscillations with the dispersion of damping factors is presented in Table 4. The maximum acceptable value of the dispersion of damping factors is 18.0%.

Table 4. Data fusion of cycle before maximum increase in moment arm oscillations with the dispersion of damping factors.

Examination Section	Time of Test [h]	Maximum of Increase in Moment Arm Oscillations [%]	Maximum of Dispersion of Damping Factors [%]
6.1	32	11.1	11.2
6.2	32	20.0 *	9.4
	66	10.7	17.8
6.3	40	9.7	10.1
6.4	40	12.7	17.1
6.5	33	18.2	18.0

* Local maximum of increase in moment arm oscillations.

As the dispersion of damping factors is determined with 0.1% tolerance, the threshold between acceptable and not acceptable technical conditions of the rim can be set at 18.2%.

7.2. Field Testing—Used Rims Type 51/2Jx14H2 of OEM

The actual task is to determine the technical condition of a wheel of unknown history and any type using non-destructive testing. The examinations of the used 14-inch rim of type 51/2Jx14H2 from OEM are presented here. One rim was used, and the other was brand new. The initial state of those rims described with the run-out standards meets the ETRTO standard but does not meet the PN-93S standard. The used rim parameters are presented in Table 5. The initial visual condition of the used rim can be described as acceptable, as there is no visible rust or cracks, and the air does not leak from the tire. In such a state, the low band vibration excited of the rim is presented in Figure 31.

Table 5. Run-out values of the rim during the fatigue experiment in natural heavy conditions.

Examination	Run-Out [mm]					
	Inner Flange		Outer Flange		Standard	
	Axial	Radial	Axial	Radial	PN-93S	ETRTO
Initial (used rim)	1.65	1.02	0.64	0.57	not met	met
After hitting the curb	4.35	2.75	4.52	2.5	not met	not met
After six months of operation on forest roads	3.53	1.77	2.92	1.12	not met	not met

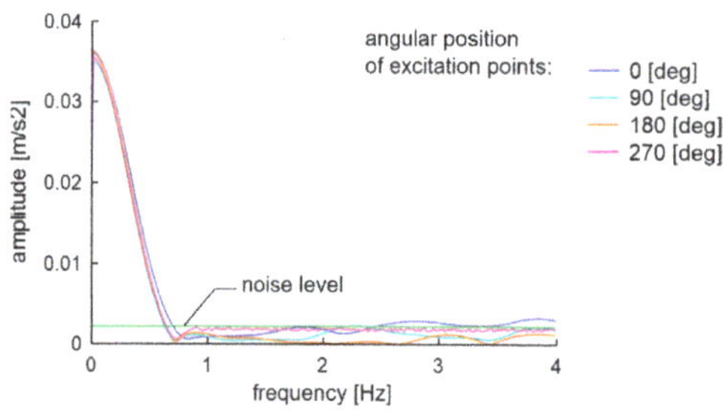

Figure 31. Low band spectra of EMA of 51/2Jx14H2 rim in an acceptable technical condition with a visual and run-out examination at the beginning of the experiment.

However, the visual and run-out examination does not involve EMA and the value of dispersion of damping factors, which in this case is 22.4%. Thus, according to the presented investigation, this rim must be classified as unfit for use. However, an experimental verification of this statement must be conducted. As experimental conditions, classic cases of mechanical damage to the rim were assumed. Hitting the curb of a nominal load vehicle with adequately inflated wheels and operation on partially paved forest roads with numerous uplifts caused by the roots of old trees were tested. For safety reasons, as the used rim was classified as unfit, this experiment was performed for a small trailer towed by a passenger car with one used and one new rim.

The first examination was hitting the curb at 20 km/h. The second examination was a six-month operation on partially paved forest roads with numerous uplifts caused by the roots of old trees. During the first experiment, the damage of the used rim was significant, as shown in the run-out values and the view presented in Figure 32, while nothing occurred to the new rim.

The run-out values of the used rim during the experiment are presented in Table 5. It is visible that the rim does not meet the run-out standards after hitting the curb. However, interestingly, the wheel passed the soapy water test for rim–tire integrity. Therefore, further operation was on non-public roads.

(a) (b)

Figure 32. View of rim: (**a**) After it hit a curb; (**b**) Further six months of operation on partially paved forest roads.

The obtained results show that during operation in the conditions of forest roads, the heavy degradation of the radial component of the geometrical dimension somehow decreases. Differences in natural frequencies and the dispersion of damping factors of the used rim are presented in Figure 33, which confirms this thesis. Still, all measured values here are much higher than in previously submitted laboratory cases of rim degradation with fatigue station use, as shown in Section 6.1, Section 6.2, Section 6.3, Sections 6.4 and 6.5.

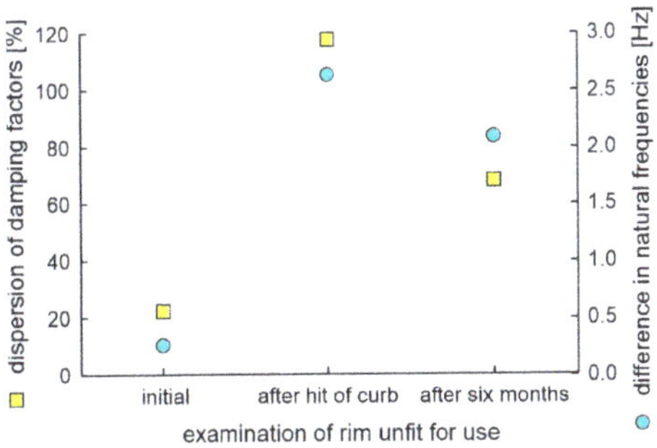

Figure 33. Differences in natural frequencies and dispersion of damping factors of unfit for use rim type 51/2Jx14H2 of OEM.

The low band signals after two experiments of unfit for use rims are presented in Figure 34. Characteristics shown in Figure 34 confirm the assumptions of the measured difference in natural frequencies. The result of frequency mixing is visible in the low band when differences in natural frequencies are greater than 0.4 Hz. The soapy water test carried out at the end of the sixth month of operation showed a loss of integrity of the unfit for use rim and tire and the examination was finished.

The postulate that the technical condition of the rim is related to the dispersion of damping factors, and the possibility of classification of technical condition based on simple threshold criteria has been proved. The 14-inch wheel with 22.4% dispersion of damping factors was unacceptably deformed when driving over a curb at 20 km/h. The rim with the initial dispersion of damping factor 8.1 passed all driving tests, including hitting the curb of a nominal load vehicle with properly inflated wheels and operation on partially paved forest roads.

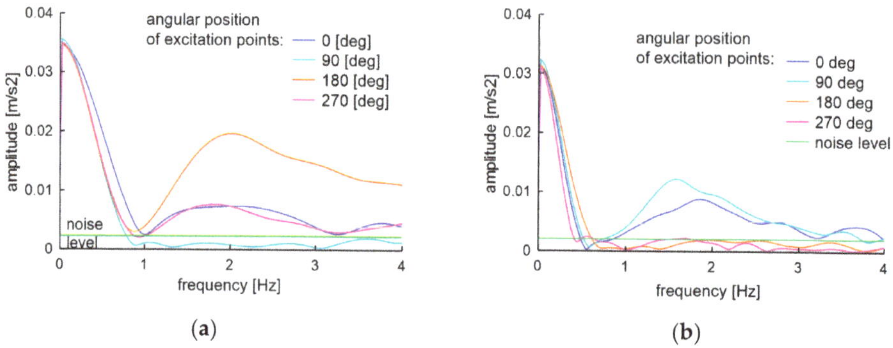

Figure 34. Low band signals of unfit for use rim: (**a**) After a hit of the curb; (**b**) Further six months of operation on partially paved forest roads.

7.3. Prediction of Remaining Time of Use

The prediction of the remaining service life of a wheel rim in relation to its unknown mileage and the unknown nature of its operation must be based on the determination of its current technical condition and signals prior to its unusable state. In Section 7.1, we have shown that the current technical condition of a wheel rim can be precisely determined by the threshold value of the damping factors dispersion. As a signal prior to the unusable state, a characteristic signal with repeating features should be selected. In almost all cases, the time course of the moment arm oscillation is similar to the shape of the μ mark. The first local maximum is lower than the second. The second maximum is in our examination always the termination of the fatigue test; for example, see Figure 10. The only exception was in Section 6.3 in the case of a rusty hub. However, the use of oscillations of moment arm to define or predict the remaining time of use is at least a semi-destructive test. Therefore, it is essential to check if other parameters measured in non-destructive tests can indicate that the threshold value mentioned is approaching, by way of the time relation analysis from the experiment termination. The time relation refers to a set of research sequences preceding the end of the experiment:

$$\{T(te-l)\}, \tag{3}$$

where T is the time before the end of the experiment (te) preceded by the number of sequences equal to l. The cyclic change of characteristic parameters can be described with the fold change $FC(P)$ according to Equation (4):

$$FC(P) = \frac{P(T(n))}{P(T(m))}, \tag{4}$$

where P is the measured parameter from the set {RO—run-out, aDF—an average of damping factors, dDF—dispersion of damping factors}, n and m are numbers of examination sequence where $n < m$. The above definitions allow for the construction of Table 6, in which the parameters indicating that the threshold value of the dissipation of damping factors is approaching are characterized.

The data in Table 6 show that there is a set of parameters that indicate the approaching end of service life. In three out of five cases, the end of the rim's service life is preceded by a fold change in the dispersion of the damping factor greater or equal to 1.46. It seems that this parameter fold change can act as a separate condition of indication. In four out of five cases, the end of the rim service life is preceded by a fold change in the average of damping factors. In three cases out of four, the fold change of the average of damping factors is correlated with an increased value of run-out greater than 2.2 mm. The fold change in the average of damping factors is coupled with the fold change of dispersion of damping factors for the experiment depicted in Section 6.2, which was conducted with different sequences. Thus, the binary indicator of approaching the end of the steel car wheel rim can be written in the form of Formula (5):

$$I = \begin{Bmatrix} 1 & FC(dDF) > 1.71 \\ 1 & FC(aDF) > 1.33 \; and \; RO > 2.43 \\ 0 & in \; other \; cases \end{Bmatrix}, \quad (5)$$

where I is the binary indicator, while the other designations are as in Equation (4). Unfortunately, calculating the proposed indicator requires at least two consecutive rim tests for the vehicle's standard annual mileage. According to our estimations implemented here, when the indicator equals one, the expected left mileage in mixed traffic conditions is about 9600 km. When the indicator is equal to zero, the rim can be used without fear.

Table 6. The set of parameters indicating that the threshold value of the dissipation of damping factors is approaching.

Examination Section	Characteristic Sign before Maximum Dispersion of Damping Factors	The Characteristic Sign and Its Value	Time Relation	Time Indicated [h]
6.1	run-out of inner side of flange > 2.2 mm	2.71 mm	$\{T(te-1)\}$	8
	local increase in the average of damping factors	$FC(aDF) = 1.76$	$\{T(te-2), T(te-1)\}$	8
6.2	local increase in the average of damping factors	$FC(aDF) = 1.89$	$\{T(te-2), T(te-1)\}$	46 *
	local increase of dispersion of damping factors	$FC(dDF) = 2.15$	$\{T(te-2), T(te-1)\}$	46 *
6.3	run-out of inner side of flange > 2.2 mm	2.43 mm	$\{T(te-1)\}$	8
	local increase in average of damping factors	$FC(aDF) = 1.33$	$\{T(te-3), T(te-2)\}$	16
6.4	run-out of inner side of flange > 2.2 mm	2.21 mm	$\{T(te-1)\}$	8
	local increase in average of damping factors	$FC(aDF) = 1.26$	$\{T(te-2), T(te-1)\}$	8
	local increase of dispersion of damping factors	$FC(dDF) = 1.53$	$\{T(te-2), T(te-1)\}$	8
6.5	local increase of dispersion of damping factors	$FC(dDF) = 1.46$ $FC(dDF) = 1.71$	$\{T(te-2), T(te-1)\}$ $\{T(te-3), T(te-1)\}$	3 6

* With the aging procedure in accordance with the standard, no aging sequence with a time equal to 8 h before the end of the experiment.

7.4. Presented Method Versus Computer Vision Methods

It should be noted that the presented method uses data fusion of modal analysis and geometrical measurements to identify the technical condition and indicate the approaching end of the service life of the steel automotive wheel rims. This method belongs to the noninvasive measurement group. However, the noninvasive measurement group is wider. Computer vision techniques are also used to analyze automotive wheel rims; see Table 7.

The detection results of the proposed method shown in Section 6 and the computer vision methods are quite different. The methods presented in Table 7 are primarily oriented to be implemented at the rim factories, while the proposed method is intended to be used at car repair stations.

Table 7. Sample of noninvasive automotive wheel rim examinations based on computer vision.

Method	Requirements	Method Application	Examined Object	Detection Result	Ref.
X-ray, digital radiography (DR)	radioscopic images taken at different positions and reference image	image segmentation analysis	aluminum castings	internal cracks in the material structure	[67]
X-ray, digital radiography (DR)	series of images with different radiation intensity	adaptive threshold and morphological reconstruction of structure	automotive aluminum rim	internal defects such as gas holes and shrinkage cavities	[68]
X-ray, NIR photography	radioscopic and NIR images of the same object	convolutional neural network application to image recognition	automotive aluminum rim	internal casting defects	[69]
VIS photography	series of photography of multiple views	feature analysis in numerous views	automotive aluminum rim	geometry imperfections	[70]

8. Conclusions

A set of conclusions can be grouped depending on the type of investigation methods: simulation, laboratory experiment, field trial case examinations, and data fusion.

The simulation results of modal analysis enable the proper construction of laboratory measurement systems and fatigue stations. Analysis of traffic conditions, service maintenance, and standards applied by rim producers of steel car wheel rims enable the proposition of cyclic examination, including initial measurement and following fatigue and measurement sequences. The presented investigation shows that sequential fatigue and rim characterization provide many more possibilities for the analysis of the character of wheel rim parameters than standard continuous tests until the end of fatigue test condition is met.

The practical side of the numerical modal analysis shows the possibility to define the degree of the rim degradation based on the variability of natural frequencies, whereby when the scale of the damage is greater, the bigger the frequency difference. According to simulations, the difference in natural frequencies for an ideal automotive rim should be lower than 0.21 Hz, while the difference may be as high as 12 Hz for a completely damaged rim. Experiments confirm that values of peaks for an ideal rim that met standards are below 0.2 Hz and for significantly degraded rim, the differences can reach 4 Hz. Experimental modal analysis also shows that the difference of natural frequencies is related to low band signals which range from 0 to 4 Hz, as these signals come from excited frequencies mixing. Thus, peaks at the low band can be a measure of geometrical rim degradation.

The experimental data shows that processes of wheel rim degradation are too complex to describe the technical condition of the rim with a single standard or one basic parameter. Our examination shows that the dispersion of damping factors of vibration excited at four points that are spread angular at the outer rim flange can act as a precise tool for binary identification of the technical condition of the rim with an unknown history of exploitation. The dispersion of damping factors acts as the threshold value for acceptance of the material health of the rim. We showed that this parameter is strongly related to the increase of momentum arm oscillations depicted in standards of precise rim factory testing with the destructive method. At the same time, the dispersion of damping factors can be measured in a non-destructive way, and it appears to work for rims with an unknown history. Moreover, the threshold value obtained in experiments for a 15-inch rim works appropriately for a 14-inch rim.

A more complex situation is related to predicting the end of the service life of the rim. This prediction may be based on natural frequencies, damping factors, and manufacturer data. However, that is only possible if its exact sequence of historical data, e.g., collected by the dealership during routine inspections, is known, and if the vehicle is in ideal condition and exploited in standard conditions. However, the technical condition of a vehicle can affect the expected remaining time of rim use.

On the other hand, the proposed examination results give an indication of the approaching end of rim service life with acceptable complexity. The indication requires two sequential examinations of the rim in standard periods related to the average annual mileage of the vehicle. For the proposed method, the data must include two sets of three parameters: run-out, average of damping factors, and dispersion of damping factors. On this basis, the indication of approaching the end of rim service life is advisable in advance, which can be described as about 10,000 km left of mileage.

The current limitation of the presented method is that it was tested only for steel car wheel rims. Given the above, the natural direction of further work is to show the technique's operation for automotive alloy rims and rims for particular purposes, e.g., for agricultural and military vehicles. Of course, expanding the area of research will force the automation of data processing and access. Furthermore, increases in data will enable statistical analysis of the methodology. However, this research will require a significant investment in time.

Author Contributions: Conceptualization, M.B. and A.R.; methodology, M.B.; software, L.Z. and A.R.; validation, M.B. and A.R.; formal analysis, M.B.; investigation, M.B. and A.R.; resources, A.R. and L.Z.; data curation, M.B. and A.R.; writing—original draft preparation, M.B.; writing—review and editing, M.B. and M.L.K.-P.; visualization, M.B. and A.R.; supervision, M.L.K.-P. All authors have read and agreed to the published version of the manuscript.

Funding: This research received no external funding.

Institutional Review Board Statement: Not applicable.

Informed Consent Statement: Not applicable.

Data Availability Statement: Data are contained within the article.

Conflicts of Interest: The authors declare no conflicts of interest.

References

1. Akay, M.E.; Ridvanogullari, A. Optimisation of machining parameters of train wheel for shrink-fit application by considering surface roughness and chip morphology parameters. *Eng. Sci. Technol. Int. J.* **2020**, *23*, 1194–1207. [CrossRef]
2. Wan, X.; Shan, Y.; Liu, X.; Wang, H.; Wang, J. Simulation of biaxial wheel test and fatigue life estimation considering the influence of tire and wheel camber. *Adv. Eng. Softw.* **2016**, *92*, 57–64. [CrossRef]
3. Zanchini, M.; Longhi, D.; Mantovani, S.; Puglisi, F.; Giacalone, M. Fatigue and failure analysis of aluminium and composite automotive wheel rims: Experimental and numerical investigation. *Eng. Fail. Anal.* **2023**, *146*, 107064. [CrossRef]
4. Somayaji, A.; Marakala, N.; Somayaji, S.; Hebbale, A.M. Studies on modal analysis of aluminium based carbon fibre reinforced optimized alloy wheel used in automotive sectors. *Mater. Today Proc.* **2022**, *52*, 274–277. [CrossRef]
5. Slavič, J.; Boltezar, M.; Mrsnik, M.; Cesnik, M.; Javh, J. (Eds.) Experimental modal analysis. In *Vibration Fatigue by Spectral Methods*; Elsevier: Amsterdam, The Netherlands, 2020; pp. 75–98. ISBN 9780128223666.
6. Ravi, S.K.; Dong, P.; Wei, Z. Data-driven modeling of multiaxial fatigue in frequency domain. *Mar. Struct.* **2022**, *84*, 103201. [CrossRef]
7. He, J.; Fu, Z.F. (Eds.) Overview of modal analysis. In *Modal Analysis*; Butterworth-Heinemann: Oxford, UK, 2001; pp. 1–11.
8. Ewins, D.J. Mode of vibration. In *Encyclopedia of Vibration*; Braun, S., Ed.; Elsevier: Amsterdam, The Netherlands, 2001; pp. 838–844. [CrossRef]
9. Liu, X.; Feng, Q. Chapter 12-Statistical energy analysis of tire/road noise. In *Automotive Tire Noise and Vibrations*; Xu, W., Ed.; Butterworth-Heinemann: Oxford, UK, 2020; pp. 271–296. [CrossRef]
10. Fazail, M.N.B.; Chazot, J.D.; Lefebvre, G.; Atalla, N. Damping loss factor characterization of complex structures using a Green's function-based model. *J. Sound Vib.* **2023**, *552*, 117642. [CrossRef]
11. Feng, S.W.; Chang, W.J.; Lin, C.T.; Lee, S.Y.; Teng, N.C.; Huang, H.M. Modal damping factor detected with an impulse-forced vibration method provides additional information on osseointegration during dental implant healing. *Int. J. Oral Maxillofac Implant.* **2015**, *30*, 1333–1340. [CrossRef]
12. Sabri, M.; Rezal, M.; Mu'az, A.; Shahril, K.; Ihsan, J. Deformation behaviour analysis of car wheel rim under different loading using finite element method. *Int. J. Eng. Technol.* **2015**, *5*, 181–184.

13. Sharma, A.; Yadav, R.; Sharma, K. Optimization and investigation of automotive wheel rim for efficient performance of vehicle. *Mater. Today Proc.* **2021**, *45*, 3601–3604. [CrossRef]
14. Gadwala, W.K.; Babu, R.G. Modeling and analysis of car wheel rim for weight optimization to use additive manufacturing process. *Mater. Today Proc.* **2022**, *62*, 336–345. [CrossRef]
15. Hawkins, G.; Kumar, V. Structural analysis of alloy wheels. *J. Phys. Conf. Ser.* **2020**, *1478*, 012007. [CrossRef]
16. Richter, R. Destructive Testing of Aluminum Wheels. *SAE Tech. Pap.* **1970**, 700444. [CrossRef]
17. ZwickRoell, Testing of Wheels, Rims, and Tires, ZwickRoell Testing Machines and Systems for the Automotive Industry, Testing of Wheels, Rims, and Tires. ZwickRoell, Ulm Austria. Available online: https://www.zwickroell.com/fileadmin/content/Files/SharePoint/user_upload/Brochures_EN/Industry_Brochures/99_389_Testing_Machines_and_Systems_for_the_Automotive_Industry_FP_EN.pdf (accessed on 12 July 2023).
18. Rychlik, A.; Kozubel, W. A method of fatigue strength testing of wheel rim fragments at production process stage. *J. KONES Powertrain Transp.* **2016**, *23*, 289–296. [CrossRef]
19. *SAE Standard J328*; Wheels-Passenger Car and Light Truck Performance Requirements and Test Procedures. SAE International: Warrendale, PA, USA, 2005; pp. 1–7.
20. Qu, D.; Wang, S.; Liu, H.; Meng, Y.A. Car-following model based on trajectory data for connected and automated vehicles to predict trajectory of human-driven vehicles. *Sustainability* **2022**, *14*, 7045. [CrossRef]
21. Thiemann, C.; Treiber, M.; Kesting, A. Estimating acceleration and lane-changing dynamics based on NGSIM trajectory data. *Transp. Res. Rec. J. Transp. Res. Board* **2008**, *2088*, 90–101. [CrossRef]
22. Yan, X.H.; Zhou, Z.W. A car-following model using online sequential extreme learning machine. *J. Phys. Conf. Ser.* **2021**, *1848*, 012095. [CrossRef]
23. Gunay, B.; Erdemir, G. Lateral analysis of longitudinal headways in traffic flow. *Int. J. Eng. Appl. Sci.* **2011**, *3*, 90–100.
24. Billal, M.K.; Oery, T.; Sankaran, R.T.; Nesarikar, A.S. *Simulation and Test Correlation of Wheel Radial Fatigue Test*; SAE Technical Paper; SAE International: Warrendale, PA, USA, 2013. [CrossRef]
25. Borase, K.H.; Deore, E.R. Review of radial fatigue test. *Int. J. Innov. Res. Multidiscip. Field* **2016**, *2*, 1–5.
26. Kusmierczak, S.; Briza, T. Degradation of coatings in car repair. In Proceedings of the Engineering for Rural Development, Jelgava, Latvia, 24–25 May 2012; pp. 262–266. Available online: https://www.tf.lbtu.lv/conference/proceedings2012/Papers/045_Kusmierczak_S.pdf (accessed on 26 July 2023).
27. He, C.G.; Song, Z.B.; Gan, Y.Z.; Ye, R.W.; Zhu, R.Z.; Liu, J.H.; Xu, Z.B. Study on the corrosion behavior and mechanism of ER8 wheel steel in neutral NaCl solution. *Coatings* **2022**, *12*, 713. [CrossRef]
28. Turner, B. Believe It or Not, Even Wheels Have Expiration Dates. Published 18 October 2018. Available online: https://driving.ca/auto-news/news/believe-it-or-not-even-wheels-have-expiration-dates (accessed on 13 June 2023).
29. Hagarová, M.; Brezinová, J.; Baranová, G.; Vináš, J.; Maruschak, P. Degradation of components in cars cue to bimetallic corrosion. *Materials* **2021**, *14*, 3323. [CrossRef]
30. Das, T.; Samantaray, S.; Thacker, B.; Danish, M.D.; Ranjan, K.R.; Pintu, K. Design and analysis of multi nut remover/tightener for car wheel. *Mater. Today Proc.* **2022**, *67*, 1170–1174. [CrossRef]
31. Wang, M.; Sun, T.; Gao, X.; Hu, J.; Kong, D.; Zan, T. Preload degradation mechanism of double-nut ball screws revealed by an innovative wear distribution model. *J. Eng. Manuf.* **2022**, *236*, 1653–1666. [CrossRef]
32. Artigas, A.; Monsalve, A.; Colás, R.; Garza-Montes-de-Oca, N.F. Failure analysis of the fasten system of wheels used in mining pickup trucks. *Case Stud. Eng. Fail. Anal.* **2017**, *8*, 28–35. [CrossRef]
33. Jayakumar, V.; Madhu, S.; Muniappan, A.; Ragul, G. Study of torque loss elimination in lug nuts used in agriculture tractor wheels. *Mater. Today Proc.* **2019**, *18*, 327–331. [CrossRef]
34. Tarkowski, S.; Nieoczym, A.; Caban, J.; Jilek, P.; Sejkorová, M. The analysis of pneumatic wheel rim deformation while hitting an obstacle. *Appl. Sci.* **2022**, *12*, 6371.
35. Ballo, F.; Previati, G.; Mastinu, G.; Comolli, F. Impact tests of wheels of road vehicles: A comprehensive method for numerical simulation. *Int. J. Impact Eng.* **2020**, *146*, 103719. [CrossRef]
36. Lee, H.; Taheri, S. Intelligent tires-a review of tire characterization literature. *IEEE Intel. Transp. Syst. Mag.* **2017**, *9*, 114–135. [CrossRef]
37. Juodvalkis, D.; Sapragonas, J.; Griškevičius, P.; Ostrowski, M. Investigation of cars' damaged deformational zones characteristics. In Proceedings of the 12th International Conference. Transport Means, Kaunas, Lithuania, 25–26 October 2008; pp. 300–304. Available online: https://www.academia.edu/29804526/Investigation_of_Cars_Damaged_Deformational_Zones_Characteristics (accessed on 10 September 2023).
38. Moskvitina, L.V. Fatigue deformations and destructions rims of wheels of the heavy dump-trucks. *Procedia Struct. Integr.* **2020**, *30*, 100–104. [CrossRef]
39. Jannifar, A.; Zubir, M.N.M.; Kazi, S. An innovative approach for conducting experimental modal analysis (EMA) in running harmonic for structural modal identification. *Measurement* **2020**, *159*, 107795. [CrossRef]
40. Borecki, M.; Rychlik, A.; Vrublevskyi, O.; Olejnik, A.; Korwin-Pawlowski, M.L. Method of non-invasive determination of wheel rim technical condition using vibration measurement and artificial neural network. *Measurement* **2021**, *185*, 110050. [CrossRef]

41. Mazzoni, A.; Solazzi, L. Experimental field test on a multipiece steel wheel and influence of the material properties on its fatigue life evaluation. *Eng. Fail. Anal.* **2022**, *135*, 106106. [CrossRef]
42. Borecki, M.; Rychlik, A.; Olejnik, A.; Prus, P.; Szmidt, J.; Korwin-Pawlowski, M.L. Application of wireless accelerometer mounted on wheel rim for parked car monitoring. *Sensors* **2020**, *20*, 6088. [CrossRef] [PubMed]
43. Cai, W.; Chi, M.; Wu, X.; Yang, N.; Huang, H.-Z. Effect of wheel initial state on the growth of polygonal wear on high-speed trains. *Wear* **2023**, *526–527*, 204894. [CrossRef]
44. Sang, H.; Zeng, J.; Qi, Y.; Mu, J.; Gan, F. Study on wheel wear mechanism of high-speed train in accelerating conditions. *Wear* **2023**, *516–517*, 204597. [CrossRef]
45. Cremona, M.A.; Liu, B.; Hu, Y.; Bruni, S.; Lewis, R. Predicting railway wheel wear under uncertainty of wear coefficient, using universal kriging. *Reliab. Eng. Syst. Saf.* **2016**, *154*, 49–59. [CrossRef]
46. Zeng, Y.; Song, D.; Zhang, W.; Zhou, B.; Xie, M.; Tang, X. A new physics-based data-driven guideline for wear modelling and prediction of train wheels. *Wear* **2020**, *456–457*, 203355. [CrossRef]
47. Shangguan, A.; Xie, G.; Fei, R.; Mu, L.; Hei, X. Train wheel degradation generation and prediction based on the time series generation adversarial network. *Reliab. Eng. Syst. Saf.* **2023**, *229*, 108816. [CrossRef]
48. Great Dane Limited Partnership, Maintenance Manual for Van and Platform Trailers, Great Dane Limited, Part No. 42101401, Savannah, GA, pp. 1–17. Available online: https://greatdane.com/wp-content/uploads/2021/03/Maintenance-Manual-2017-web.pdf (accessed on 13 June 2023).
49. Kubiak, P.; Wozniak, M.; Zakrzewski, S.; Siczek, K.; Rylkski, A.; Mrowicki, A.; Matej, J.; Deda, J.; Knap, L. Research of damage to the rear wheel rim in a tipper-truck. *Bull. Pol. Acad. Sci. Tech. Sci.* **2022**, *70*, e141000. [CrossRef]
50. ETRTO. Standards Manual. In *The European Tyre and Rim Technical Organisation 32/2*; Avenue Brugmann–B-1060; ETRTO: Brussels, Belgium, 2002; pp. G1–V50.
51. *PN-93S-912440/03*; Wheels with Pneumatic Tyres. Disc Wheels. Requirements and Tests. Polski Komitet Normalizacyjny: Warsaw, Poland, 2003; pp. 1–6.
52. Rychlik, A.; Borecki, M.; Korwin-Pawlowski, M.L. Non-invasive method of car wheel rim examination. *Proc. SPIE* **2018**, *10808*, 108085S. [CrossRef]
53. *ISO 3006:2005*; Road Vehicles—Passenger car Wheels for Road Use—Test Methods. ISO Copyright Office: Geneva, Switzerland, 2005; pp. 1–5.
54. *SAE J267*; Surface Vehicle Recommended Practice, Wheels/Rims-Truck and Bus—Performance Requirements and Test Procedures for Radial and Cornering Fatigue. SAE International: Warrendale, PA, USA, 2014; pp. 1–12.
55. Hasbollah, R. Do Car Rims Wear Out? Car Parts Australia October 13, 2021. Available online: https://carpart.com.au/blog/do-car-rims-wear-out (accessed on 14 November 2023).
56. Miren Kisshan, J.L.; Sankara Narayanan, K.; Augustine, B.M.; Vijayaganapathy, D. Design and analysis of wheel rim. *Int. J. Pure Appl. Math.* **2018**, *120*, 3933–3943. Available online: https://acadpubl.eu/hub/2018-120-6/3/286.pdf (accessed on 10 September 2023).
57. Luo, S.-J.; Fu, Y.-T.; Zhou, Y.-X. Perceptual matching of shape design style between wheel hub and car type. *Int. J. Ind. Ergon.* **2012**, *42*, 90–102. [CrossRef]
58. Corrigendum to Regulation No 124 of the Economic Commission for Europe of the United Nations. Uniform Provisions Concerning the Approval of Wheels for Passenger Cars and Their Trailers. Official Journal of the European Union 2007, L 70, 413-438. Available online: https://eur-lex.europa.eu/legal-content/EN/TXT/PDF/?uri=OJ:L:2007:070:FULL&from=EN (accessed on 10 September 2023).
59. Vakakis, A.F. Nonlinear normal modes. In *Encyclopedia of Vibration*; Braun, S., Ed.; Elsevier: Amsterdam, The Netherlands, 2001; pp. 918–924. [CrossRef]
60. Inman, D.J. *Engineering Vibration*; Pearson Education, Inc.: Upper Saddle, NJ, USA, 2008; pp. 43–48. ISBN 978-0-13-228173-7.
61. Wang, X.; Zhang, X. Simulation of dynamic cornering fatigue test of a steel passenger car wheel. *Int. J. Fatigue* **2010**, *32*, 434–442. [CrossRef]
62. Zheng, Z.-G.; Sun, T.; Xu, X.-Y.; Pan, S.-Q.; Yuan, S. Numerical simulation of steel wheel dynamic cornering fatigue test. *Eng. Fail. Anal.* **2014**, *39*, 124–134. [CrossRef]
63. Kocabicak, U.; First, M. Numerical analysis of wheel cornering fatigue tests. *Eng. Fail Anal.* **2001**, *8*, 339–354.
64. Rakesh, B.T. Stress analysis in wheel rim by using dynamic cornering fatigue test under different conditions. *IJARIIE* **2017**, *3*, 4863–4868.
65. Odyssee-Mure. Available online: https://www.odyssee-mure.eu/publications/efficiency-by-sector/transport/transport-eu.pdf (accessed on 14 November 2023).
66. *PN-EN ISO 6789-2:2017-03*; Assembly Tools for Screws and Nuts—Hand Torque Tools—Part 2: Requirements for Calibration and Determination of Measurement Uncertainty. Polski Komitet Normalizacyjny: Warsaw, Poland, 2017; pp. 1–54.
67. Mery, D.; Filbert, D. Automated flaw detection in aluminum castings based on the tracking of potential defects in a radioscopic image sequence. *IEEE Trans. Robot. Autom.* **2002**, *18*, 890–901. [CrossRef]
68. Zhang, J.; Guo, Z.; Jiao, T.; Wang, M. Defect detection of aluminum alloy wheels in radiography images using adaptive threshold and morphological reconstruction. *Appl. Sci.* **2018**, *8*, 2365. [CrossRef]

69. Wu, J.-D.; Huang, Y.-H. Enhanced identification of internal casting defects in vehicle wheels using YOLO object detection and X-ray inspection. *Trait. Du Signal* **2023**, *40*, 1909. [CrossRef]
70. Carrasco, M.; Mery, D. Automatic multiple view inspection using geometrical tracking and feature analysis in aluminum wheels. *Mach. Vis. Appl.* **2011**, *22*, 157–170.

materials

MDPI

Article

Numerical Investigation on Anti-Explosion Performance of Non-Metallic Annular Protective Structures

Xiaobing Bian [1], Lei Yang [1], Tao Wang [1] and Guangyan Huang [1,2,*]

[1] State Key Laboratory of Explosion Science and Technology, Beijing Institute of Technology, Beijing 100081, China; 3220205032@bit.edu.cn (X.B.); yangl@bit.edu.cn (L.Y.); wang_tao@bit.edu.cn (T.W.)
[2] Beijing Institute of Technology Chongqing Innovation Centre, Chongqing 401120, China
* Correspondence: huanggy@bit.edu.cn

Abstract: Explosive shock wave protection is an important issue that urgently needs to be solved in the current military and public security safety fields. Non-metallic protective structures have the characteristics of being lightweight and having low secondary damage, making them an important research object in the field of equivalent protection. In this paper, the numerical simulation was performed to investigate the dynamic mechanical response of non-metallic annular protective structures under the internal blast, which were made by the continuous winding of PE fibers. The impact of various charges, the number of fiber layers, and polyurethane foam on the damage to protective structures was analyzed. The numerical results showed that 120 PE fiber layers could protect 50 g TNT equivalent explosives. However, solely increasing the thickness of fiber layers cannot effectively enhance the protection efficiency. By adding polyurethane foam in the inner layer, the stress acting on the fiber could be effectively reduced. A 30 mm thick polyurethane layer can reduce the equivalent stress of the fiber layer by 41.6%. This paper can provide some reference for the numerical simulations of non-metallic explosion protection structures.

Keywords: blast protection; non-metallic annular structures; numerical simulation; UHMWPE

1. Introduction

In today's world, the spread of explosive terrorist incidents poses a serious challenge to human civilization, regional peace and national security [1]. The disposal of various types of military explosives, homemade explosives, unexploded ordnance and other explosives is complex and arduous. The safety protection of explosives is the most important part of the explosive ordnance disposal process, which usually involves the use of explosion-proof containers of a certain strength to restrain the hazards generated by explosives [2,3].

Unlike traditional metal or concrete materials [4–7], the structurally weak materials represented by fibers have shown great promise for internal blast mitigation. The structurally weak materials can convert the blast energy into their own kinetic and internal energy and are almost completely converted into small soft particles after loading, which have virtually no secondary damage compared to rigid materials such as metals [8].

In recent years, the structurally weak materials such as water, fibers and foams have been widely used in the field of blast mitigation [9–11]. Zhou et al. [9] showed that the annular foam and liquid protective structure, which is a protective structure with promising applications, can significantly reduce the peak pressure of the explosion at a certain distance. Batra et al. [10] investigated the three-dimensional transient deformation of unidirectional fiber laminates subjected to blast loading. Ply splitting was found to be the dominant damage mode, absorbing 80% of the shock wave energy. Fallah et al. [11] compared the deformation of mild steel and ultra-high molecular weight polyethylene (UHMWPE, referred to as PE) fiber with the same face density under blast loading. It was found that the mild steel had ruptured after loading, and PE fibers could reduce the local deformation by 30%.

PE fiber is one of the best ballistic performance fibers available, which has been widely used in bulletproof vests, bulletproof helmets, bulletproof panels, and other equipment designs [12,13]. The common PE fiber structures mainly include fiber filaments, fiber cloths, and fiber plates. The single-layer PE fiber cloth is generally made of four layers of PE fibers orthogonally hot-pressed, and the ratio of fiber to resin is about 4:1 [14].

However, according to our previous blast tests (Figure 1), visible damage or even disintegration occurred in the PE fiber after loading. The structure of the PE fiber after loading is shown in Figure 2, which included various failure modes such as ply splitting, bucking, and fiber matrix de-bonding, etc. Hence, adding a cushioning material in front of the fiber layer was a viable way to reduce structural damage [15–19]. Karagiozova et al. [15] analyzed the deformation mechanism of FML (Fiber-metal laminate) based on the composition of glass fiber panels and aluminum panels under blast loading. Sitnikova et al. [16] conducted a series of experiments and numerical simulations to analyze the dynamic response of FML under blast loading. The fiber layer was found to have fractured locally and a petal-shaped perforation was formed in the middle of the target plate. Although the FML sandwich composite structure was effective in reducing fiber deformation, the secondary damage could be caused by the broken pieces when the aluminum plate was shattered. Thus, it was necessary to design a weak material such as foam to act as a buffer layer for the fibers and evaluate their blast mitigation performance, which has hardly been studied before.

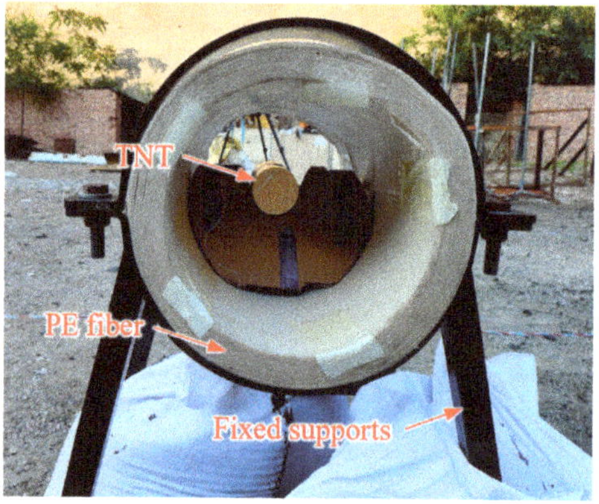

Figure 1. Photograph of the previous experimental setup.

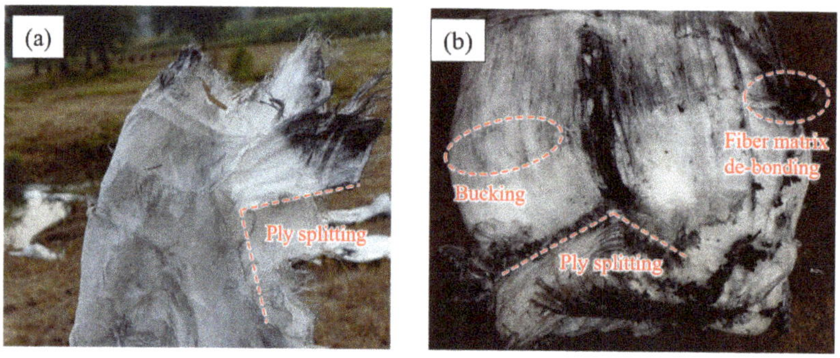

Figure 2. Photograph of the PE fibers after loading: (**a**) overall structure, (**b**) outermost layer.

This paper carried out a series of numerical simulations of the non-metallic composites under blast loading. The blast mitigation effect of PE fibers was comparatively investigated before and after the addition of polyurethane (PU) foam. The results of this investigation showed that the added PU foam on the inside of the PE fibers can effectively reduce the stress of the fiber, avoiding the direct disintegration of the fiber structure under the internal blast loading. The main innovation of this paper is the use of foam "weak material" as a cushioning layer for fibers, and greatly improving the protection of fiber materials against blast shock waves, which has been little investigated before. It is expected to be able to provide a certain reference for the engineering design of explosion-proof structures.

2. Numerical Simulation Description

The numerical simulation conditions in this paper are shown in Table 1. The inner diameter and height of the annular PE fiber were 200 mm, and 400 mm, respectively. Working conditions 1–3 were mainly used to compare the protective effect of the same number of fiber layers under different trinitrotoluene (TNT) equivalents. The working conditions 1, 4 and 5 were to compare the effects of different fiber layers on protection under the same charge (200 g TNT). The working conditions 1, 6 and 7 were to compare the effects of the increased PU foam energy-absorbing layer on the protective structure under the same charge (200 g TNT) and the same number of fiber layers (120 layers of PE fibers).

Table 1. Typical numerical simulation conditions.

Working Conditions	Protective Structure	Mass of TNT Charge
1	120 layers of PE fiber	200 g (50 mm diameter, 63 mm height)
2	120 layers of PE fiber	100 g (40 mm diameter, 50 mm height)
3	120 layers of PE fiber	50 g (30 mm diameter, 44 mm height)
4	80 layers of PE fiber	200 g (50 mm diameter, 63 mm height)
5	160 layers of PE fiber	200 g (50 mm diameter, 63 mm height)
6	120 layers of PE fiber (with 20 mm PU)	200 g (50 mm diameter, 63 mm height)
7	120 layers of PE fiber (with 30 mm PU)	200 g (50 mm diameter, 63 mm height)

2.1. Numerical Model and Simplification

To further investigate the buffering effect of PU foam on PE fiber, a series of numerical simulations were conducted using explicit dynamics solvers in LS-DYNA. The Structured Arbitrary Lagrange-Euler (S-ALE) method was used to define fluid-structure coupling contact, which referenced the immersed boundary method (IBM) [20,21]. Compared with the traditional ALE method, the S-ALE method can avoid leakage and improve computing efficiency [22]. The pinball segment-based contact penalty formulation has been used to define the contact force between the foam and the PE fibers without tangential friction.

As shown in Figure 3, the overall model was constructed using a 1/8 symmetric model, with the detonation point set at the center. Corresponding to this, three symmetry boundary conditions were applied to the $X = 0$, $Y = 0$, $Z = 0$ planes and no reflection boundary conditions were applied to other planes (Figure 3b,c).

The S-ALE calculation area was divided using 2 mm structured orthogonal meshes with a side length of 140 mm and a height of 280 mm. The sizes of TNT, PU foam and PE fibers varied under different working conditions, and Figure 3a showed only one of them. The PE fiber bulk density and surface density were 970 g/m^3 and 150 g $\pm$ 5 g/cm^2, respectively. The thickness of a single fiber layer was 0.15 mm. According to the relevant literature, combining four layers of 0.15 mm thickness PE fibers into one layer of 0.6 mm thickness fibers had little effect on the deformation of the structure at a high strain rate. In order to further simplify the calculations, four layers of fibers were equated to one layer [23].

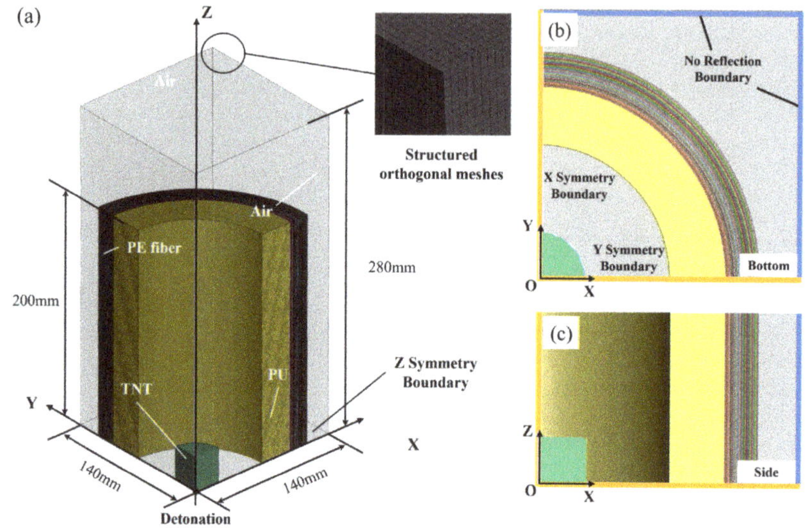

Figure 3. Numerical simulation model of the PE fiber and PU. (**a**) 1/8 of 3D model. (**b**) Bottom view of the numerical model (X-Y plane) with the boundary conditions. (**c**) Side view of the numerical model (X-Z plane).

The PU foam, PE fiber and Euler regions were meshed in numerical simulations. The total number of PU foam (30 mm thickness) elements was 100,500, and the total number of nodes was 109,888. The total number of PE fiber (30 layers) elements was 256,200 and the total number of nodes was 524,172. The element size of the Euler region was 2 mm, and the total number of elements was 70 × 70 × 140 = 686,000. The mesh was automatically generated using Hypermesh meshing software version 2019, with all elements being hexahedral and having a size of 2 mm. The thickness of the PE section was 0.6 mm, and the thickness direction was meshed as a single layer. The official LS-DYNA theory manual states that the Lagrange elements and Euler elements with the same size were recommended to avoid S-ALE algorithm crashes [24]. Thus, the TNT and air domains were adopted as S-ALE meshed with a size of 2 mm in this model.

2.2. Material Model

The standard Jones Wilkins Lee (JWL) equation of state was used to describe the mechanical behavior of TNT detonation products, with parameters derived from Lawrence Livermore National Laboratory [25]:

$$P = A\left(1 - \frac{\omega}{R_1 V}\right)e^{-R_1 V} + B\left(1 - \frac{\omega}{R_2 V}\right)e^{-R_2 V} + \frac{\omega E}{V'} \quad (1)$$

where A, B, w, R_1, R_2 are the equation of state constants of the explosives.

The Linear Polynomial equation of state was used to describe the impact response behavior of the air [26]:

$$P = C_0 + C_1\mu + C_2\mu^2 + C_3\mu^3 + (C_4 + C_5\mu + C_6\mu^2)E_0 \quad (2)$$

where $\mu = \rho/\rho_0 - 1$ is the ratio of the current density to the initial density. $C_0 = C_1 = C_2 = C_3 = C_6 = 0$, $C_4 = C_5 = \gamma - 1$. For an ideal gas, $\gamma = 1.4$. E_0 is the internal energy per unit volume. The specific parameters are shown in Table 3.

The PU foam material model was adopted as Crushable Foam with the parameters as shown in Table 3 [9]. It should be noted in particular that a failure criterion based on maximum principal strain (ε_{MXEPS}) was used in the simulations. The volumetric strain and stress relationship is shown in Figure 4.

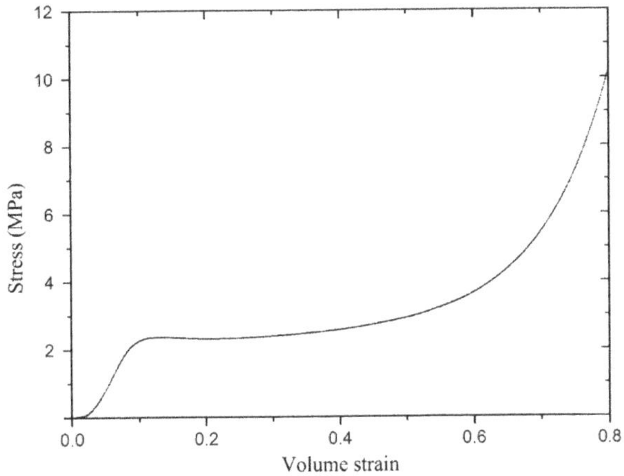

Figure 4. Volume strain and stress curve of the PU foam.

The PE fiber material was various anisotropic material. Considering that the fibers were wound molding, there was no connection between faces. Hence, there was no need to set up face-to-face bound contact. The composite failure material model was used in the LS-DYNA. The linear relationship of stress-strain was expressed as [27]:

$$\begin{pmatrix} \varepsilon_a \\ \varepsilon_b \\ \varepsilon_c \\ \gamma_{bc} \\ \gamma_{ca} \\ \gamma_{ab} \end{pmatrix} = \begin{pmatrix} \frac{1}{E_a} & -\frac{v_{ba}}{E_b} & -\frac{v_{ca}}{E_c} & 0 & 0 & 0 \\ -\frac{v_{ab}}{E_a} & \frac{1}{E_b} & -\frac{v_{cb}}{E_c} & 0 & 0 & 0 \\ -\frac{v_{ac}}{E_a} & -\frac{v_{bc}}{E_b} & \frac{1}{E_c} & 0 & 0 & 0 \\ 0 & 0 & 0 & \frac{1}{G_{cb}} & 0 & 0 \\ 0 & 0 & 0 & 0 & \frac{1}{G_{ca}} & 0 \\ 0 & 0 & 0 & 0 & 0 & \frac{1}{G_{ab}} \end{pmatrix} \begin{pmatrix} \sigma_a \\ \sigma_b \\ \sigma_c \\ \tau_{bc} \\ \tau_{ca} \\ \tau_{ab} \end{pmatrix} \quad (3)$$

where E, G and v are the modulus of elasticity, shear modulus, and Poisson's ratio of the material, respectively. ε, γ, σ and τ are the strain, shear strain, positive stress and shear stress of the fiber, respectively. The following table a, b and c represent longitudinal, transverse and normal directions, respectively.

Five fiber material failure models and nine strengths were presented in Table 2, including three shear strengths (S_{ab}, S_{ac}, S_{bc}), three compressive strengths (C_a, C_b, C_c) and three tensile strengths (T_a, T_b, T_c) [28]. The corresponding parameters are shown in Table 3.

Table 2. Failure model and criterion of the PE fiber.

Failure Model	Criterion
In-plane tensile failure	$\left(\frac{\sigma_a}{T_a}\right)^2 + \left(\frac{\tau_{ab}}{S_{ab}}\right)^2 + \left(\frac{\tau_{ac}}{S_{ac}}\right)^2 \geq 1$ $\left(\frac{\sigma_b}{T_b}\right)^2 + \left(\frac{\tau_{ab}}{S_{ab}}\right)^2 + \left(\frac{\tau_{ac}}{S_{ac}}\right)^2 \geq 1$
Through-thickness tensile failure	$\left(\frac{\sigma_c}{T_c}\right)^2 + \left(\frac{\tau_{ac}}{S_{ac}}\right)^2 + \left(\frac{\tau_{bc}}{S_{bc}}\right)^2 \geq 1$
Through-thickness shear failure	$\left(\frac{\sigma_a}{T_b}\right)^2 + \left(\frac{\tau_{ac}}{S_{ac}}\right)^2 \geq 1$, $\left(\frac{\sigma_b}{T_b}\right)^2 + \left(\frac{\tau_{bc}}{S_{bc}}\right)^2 \geq 1$
Longitudinal compression failure	$\left(\frac{\sigma_a}{C_a}\right)^2 \geq 1$

Table 2. *Cont.*

Failure Model	Criterion
Through-thickness and transverse compressive failure	$\left(\frac{\sigma_b}{S_{ab}+S_{bc}}\right)^2 + \left[\left(\frac{C_b}{S_{ab}+S_{bc}}\right)^2 - 1\right]\frac{\sigma_b}{\lvert C_b \rvert} + \left(\frac{\tau_{ab}}{S_{ab}}\right)^2 + \left(\frac{\tau_{bc}}{S_{bc}}\right)^2 \geq 1$ $\left(\frac{\sigma_c}{S_{ac}+S_{bc}}\right)^2 + \left[\left(\frac{C_c}{S_{ac}+S_{bc}}\right)^2 - 1\right]\frac{\sigma_c}{\lvert C_c \rvert} + \left(\frac{\tau_{ac}}{S_{ac}}\right)^2 + \left(\frac{\tau_{bc}}{S_{bc}}\right)^2 \geq 1$

Table 3. Material model parameters used for TNT, Air, Pu foam and PE fiber.

TNT material parameters [25]					
ρ (kg/m^3)	A (Pa)	B (Pa)	R_1	R_2	w
1.63	3.738×10^{11}	3.747×10^9	4.15	0.9	0.35
D (m/s)	P_{CJ} (Pa)				
6930	2.1×10^{10}				
Air material parameters [26]					
ρ (kg/m^3)	C_4		C_5		E_0 (Pa)
1.225	0.4		0.4		2.5×10^5
Foam material parameters [9]					
ρ (kg/m^3)	E (Pa)	v_r	T_{sc} (Pa)	Damp	ε_{MXEPS}
200	3.195×10^8	0.001	5.3×10^6	0.1	0.05
PE material parameters [27,28]					
E_a (Pa)	E_b (Pa)	E_c (Pa)	v_{ba}	v_{ca}	v_{cb}
34.257×10^9	34.257×10^9	5.1×10^9	0	0.013	0.013
G_{ab} (Pa)	G_{ca} (Pa)	G_{cb} (Pa)	T_a (Pa)	T_b (Pa)	C_c (Pa)
0.1738×10^9	0.5478×10^9	0.5478×10^9	1.25×10^9	1.25×10^9	1.9×10^9

3. Numerical Calculations and Analysis of Results

3.1. Effect of Different Explosive Equivalents

As shown in Figure 5, under the 50 g TNT internal blast loading, the fiber protection structure had no significant deformation. There was only a little damage to the innermost layer, which mainly concentrated in the center of the explosion. It could be basically considered that the structure was able to effectively protect against 50 g TNT explosives. Under the 100 g TNT internal blast loading, the inner and outer layers of fiber protection incurred some damage, which mainly concentrated in the center of the explosion. Hence, the structure under the 100 g TNT charge had a certain risk. Under the conditions of the 200 g TNT, the protective structure incurred a large fragmentation. The outermost layer was torn, which has been completely ineffective.

Numerical simulation showed that under the action of the explosion shock wave, the height of the protective structure could be reduced to the middle, as shown in Figure 6. For the 200 g TNT charge, its height before the explosion was 400 mm. After the explosion, the height was reduced to about 360 mm, a 10% height reduction. This was due to the large outward deformation of the middle protective layer under the action of the shock wave, as shown in Figure 7.

The time history of the blast shock wave and fiber interaction was shown in Figure 8. At 0.02 ms, the blast shock wave reached the fiber layer. At 0.07 ms, some overall deformation occurred, and some fibers were broken. At 0.2 ms, a large number of fibers failed and broke off. At 0.5 ms, the middle part has been completely broken and the protective structure has lost its protective ability.

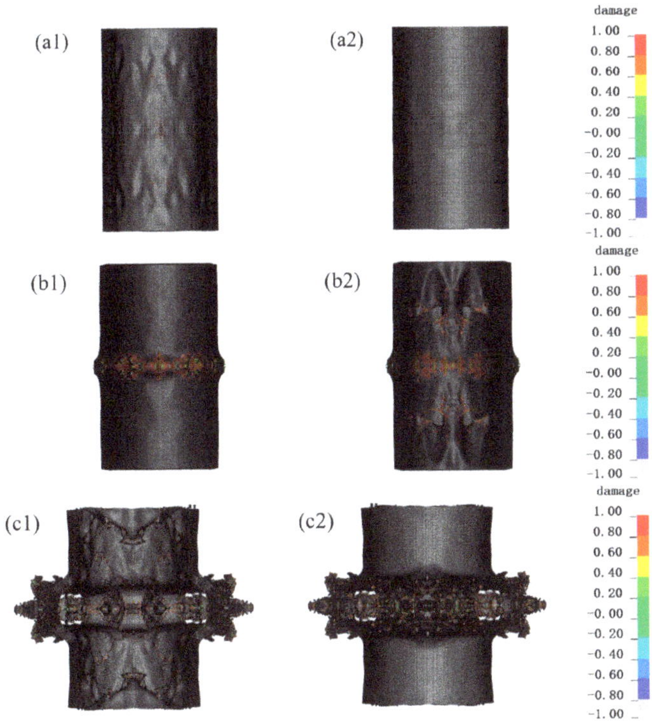

Figure 5. The damage of protective structure under different equivalent explosives. The legend represents the extent of damage to the material, where red (1) represents complete damage to the material and white (−1) represents no damage. (**a1**) Inner side, (**a2**) outer side of the PE fiber within 50 g TNT; (**b1**) Inner side, (**b2**) outer side of the PE fiber within 100 g TNT; (**c1**) Inner side, (**c2**) outer side of the PE fiber within 200 g TNT.

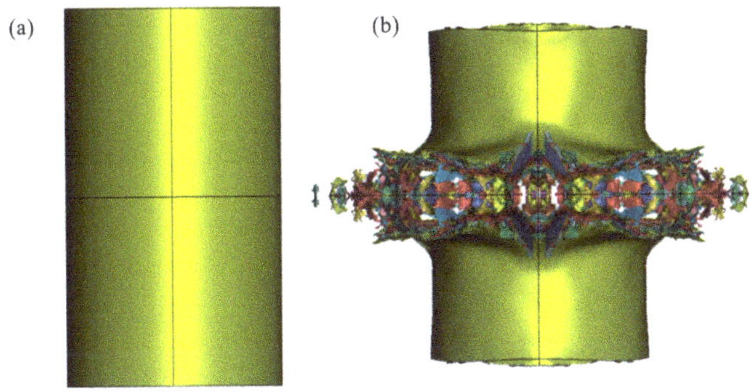

Figure 6. Comparison of protective structure heights before and after the explosion. Height of protective structure (**a**) before explosion, (**b**) after explosion.

Figure 7. Crushing condition of the middle structure.

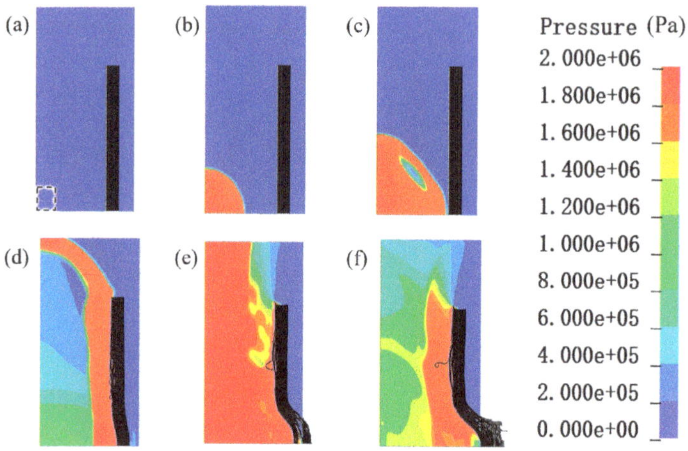

Figure 8. Time history of the interaction between shock wave and PE fiber: (**a**) Initial state at 0 ms, (**b**) blast wave propagation at 0.01 ms, (**c**) shock waves on fibers at 0.02 ms, (**d**) partial fiber failure at 0.07 ms, (**e**) massive fiber failure at 0.2 ms, (**f**) outermost layer destruction at 0.5 ms.

3.2. Effect of Different Fiber Layers

The dynamic mechanical response of 80-layer, 120-layer and 160-layer PE fibers under 200 g TNT charge is compared in Figure 9. And Figure 10 shows the damage of different fiber layers under blast loadings. Under the action of 200 g TNT charge, 80 layers of PE fibers were obviously deformed and broken. Of PE fibers, 120 layers were broken to a relatively small extent, and 160 layers were broken to the smallest extent, but the inner layer and the outermost layer were also broken. It can be seen that increasing the number of fiber layers can improve the protective effect to a certain extent, but cannot completely realize the explosion protection.

The shock wave generated by the explosion first acted in the innermost layer of PE, and the stress waves continued to propagate. In the continuous winding of the dense structure, the stress wave attenuation was small. A shock wave in the outermost layer of PE formed a reflected stress wave, resulting in the outermost fiber structure two times the stress state. Therefore, the damage generally occurred in the innermost and outermost layers of the fiber structure, and in some cases, the damage in the middle layer was smaller.

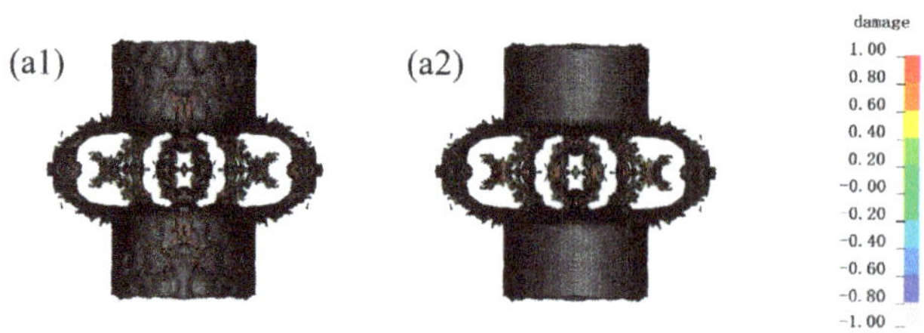

Figure 9. *Cont.*

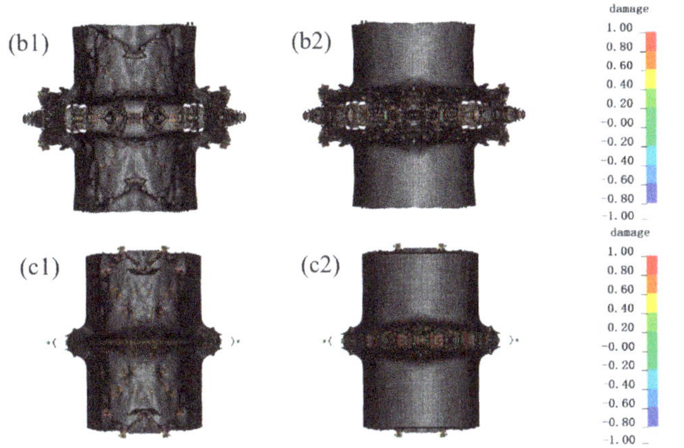

Figure 9. The effect of different fiber layers on protection. The legend represents the extent of damage to the material, where red (1) represents complete damage to the material and white (−1) represents no damage. (**a1**) Inner side, (**a2**) outer side of the 80-layer PE fiber; (**b1**) Inner side, (**b2**) outer side of the 120-layer PE fiber; (**c1**) Inner side, (**c2**) outer side of the 160-layer PE fiber.

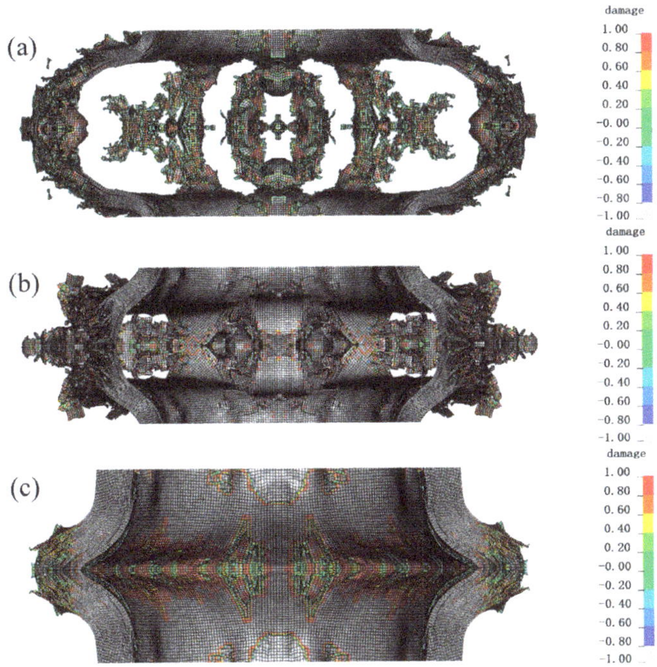

Figure 10. Damage situation of central parts with different fiber layers under blast loadings: (**a**) 80 layers of PE fibers, (**b**) 120 layers of PE fibers, (**c**) 160 layers of PE fibers. The legend represents the extent of damage to the material, where red (1) represents complete damage to the material and white (−1) represents no damage.

3.3. *Effect of the PU Foam*

Considering that increasing the number of PE layers did not effectively improve the stress state of the overall protective structure, the pressure peak of the inner PE fiber layer must be reduced to effectively reduce the damage. The PU foam was an excellent shock wave absorbing material, which could be placed inside the fiber structure to provide a cushioning effect.

As shown in Figure 11, there was a significant reduction in the deformation of PE fiber after adding 20 mm PU foam, with only a little damage in the inner and outer layers. After adding 30 mm PU foam, there was almost no damage in the outermost layer of fibers.

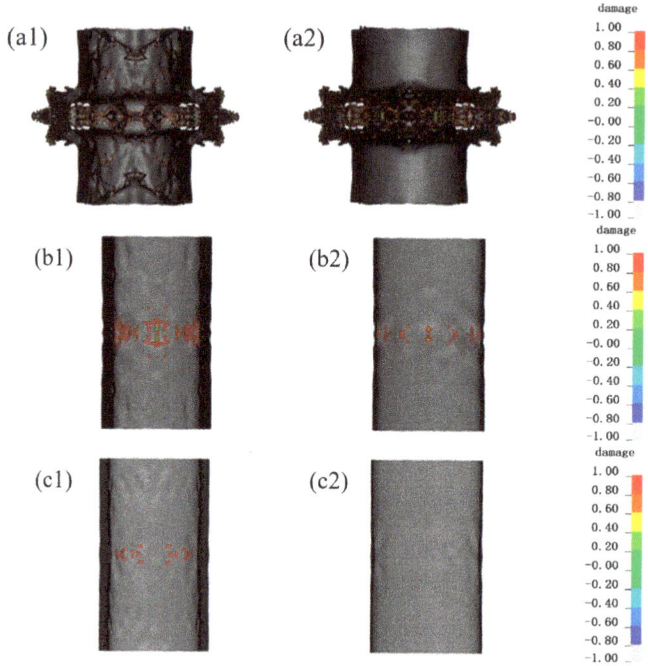

Figure 11. Damage situation of PE fiber structure after adding PU foam protective layer. The legend represents the extent of damage to the material, where red (1) represents complete damage to the material and white (−1) represents no damage. (**a1**) Inner side, (**a2**) outer side of the PE fiber without PU foam; (**b1**) Inner side, (**b2**) outer side of the PE fiber with 20 mm PU foam; (**c1**) Inner side, (**c2**) outer side of the PE fiber with 30 mm PU foam.

The protective conditions of single fibers and 30 mm PU foam/fibers are compared in Figure 12. The yellow area in Figure 12b exclusively represents the PU foam material, excluding any indication of pressure. In the no PU foam condition, at 0.02 ms, the shock wave acted directly on the PE fibers, creating a large stress and a large number of fibers failing. At 0.03 ms, the stress wave continued to propagate to the outer layer, forming a reflection in the outermost layer. In contrast, under the 30 mm PU foam protection condition, at 0.02 ms, the shock wave acted on the PU foam layer and transmitted inward to a certain extent, and there was no crushing of fibers in the outermost layer. At 0.03 ms, a large amount of PU foam crushing occurred, absorbing a large amount of energy, while the foam in contact with the fiber generated a certain amount of stress wave propagation to the outer layer.

Comparing Figures 13 and 14, it can be found that at 0.05 ms, under the condition of no PU foam protection, the area of higher effective stresses in the fiber layer (≥500 MPa) was much larger than that of the 30 mm PU foam protection layer. Under the condition of no protection, the effective stress in the outermost layer of the fiber was greater than that in the innermost layer.

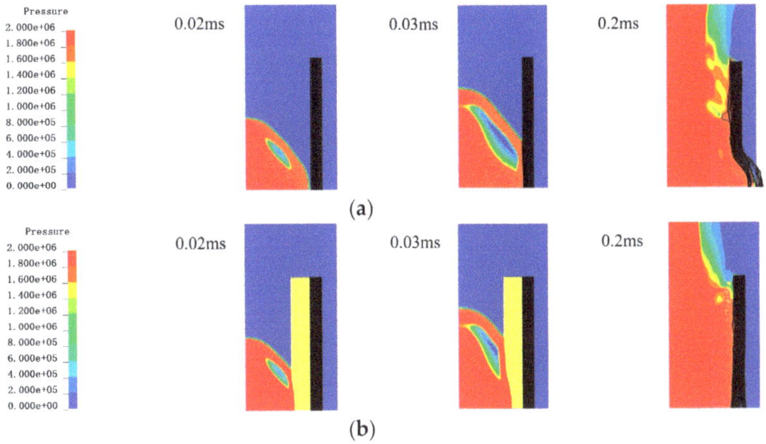

Figure 12. Comparison of the pressure contours in the Euler region (including air and TNT) in different conditions. The figure only showed the pressure of air and TNT without showing the fiber and foam. The unit of the pressure is Pa. (**a**) The PE fiber without PU foam at 0.02 ms, 0.03 ms and 0.2 ms; (**b**) The PE fiber with 30 mm PU foam at 0.02 ms, 0.03 ms and 0.2 ms.

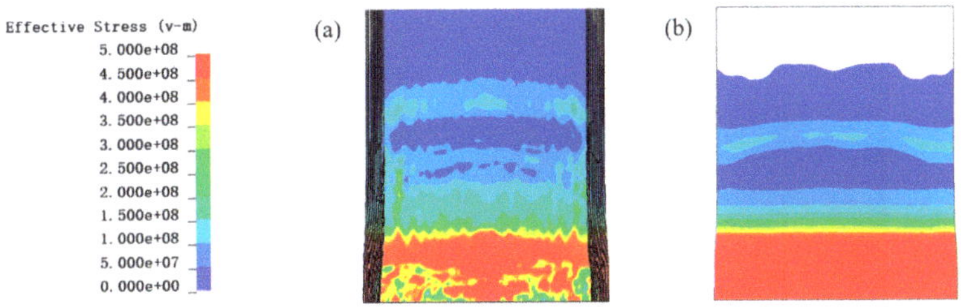

Figure 13. Contour diagram of effective stress of PE fiber without PU foam at 0.5 ms: (**a**) inner layer, (**b**) outer layer. The unit of the effective stress is Pa.

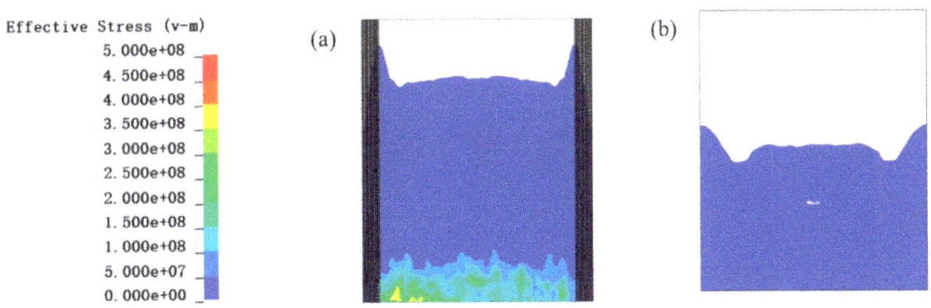

Figure 14. Contour diagram of effective stress of PE fiber with 30 mm PU foam at 0.5 ms: (**a**) inner layer, (**b**) outer layer. The unit of the effective stress is Pa.

As shown in Figures 15 and 16, comparing the PU foam (PU) and PE-only protection conditions at a thickness of 30 mm, the PU foam absorbed a significant amount of energy during the period of 0 to 0.05 ms. Thus, the total energy acting on PE fibers was significantly reduced.

Comparing no PU (no_PU), 20 mm thickness PU (20 mm PU), and 30 mm thickness PU (30 mm PU), the effective stress in the PE fiber layer is shown in Figure 17. It is important to note that Figure 17 shows the effective stress in the PE fiber, while Equation (2) calculates the shock wave overpressure in the air. Thus, it is reasonable that there is a large difference between them. The probed points were chosen to be three cells at the bottom, and the equivalent stress data from the three cells were averaged for comparison.

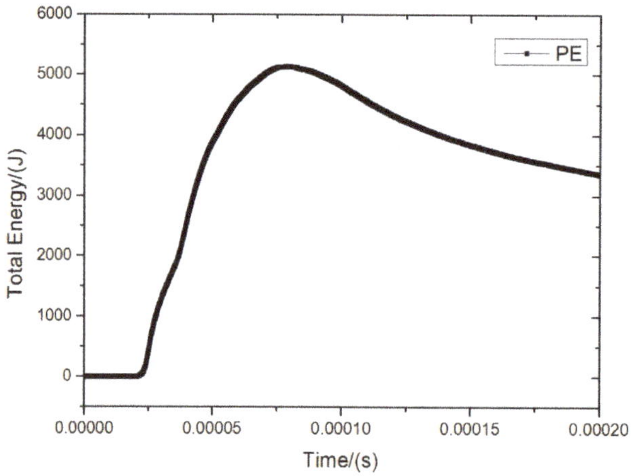

Figure 15. Total energy history curve of PE fiber layers without PU protection.

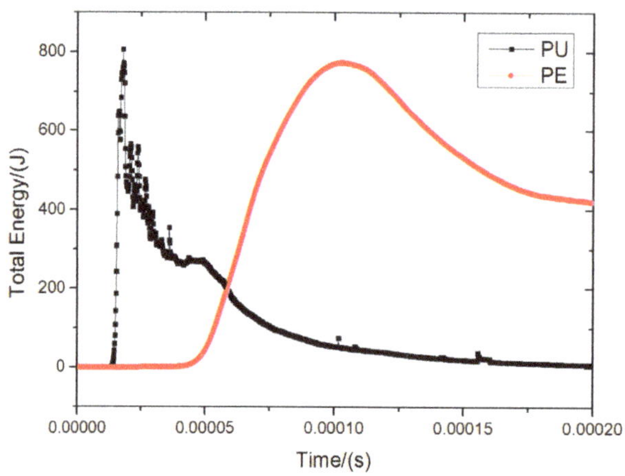

Figure 16. Energy history curve of 30 mm PU foam protective structure.

For the no PU foam, 20 mm PU foam, and 30 mm PU foam condition, the peak equivalent stress was 970 MPa, 727 MPa, and 566 MPa, respectively. As shown in Table 4, the 30 mm thickness of PU was able to significantly reduce the peak equivalent stress in the fiber layer by 41.6%.

Table 4. Comparison of equivalent stress peaks with different protective layers.

Protective Structure	Peak Equivalent Stress (MPa)	Comparison
No PU foam	970	0
20 mm PU foam	727	−25.1%
30 mm PU foam	566	−41.6%

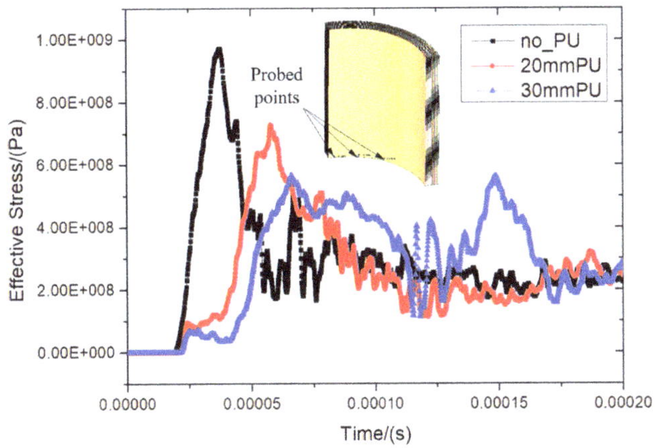

Figure 17. Effective stress history curves of fibers with different foam buffer layers.

4. Conclusions

In this paper, the numerical simulation of the non-metallic annular structures was investigated. By comparing the effects of different explosive equivalents, fiber layer thicknesses, and the PU foam on the protective structure, the numerical simulation analysis showed that:

(1) With the increase of explosive equivalents, the fiber protective structure under the same conditions was more likely to be broken. For this condition (inner diameter of 200 mm, length of 400 mm ring fiber structure, PE fiber layer number of 120 layers), the ring fiber could protect 50 g TNT bare explosives. For 100 g and above TNT charge, the annular fiber inner and outer layers would be torn, and there was a protection gap;

(2) The fiber thickness for bare explosives protection had less impact, and simply increasing the thickness of the fiber did not significantly improve the tearing situation;

(3) The inner layer increased 30 mm thickness PU foam material, and the PU foam can effectively absorb the shock wave energy. It reduced the inner layer of fiber on the equivalent stress of 41.6%, thus significantly reducing the tearing of PE fibers by shock waves.

The work in this paper for the non-metallic protective structures under the internal blast provided a numerical method reference. The next step of the work will focus on the protection structure for the blast test research.

Author Contributions: Conceptualization, X.B. and L.Y.; methodology, T.W.; software, G.H.; validation, X.B. and L.Y.; formal analysis, X.B.; investigation, X.B. and L.Y.; resources, G.H.; data curation, T.W.; writing—original draft preparation, X.B.; writing—review and editing, L.Y.; visualization, X.B.; supervision, T.W.; project administration, G.H.; funding acquisition, G.H. All authors have read and agreed to the published version of the manuscript.

Funding: The work in this investigation was supported by the National Key Research and Development Programs of China (Grant number 2022YFC3320504 and 2020YFC0826804).

Institutional Review Board Statement: Not applicable.

Informed Consent Statement: Not applicable.

Data Availability Statement: Considering that there is still more work to be done in the future, we have not chosen to publish all the data in this article.

Conflicts of Interest: The authors declare no conflict of interest.

References

1. Lerner, E.B.; O'Connor, R.E.; Schwartz, R.; Brinsfield, K.; Ashkenazi, I.; Degutis, L.C.; Dionne, J.-P.; Hines, S.; Hunter, S.; O'Reilly, G.; et al. Blast-related injuries from terrorism: An international perspective. *Prehospital Emerg. Care* **2007**, *11*, 137–153. [CrossRef] [PubMed]
2. Dong, Q.; Li, Q.M.; Zheng, J. Strain growth in a finite-length cylindrical shell under internal pressure pulse. *J. Press. Vessel Technol.* **2017**, *139*, 021213. [CrossRef]
3. Zhu, W.; Huang, G.Y.; Liu, C.M.; Feng, S.S. Experimental and numerical investigation of a hollow cylindrical water barrier against internal blast loading. *Eng. Struct.* **2018**, *172*, 789–806. [CrossRef]
4. Shi, Y.; Xiong, W.; Li, Z.X.; Xu, Q. Experimental studies on the local damage and fragments of unreinforced masonry walls under close-in explosions. *Int. J. Impact Eng.* **2016**, *90*, 122–131. [CrossRef]
5. Eamon, C.D.; Baylot, J.T.; O'Daniel, J.L. Modeling concrete masonry walls subjected to explosive loads. *J. Eng. Mech.* **2004**, *130*, 1098–1106. [CrossRef]
6. Li, H.J.; Shen, C.J.; Lu, G.; Zhang, Z.H. Response of cylindrical tubes subjected to internal blast loading. *Eng. Struct.* **2022**, *272*, 115004. [CrossRef]
7. He, Y.; Liu, Z.; Li, M.; Li, P.; Zhao, Y.; Liu, Q.; Liu, C.; Ye, P. The damage to thick steel plates by local contact explosions. *Materials* **2023**, *16*, 2966. [CrossRef]
8. Zhu, W.; Huang, G.Y.; Guo, Z.W.; Lai, W.; Feng, S.-S. Ballistic penetration of bi-layer structures with water container and fiber composite: Effects of the container position. *Compos. Struct.* **2019**, *227*, 111320. [CrossRef]
9. Zhou, Y.; Wang, T.; Zhu, W.; Bian, X.-B.; Huang, G.-Y. Evaluation of blast mitigation effects of hollow cylindrical barriers based on water and foam. *Compos. Struct.* **2022**, *282*, 115016. [CrossRef]
10. Batra, R.C.; Hassan, N.M. Blast resistance of unidirectional fiber reinforced composites. *Compos. Part B Eng.* **2008**, *39*, 513–536. [CrossRef]
11. Fallah, A.S.; Micallef, K.; Langdon, G.S.; Lee, W.C.; Curtis, P.T.; Louca, L.A. Dynamic response of Dyneema® HB26 plates to localised blast loading. *Int. J. Impact Eng.* **2014**, *73*, 91–100. [CrossRef]
12. Shah, I.A.; Khan, R.; Koloor, S.S.R.; Petrů, M.; Badshah, S.; Ahmad, S.; Amjad, M. Finite element analysis of the ballistic impact on auxetic sandwich composite human body armor. *Materials* **2022**, *15*, 2064. [CrossRef]
13. Chen, Z.; Xu, Y.; Li, M.; Li, B.; Song, W.; Xiao, L.; Cheng, Y.; Jia, S. Investigation on residual strength and failure mechanism of the ceramic/UHMWPE armors after ballistic tests. *Materials* **2022**, *15*, 901. [CrossRef]
14. Zhao, Y.; Wang, D.J.; Yu, J.R. *Ultra-High Molecular Weight Polyethylene Fiber*; National Defense Industry Press: Beijing, China, 2018. (In Chinese)
15. Karagiozova, D.; Langdon, G.S.; Nurick, G.N.; Yuen, S.C.K. Simulation of the response of fibre–metal laminates to localised blast loading. *Int. J. Impact Eng.* **2010**, *37*, 766–782. [CrossRef]
16. Sitnikova, E.; Guan, Z.W.; Schleyer, G.K.; Cantwell, W.J. Modelling of perforation failure in fibre metal laminates subjected to high impulsive blast loading. *Int. J. Solids Struct.* **2014**, *51*, 3135–3146. [CrossRef]
17. Lin, G.; Li, F.; Zhang, Q.; Chen, P.; Sun, W.; Saikov, I.; Shcherbakov, V.; Alymov, M. Dynamic instability of fiber composite cylindrical shell with metal liner subjected to internal pulse loading. *Compos. Struct.* **2022**, *280*, 114906. [CrossRef]
18. Cai, S.; Liu, J.; Zhang, P.; Li, C.; Cheng, Y.; Chen, C. Experimental study on failure mechanisms of sandwich panels with multi-layered aluminum foam/UHMWPE laminate core under combined blast and fragments loading. *Thin-Walled Struct.* **2021**, *159*, 107227. [CrossRef]
19. Wen, Z.; Li, M. Numerical study of low-velocity impact response of a fiber composite honeycomb sandwich structure. *Materials* **2023**, *16*, 5482. [CrossRef] [PubMed]
20. Xin, C.L.; Tu, J.; Wang, J.L.; Sun, F.T.; Liu, A.Y. *Mastering LS-DYNA from Shallow to Deep*; Water Resources and Hydropower Press: Beijing, China, 2019. (In Chinese)
21. Mittal, R.; Iaccarino, G. Immersed boundary methods. *Annu. Rev. Fluid Mech.* **2005**, *37*, 239–261. [CrossRef]
22. Yang, L.; Wang, T.; Bian, X.B.; Huang, G.-Y. Evaluation of blast mitigation performance of cylindrical explosion containment vessels based on water containers. *Int. J. Impact Eng.* **2023**, *181*, 104729. [CrossRef]
23. Henrych, J.; Abrahamson, G.-R. The dynamics of explosion and its use. *J. Appl. Mech.* **1980**, *47*, 218. [CrossRef]
24. LS-DYNA. *Theory Manual Revision R12.0*; Livermore Software Technology Corporation (LSTC): Livermore, CA, USA, 2020.
25. Lee, E.; Finger, M.; Collins, W. *JWL Equation of State Coefficients for High Explosives*; Lawrence Livermore National Laboratory (LLNL): Livermore, CA, USA, 1973.
26. Varas, D.; Zaera, R.; López-Puente, J. Numerical modelling of the hydrodynamic ram phenomenon. *Int. J. Impact Eng.* **2009**, *36*, 363–374. [CrossRef]

27. Zhang, R.; Han, B.; Zhong, J.Y.; Qiang, L.S.; Ni, C.Y.; Zhang, Q.; Zhang, Q.C.; Li, B.-C.; Lu, T.J. Enhanced ballistic resistance of multilayered cross-ply UHMWPE laminated plates. *Int. J. Impact Eng.* **2022**, *159*, 104035. [CrossRef]
28. Zhang, R.; Qiang, L.S.; Han, B.; Zhao, Z.-Y.; Zhang, Q.-C.; Ni, C.-Y.; Lu, T.J. Ballistic performance of UHMWPE laminated plates and UHMWPE encapsulated aluminum structures: Numerical simulation. *Compos. Struct.* **2020**, *252*, 112686. [CrossRef]

materials

MDPI

Article

Coupled Modal Analysis and Aerodynamics of Rotating Composite Beam

Grzegorz Stachyra, Lukasz Kloda * and Zofia Szmit

Department of Applied Mechanics, Faculty of Mechanical Engineering, Lublin University of Technology, 20-618 Lublin, Poland

* Correspondence: l.kloda@pollub.pl; Tel.: +48-81-538-45-72

Abstract: This study primarily focuses on conducting, both experimentally and numerically, a modal analysis of a cantilever composite beam. Through extended numerical simulations, we investigate Campbell diagrams, which, depending on the rotation speed of the structure, comprise natural frequencies and their corresponding modal shapes. Our results are categorized into two main aspects: the classical single-mode behavior and an innovative extension involving linearly coupled modal analysis. One key novelty of our research lies in the introduction of an analytical description for coupled mode shapes, which encompass various deformations, including bending, longitudinal deformations, and twisting. The most pronounced activation of dynamic couplings within the linear regime for a 45° preset angle is observed, though the same is not true of the 0° and 90° preset angles, for which these couplings are not visible. In addition to the modal analysis, our secondary goal is to assess the lift, drag forces, and moment characteristics of a rectangular profile in uniform flow. We provide insights into both the static and dynamic aerodynamic responses experienced by the beam within an operational frequency spectrum. This study contributes to a deeper understanding of the dynamics of composite rotating beams and their aerodynamic characteristics.

Keywords: composite material; rotating structure; modal interactions

1. Introduction

Beams are among the most popular construction elements in engineering. Therefore, it is crucial to have a solid understanding of both the basic and advanced theories related to beam modeling. This knowledge is not only applied but also enhanced in the case of rotating beams, which find widespread use in various industrial applications. Some of the most common applications include wind turbines, helicopter rotors, and airplane propellers.

In their work [1], the authors provided a comprehensive review of the most common theories pertaining to beams, which have been utilized by scientists over the past few decades. They examined classical approaches, such as those by Da Vinci, Euler-Bernoulli, and Timoshenko, in addition to the Generalized Beam Theory. Special emphasis was placed on the Carrera Unified Formulation (CUF) in one dimension, and the authors presented numerical examples illustrating its application in static, dynamic, and aeroelastic problems. Furthermore, they conducted an overview of two recently developed methods: axiomatic/asymptotic and component-wise approaches. The primary conclusion drawn from this critical review is that beam theories are still in need of further development and improvement. In another study by Wang et al. [2], a reduced model for vortex-induced vibrations (VIVs) in turbine blades is derived. In this study, the authors modeled the blades as uniform cantilever beams and employed the multiple scale method to investigate nonlinear dynamics. Subsequently, they calculated frequency–response curves and identified two types of bifurcation. The results presented underscored the necessity of employing coupled models to analyze the rich dynamics of VIVs.

The asymptotic development method is employed in [3] to investigate the free nonlinear oscillations of initially straight Timoshenko beams. The authors focused on two different definitions of curvature: one with respect to the deformed length and the other with respect to the undeformed length. The comparison of these two methods was the primary objective of their study, and the authors demonstrated that the results for slender beams are very similar when using both approaches. Furthermore, in [4], the authors analyzed the model of a geometrically exact nonlinear Timoshenko beam. They derived the equations of motion for the structure but primarily concentrated on one-dimensional constitutive equations. The paper presents basic numerical results. A similar approach is applied in [5] to analyze the dynamics of an elastic isotropic rotating beam. The eigenvalues and mode shapes are obtained for the linear problem, and the coupling between flapping, lagging, axial, and torsional components is studied. In the second part of the paper [6], the authors focused on analytical calculations. They applied the multiscale method directly to the partial differential equations of motion and drew backbone curves. Additionally, they analyzed three flapping modes as the angular speed varied from low to high. The scientists demonstrated that the nonlinearities of the flapping modes are strongly correlated with angular speed and can transition from hardening to softening and vice versa. In addition, Thomas et al. [7] conducted a study on the influence of rotation speed on the nonlinear vibrations of a cantilever beam. They focused on the phenomena of hardening/softening and jump effects, particularly when dealing with large amplitudes. To analyze these phenomena precisely, they employed three different models: two analytical models and one original model based on finite-element discretization. On a related note, the nonlinear vibrations of a rotating Timoshenko beam were investigated using the p-version finite element (FE) method in [8]. This study considered two types of nonlinearities: the strain–displacement relationship and the inertia force resulting from the rotation speed. Nonlinear forced vibrations were analyzed in the time domain, with consideration for both constant and non-constant rotation speeds. Carrera et al. [9] examined the free vibrations of a rotating composite blade. They employed the Carrera Unified Formulation (CUF) and the FE method to solve the governing equations. The authors placed their focus on both flapwise and lagwise motion, and they also accounted for the Coriolis force in their analysis. In a related study, presented in [10], the authors delved into the nonlinear vibration of a rotating beam with variable angular velocity. They concentrated on the coupling between longitudinal and bending vibrations. The authors derived the governing equations of motion using Hamilton's principle and the Galerkin method. They then applied the multiscale method to obtain a first-order approximate solution. Their results were compared to those obtained through numerical integration, demonstrating a very good agreement. In the work of [11], the same methods were applied to derive the equations of motion for a rotating composite Timoshenko beam with both open and closed box-beam cross-sections. The authors stated that the change in pitch angle significantly influences the coupling between flapwise bending and chordwise bending motions, which is associated with the centrifugal force. The presented results take into account nonconstant angular speed as well as a nonzero pitch angle. Given the practical applications of rotating beams and structures, it is of paramount importance to consider the significant impact of aerodynamic loads on their dynamics. In the paper by DiNino et al. [12], an in-depth analysis of a homogeneous viscoelastic beam was conducted under the influence of uniformly distributed turbulent wind flow. This study encompasses an examination of both the steady and turbulent components of the wind, with a particular focus on their roles in Hopf bifurcation and parametric excitation. The authors also emphasized the interaction between bifurcation phenomena and the critical and post-critical behavior of the beam. Meanwhile, in [13], Elmiligui et al. present results obtained from numerical simulations of flow past a circular cylinder. Two distinct approaches are employed to prepare the model for simulations, and the resulting data are compared with previously published experimental findings. Nonlinear vibrations of the blade under high-temperature supersonic gas flow and varying angular speed are presented in [14]. The authors assume that the blade is pre-twisted, presetting, and a thin-walled rotating cantilever beam is used. The equations of motion are derived using Hamilton's principle and the Galerkin method, revealing the presence of

1:1 internal resonance as well as primary resonance. The numerical results presented in the paper show that not only periodic motions but also chaotic motions can occur in the nonlinear vibrations of the rotating blade when the angular speed varies. Furthermore, in [15], a bifurcation analysis of a rotating pre-twisted beam is presented, taking into account varying speed and aerodynamic forces. The model is analyzed in both the chordwise and flapwise directions, revealing phenomena such as jumps, saturation, and double jumps. Additionally, in [16], a model of an Euler–Bernoulli beam with nonlinear curvature and coupled transversal–longitudinal deformation is introduced. The authors applied Hamilton's principle to derive the equations of motion, with a focus on time delay control as the primary task. They presented the influence of linear and cubic control methods on vibration reduction for different rotating speeds. Meanwhile, nonlinear vibrations of a slowly rotating beam with a tip mass are studied in [17]. The authors applied the extended Euler–Bernoulli theory to analyze longitudinal–bending–twisting vibrations. They utilized the multiple time-scale method to solve partial differential equations and demonstrate the influence of angular speed, tip mass, and hub on nonlinear vibrations. Furthermore, the free vibrations of the beam model with a tip mass are explored in [18]. The authors focused on cross-sectional rotations, lateral bending, and transverse bending. The numerical simulations illustrate the effects of tip mass, rotary inertia, viscoelastic damping, and the beam inertia ratio on the stability of the system, as well as on natural frequencies. In their work, Huang et al. [19] presented fascinating results from experimental studies on slowly rotating cantilever beams. They employed Digital Image Correlation, the Phase Mapping Method, and direct measurements under operational conditions to analyze three beams, subjecting them to twenty different angular velocities. Their findings revealed centrifugal hardening behaviors in the flap-wise direction, confirming the accuracy of their chosen model. Notably, they achieved excellent agreement between experimental data and numerical calculations for hardening frequency. Another study of rotating composite beams is discussed in the paper by Gawryluk et al. [20]. In this research, the authors assumed a constant angular velocity for the rotating beam and utilized a Macro Fiber Composite™ (MFC) actuator for excitation. They employed numerical solutions via the FE method, which were subsequently validated through experimental testing. Additionally, Rafiee et al. [21] provided a critical review of scientific papers focused on rotating beams. The authors examined various approaches to calculations, including analytical, semi-analytical, and numerical methods, and discussed different beam theories. This paper offers a comprehensive overview of research on beam vibrations that has been conducted in recent years.

In a study by Teter et al. [22], modal analysis of a rotor composed of three active composite beams is presented. They compare experimental results obtained from a laser vibrometer and a LMS Test.Lab analyzer® with modal hammer to numerical simulations performed using Abaqus® software. The authors achieved excellent agreement among all methods, not only for natural frequencies but also for mode shapes. In the subsequent paper authored by Mitura et al. [23], the dynamics analysis of the rotor operating at a constant angular velocity is presented. The authors employed a Digital Signal Processing (DSP) system to excite vibrations in the beams and control angular speed. The authors investigated the influence of the piezoelectric effect and the hub's speed on the rotor's dynamic behavior. An analysis of force vibrations in a mistuned three-bladed rotor is presented in Warminski et al. work [24]. They assumed that beam mistuning in the rotor results from manufacturing processes in composite production. The rotor was excited by harmonic torque, or by chaotic oscillations. This study revealed the localization phenomenon. Furthermore, the localization and synchronization in a rotor with three beams were studied by Szmit in [25]. The model was analyzed numerically based on equations of motion and through numerical simulations using Abaqus® software. Additionally, the paper presented results from experimental studies, including natural and force vibrations. Finally, Szmit et al. [26] conducted fully experimental studies on a three-bladed rotor. They used high-speed cameras during constant angular speed rotation to analyze the aerodynamic loads at different preset angles. The results include polynomials describing aerodynamic loads based on camera images.

Despite the extensive literature on rotating beams' vibrations, in which the single-mode linear behaviour of the eigenvalue problem is corrected through nonlinear effects, the mechanical coupling, which already occurs in the linear problem between two distinct orthogonal modes, appears to be overlooked in the analytical/numerical models. This provided motivation for conducting a numerical modal analysis within detailed inspection of interactions already in the the linear scope. Furthermore, the linear dynamics of beams is supplemented with aerodynamic characteristics that are closely dependent on the beam's geometry. Research on this aspect is lacking in the majority of studies on rotating beams.

The paper is organized as follows. In Section 2 linear modal analysis of the rotating structure are presented. Graphs illustrating the change in natural frequencies and associated linear mode shapes with rotor rotational speed are discussed, and linear mode couplings of bending, longitudinal motion and torsion in the spatial coordinate system are explored. The aerodynamic characteristics of static and dynamic lift/drag forces, together with aerodynamic momentum, are investigated in Section 3. The article concludes with final remarks and a description of future scientific research directions in Section 4.

2. Dynamic Response of Composite Structure

Let us consider a composite beam attached to a rigid hub with a radius R; see Figure 1. The beam is made of highly elastic ThinPregTM 120EP-513/CF resin and M4JB-12000-50B (TORAY) carbon fibers. Moreover, a specific stacking sequence $[0/-60/60/0/-60/60_3/-60_2/0_2/-60/0_2/60_2/-60]$ ensures isotropic properties of material in the linear elastic regime, as defined by Hook's law [27]. Uniform distribution of the material along the specimen's length $L = 595$ mm and cross-sectional area $b \times h = 35$ mm $\times$ 0.9 mm is assumed [25]. The effective mechanical properties of the composite structure are gathered in Table 1. In Figure 1a, only one coordinate system exists that rotates with the rotating beam–hub structure. The x-axis aligns with the longitudinal axis of the undeformed beam, the z-axis coincides with the hubs' rotation axis and the y-axis completes the right-handed Cartesian coordinate system. Additionally, in Figure 1b, an angle Θ is measured from xy-plane positively defined in accordance with the right-hand rule about the x-axis. A preset angle Θ can be varied from $0°$ to $90°$, and describes the orientation of the blade attached to the hub. The system rotates with a constant speed $\dot{\varphi}$. The hub's mass moment of inertia is infinite; hence, the rotating imbalance and inertial coupling between successive beams are not taken into account. The target of this assumption is to eliminate additional interactions between consecutive beams. The attention is devoted only to the beam as a 3D continuous structure, which can be deformed out of plane (outplane bending i), in-plane (inplane bending j), along the main of the beams' axis (longitudinal k) as well as twist (torsion l). Note that, since the xyz coordinate system is embedded in the rotating hub, the directions of introduced deformations i and j are not aligned with the xy and yz planes. Only longitudinal k and twisting l motions can be referenced relative to the x-axis.

Table 1. Effective mechanical properties of the composite beam: density, mass per unit length, Young's modulus, shear modulus, Poisson's ratio [16].

ρ	μ	E	G	ν
kg/m^3	**kg/m**	**GPa**	**GPa**	**(-)**
1350	0.042525	55.7225	20.4862	0.36

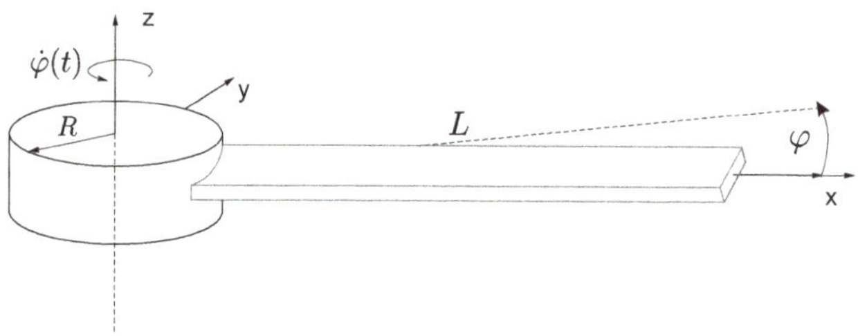

(**a**)

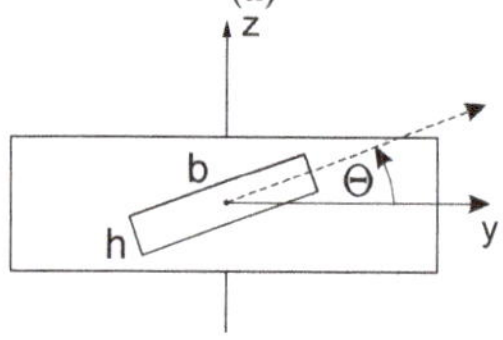

(**b**)

Figure 1. The beam–hub structure: (**a**) an isometric top view, and (**b**) a viewpoint orthogonal to the primary axis of the beam.

Commercial Ansys® software was used for all numerical simulations presented in the paper. In the first step, the natural frequencies and associated mode shapes were validated in accordance with analytical calculations, simulations of competitive commercial software [17,28], as well as experimental studies performed in the absence of rotation ($\dot{\varphi} = 0$) e.g., by neglecting the centrifugal force [16]. Nevertheless, the experimental investigations were restricted solely to the first two modes of natural vibrations. This prompted the authors to explore higher frequencies of the system's natural vibrations within the frequency range that aligns with forthcoming numerical analyses.

Experimental measurements were conducted in the laboratory of the Department of Applied Mechanics at the Lublin University of Technology. An advanced PSV 500 laser scanning vibrometer and an electromechanical exciter SmartShaker K2007E01 were used for the measurements [22]. The measurement system setup is illustrated in Figure 2. The experiment was conducted based on a periodic chirp excitation in the frequency range of 0–25,132.7 rad/s (0–4 kHz), with the excitation applied at the base of the beam using the head of the electromechanical shaker. The calibrated scanning head performed three measurements for each of the 385 predefined points. Fast Fourier Transform (FFT) was then applied to the recorded time-domain signals for each point to identify resonant peaks and their associated vibration modes. The results of the vibration tests are presented as the first twenty detected modes of vibrations shown in Figure 3, and the corresponding magnitude–frequency plots are presented in Figure 4.

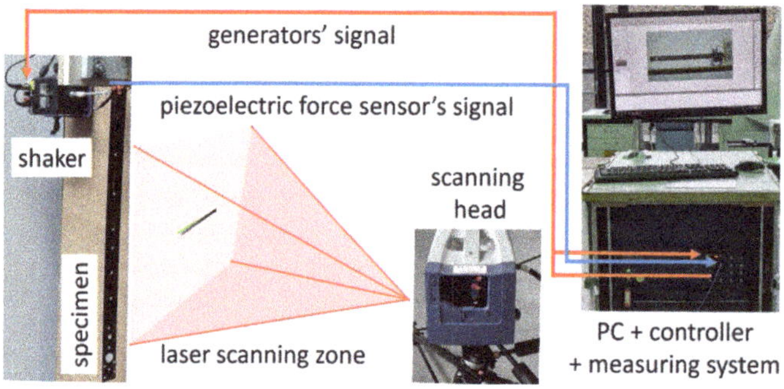

Figure 2. The scheme of the experimental setup.

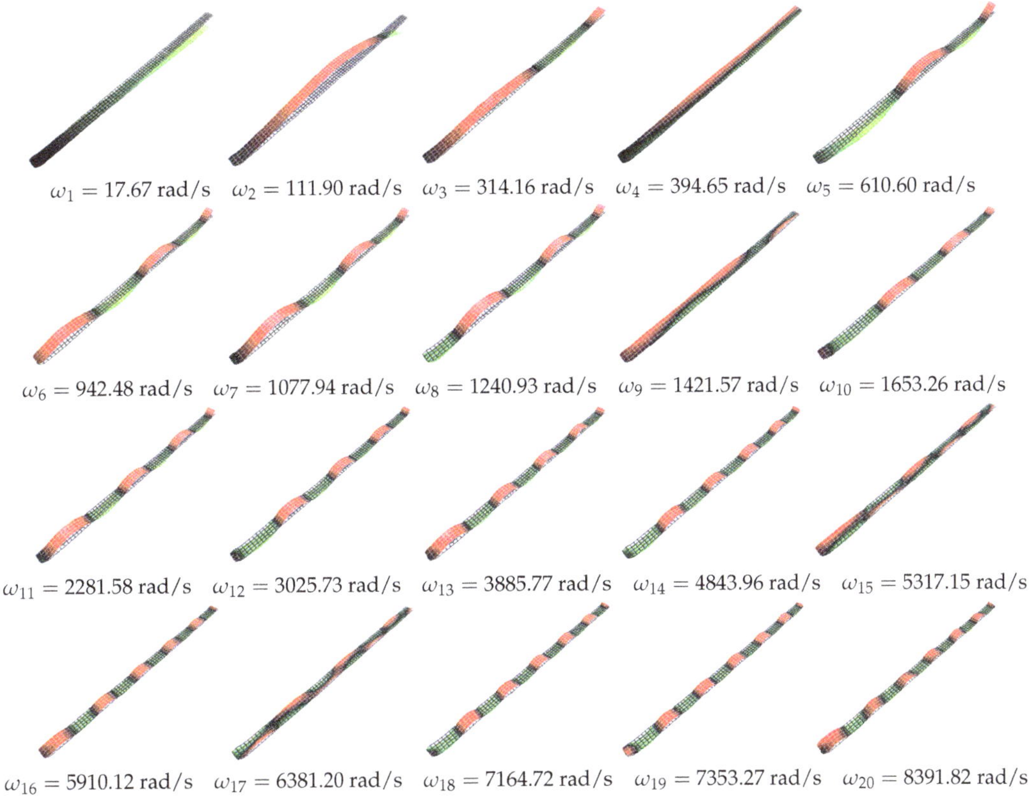

Figure 3. Experimental linear (single) mode shapes.

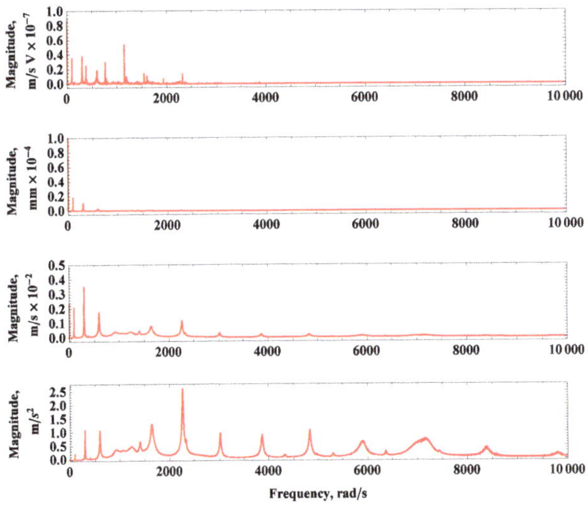

Figure 4. Frequency magnitude curves of dynamic tests for measured signals: velocity multiplied by force sensors' voltage, displacement, velocity, and acceleration (from **top** to **bottom**).

It is worth noting at this stage that in the absence of angular velocity, despite the use of a composite structure and a very broad frequency spectrum, no dynamic couplings were observed in the linear range of the dynamic response. The detected natural frequencies align with the numerical calculations reported in the Section 2.1. The authors regret that, due to technical constraints, they were unable to perform modal analysis considering a rotating structure. Consequently, experimental measurements were complemented only with numerical simulations using the finite element method.

2.1. Campbell Diagram

In this Section, the rotating system is axially pre-stressed due to centrifugal forces and then subjected to linear modal analysis. The distribution of centrifugal forces depends on angular velocity $\dot{\varphi}$ as well as the dimensions of the beam and the radius of the hub. It interferes with inertia and stiffness matrices and has a significant impact on eigenvalue problems such as linear eigenfrequencies and associated modes shapes. Natural frequencies as a function of rotational speed, e.g., Campbell diagrams, for five preset angles are presented in Figure 5. The linear natural frequencies up to 3000 rad/s are reported, and extended analysis for higher-order modes with logarithmic scale on ordinate are gathered in Appendix A.

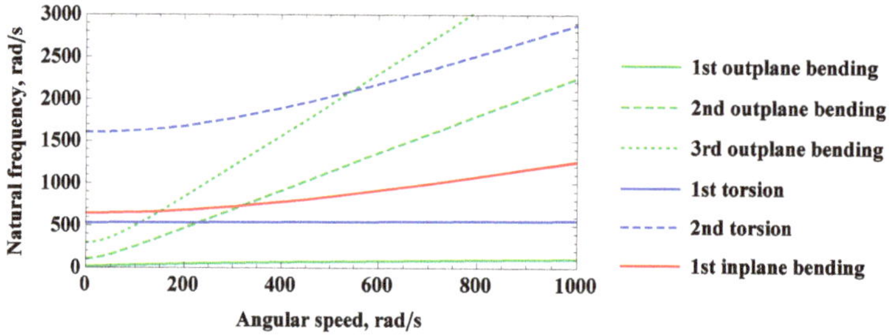

(a)

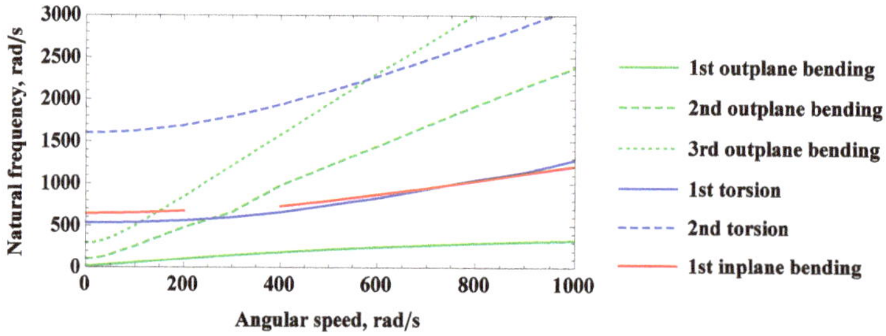

(b)

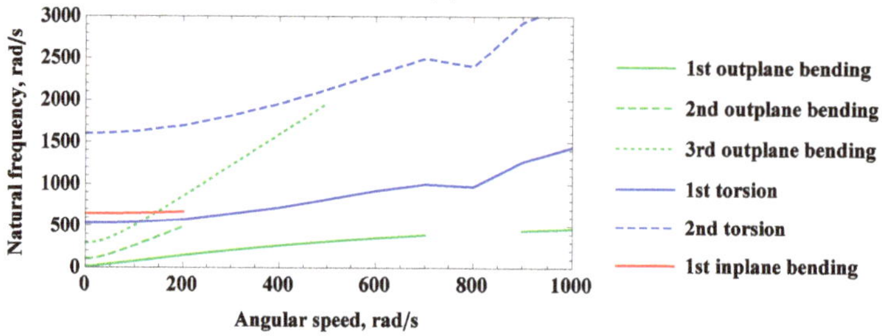

(c)

Figure 5. *Cont.*

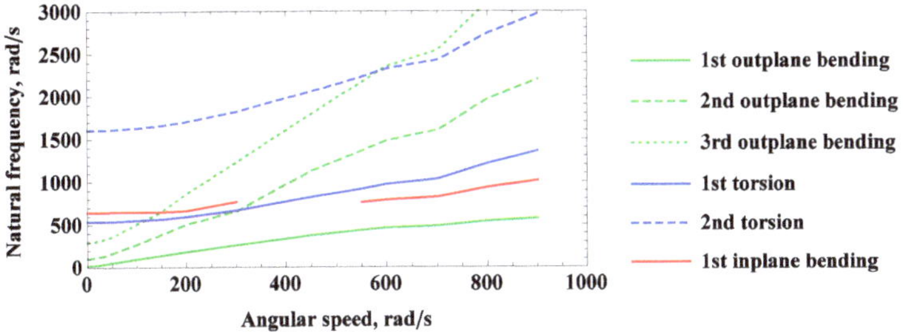

(d)

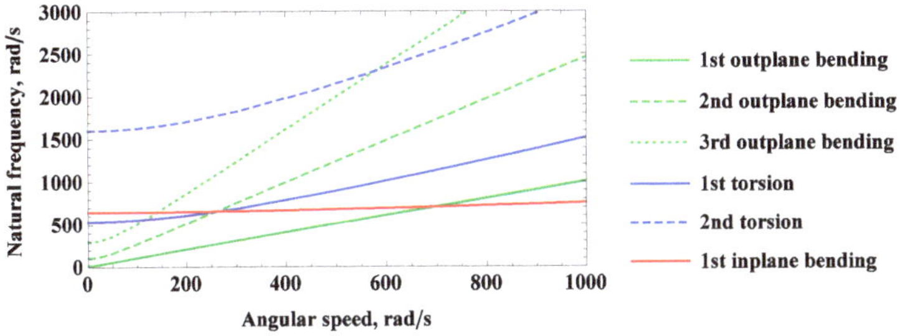

(e)

Figure 5. Campbell diagram of the rotating beam for predefined preset angle: (**a**) $\Theta = 0°$, (**b**) $\Theta = 30°$, (**c**) $\Theta = 45°$, (**d**) $\Theta = 60°$, (**e**) $\Theta = 90°$.

In the angular velocity absence $\dot{\varphi} = 0$, the natural frequencies are the same regardless of the radius of the hub and the preset angle of the blade. When rotation is activated, the two boundary angles $\Theta = 0°$ and $\Theta = 90°$ represent the dynamics of clear single-modes of vibration. In Figure 5a,b, despite numerous intersections of natural frequencies, no couplings occur. It is very interesting that for a rotational speed of 260 rad/s, three curves intersect at 650 rad/s. In general, values of natural frequencies increase with increasing rotational speed, but the slope trends are different. In contrast to the other curves, only the first torsion mode at $\Theta = 0°$ has a constant value of natural frequency. In the scenario in which linear modal couplings occur, indications on Campbell charts are not reported because they cannot be assigned to the conventional vibration modes included in the legend of the graphic. The individual interpretation will be performed in Section 2.3. Therefore, for preset angle $\Theta = 30°$, $\Theta = 45°$ and $\Theta = 60°$ strong linear modal interactions are observed in Figure 5b–d. In the first and third cases, the 1st inplane bending mode is lacking only for 300 rad/s and 400 rad/s, respectively. The preset angle $\Theta = 45°$ seems to be the critical one, for the sake of only torsion modes, and the 1st outplane modes were matched in the studied angular speed interval. This means that in the linear range, there is already a strong coupling or multiple instabilities in simulations on the beam.

2.2. Linear Mode Shapes

Campbell diagrams display only natural frequencies and lack information about the deformation of the shape. Based on three selected angular speeds, the change in linear mode shapes necessitates a proper discussion on deflection half-waves and the modal nodes location. To facilitate the observed changes, selected higher-order modes are presented in Figure 6. The third out-of-plane bending mode is very susceptible to angular velocity in the range of up to 1000 rad/s. Firstly, the natural frequency varies from 295.591 rad/s to 3917.31 rad/s. Secondly, the two modal nodes shift at approximately 1.5% and 0.84% at 100 rad/s. The changes become more prominent at approximately 3.7% and 3.5% at 100 rad/s. The mentioned shifts in % refer to nodals' displacements over the length of the beam in the free rest configuration. Standardization of the results to the maximum beam deflection also shows that with the increase in rotational speed, the amplitude of the first two half-waves decays with respect to the free end. Moreover, the deflection arrows of the half-waves are inclined to the right. The second flexural in-plane mode shape remains constant for increasing rotational speed, while its natural frequency changes from 3968.35 rad/s to 4569.89 rad/s, which provides an increase of about 15% (see Figure A1). Analogously to the bending mode in the susceptible direction, torsional vibrations display a shift in the modal node by approximately 0.5% and 5.9% for 100 rad/s and 1000 rad/s, respectively. The second analogy is the reduction in the maximum twist in the first half-wave. We note that standard linear mode shape projections (Φ_i, Φ_j, Φ_k and Φ_l) consisting of amplitudes (A_1–A_4, B_1–B_4, C_1–C_2 and D_1–D_2) and characteristic coefficients (λ_1, λ_2, λ_3 and λ_4) can be described in the form

$$\Phi_i(x) = A_1 \sin(\lambda_1 x) + A_2 \cos(\lambda_1 x) + A_3 \sinh(\lambda_1 x) + A_4 \cosh(\lambda_1 x), \quad (1)$$

$$\Phi_j(x) = B_1 \sin(\lambda_2 x) + B_2 \cos(\lambda_2 x) + B_3 \sinh(\lambda_2 x) + B_4 \cosh(\lambda_2 x), \quad (2)$$

for outplane/inplate bedning, and

$$\Phi_k(x) = C_1 \sin(\lambda_3 x) + C_2 \cos(\lambda_3 x), \quad (3)$$

$$\Phi_l(x) = D_1 \sin(\lambda_4 x) + D_2 \cos(\lambda_4 x), \quad (4)$$

for longitudinal and twisting are sufficient. Hoverer, the indicated amplitudes and characteristic coefficients must satisfy the sclerotic boundary conditions at $x = 0$ and rheonomic constraints $x = L$ by balancing internal forces, centrifugal forces, Coriolis forces and inertia terms. To date, finding an analytical solution to such a complex problem remains challenging.

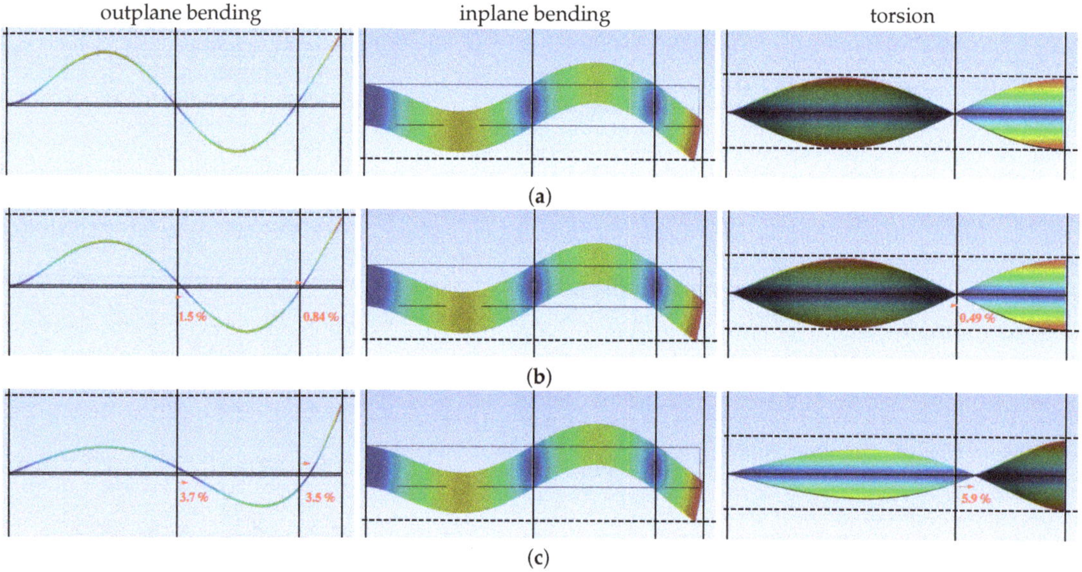

Figure 6. Linear mode shapes for preset angle $\Theta = 90°$ and varying angular velocity: (**a**) $\dot{\varphi} = 0$ rad/s, (**b**) $\dot{\varphi} = 100d$ rad/s and (**c**) $\dot{\varphi} = 1000$ rad/s.

2.3. Linear Mode Couplings

In this section, we devote attention to more sophisticated mode shapes, which involves combining at least two linear *unidirectional* mode shapes [29]. The introduced i, j, k, l notations can be extended to combined mode shapes $\Phi(i, j, k, l)$ in the linear regime, e.g., the third inplane bending mode interacting with the first longitudinal mode and second twist mode can be classified as $\Phi(0,3,1,2)$. In proposed notation, the frequency dependence is omitted for simplicity. Referring to Campbell charts of Figure 5c, a set of the most interesting solutions of numerical simulations for preset angle $\Theta = 45^\circ$ and $\dot{\varphi} = 800$ rad/s is presented in Figure 7. This is a particularly complicated case, in which twin modes of vibrations $\Phi(4,2,0,0)$ for $\omega_7 = 4164.62$ rad/s and $\Phi(4,2,0,0)^*$ for $\omega_9 = 4662.44$ rad/s are obtained. Despite the fact that both consist of the fourth flexible mode and the second flexible mode with greater stiffness, their natural frequencies differ. Moreover, in Figure 7a the i-type mode is dominant, while in Figure 7b, the j-type mode is more exposed. Since the notation counts only the dominant modes, it is conventional to implement weights ($\bar{A}$, $\bar{B}$, $\bar{C}$ and $\bar{D}$) for each mode of vibration

$$\bar{\Phi}(t,x) = \sin(\omega_n t)\Phi(i,j,k,l) = \sin\omega_n t\left[\bar{A}\Phi_i(x) + \bar{B}\Phi_j(x) + \bar{C}\Phi_k(x) + \bar{D}\Phi_l(x)\right], \quad (5)$$

$$\bar{A} + \bar{B} + \bar{C} + \bar{D} = \mathbf{1}, \quad (6)$$

where ω_n corresponds to the nth natural frequency of a given $\dot{\varphi}$, while t is the time.

Other detected modal couplings include the combination of the 9th outplane bending with the 3rd in plane bending and 1st longitudinal with 10th torsion, which are presented in Figure 7c and Figure 7d, respectively. We have observed that neither the first in-plane bending singular mode nor any coupled modes are detectable; therefore, this mode can be subtly incorporated into other coupled modes of vibration.

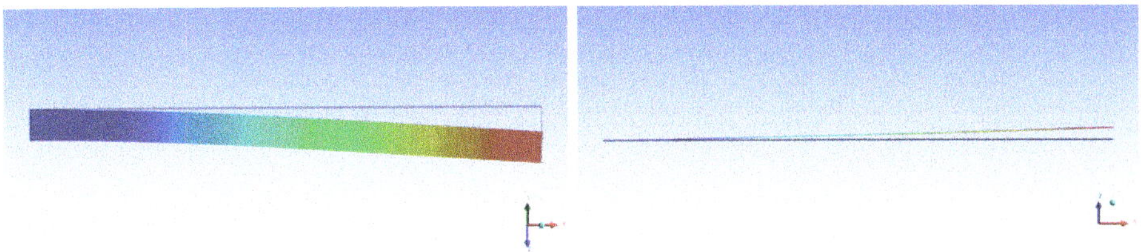

(**a**) $\Phi(1,1,0,0)$, $\omega_7 = 157.45$ rad/s

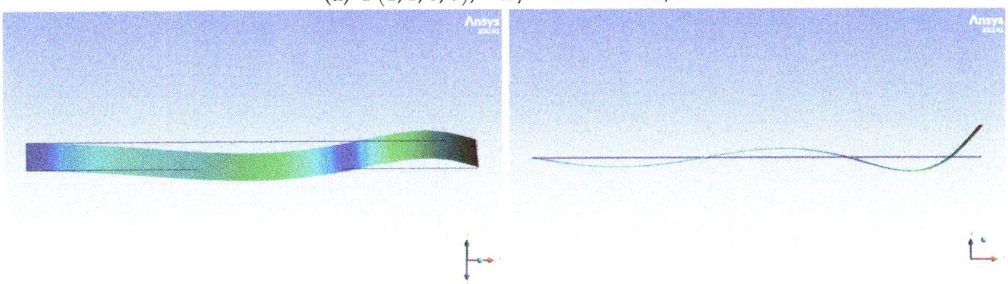

(**b**) $\Phi(4,2,0,0)$, $\omega_7 = 4164.62$ rad/s

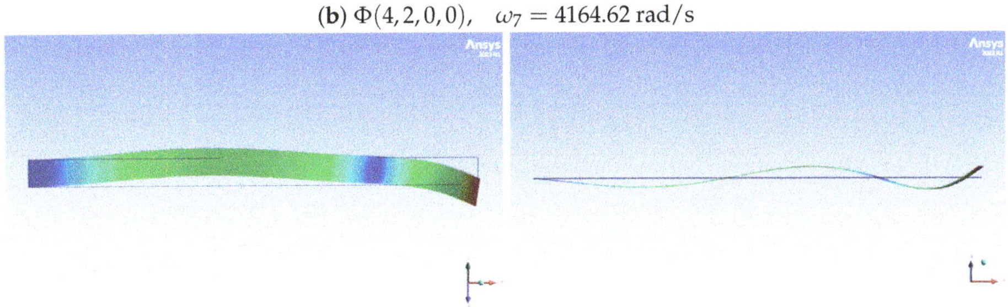

(**c**) $\Phi(4,2,0,0)*$, $\omega_9 = 4662.44$ rad/s

Figure 7. *Cont.*

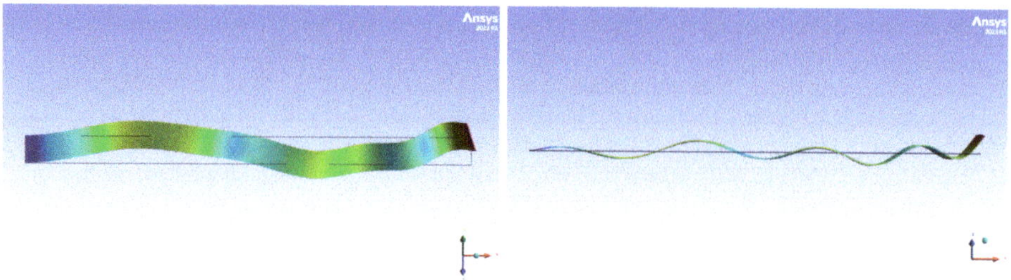

(**d**) $\Phi(9,3,0,0), \quad \omega_{17} = 11332.35$ rad/s

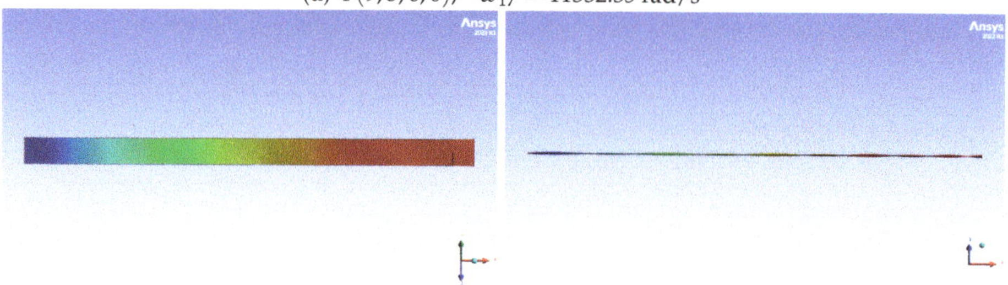

(**e**) $\Phi(0,0,1,10), \quad \omega_{25} = 16904.28$ rad/s

Figure 7. Linear mode couplings for preset angle $\Theta = 45^\circ$ and $\dot{\varphi} = 800$ rad/s.

Extended results for fixed preset angle $\Theta = 45^\circ$ and gradually varied rotational speed $\dot{\varphi}$ for 200 rad/s, 300 rad/s, 400 rad/s, 500 rad/s, 600 rad/s, 700 rad/s, 900 rad/s and 1000 rad/s are reported in Figures A2–A8. These results provide a solid basis for further analyses using analytical methods, indicating the level of complexity of the issue in 4D space and in the time/frequency domain.

3. Aerodynamic Simulations

3.1. Lift/Drag Forces and Momentum

In this section, we focus on the aerodynamic aspects of the 2D blade in the flow of uniform air. The rectangular cross-section of the beam is placed at a given preset angle, Θ. In the studied case of a non-deformable structure, the preset angle is consistent with the angle of attack. Geometric details of the Computational Fluid Dynamics (CFD) simulations are presented in Figure 8. During the simulations, aerodynamic forces were recorded over time. In post-processing, the maximum and minimum magnitudes, as well as the mean values, of steady-state motion time histories were grouped according to the angle of attack Θ and varying airflow conditions. The translational airflow to rotation of the beam–hub structure is converted as follows

$$\dot{\varphi} = \frac{v}{R+L} \tag{7}$$

where v corresponds to the air flow velocity at the tip of the blade and $R + L$ describe the distance between the main axis of rotation and the beam tip in a free and undeformed configuration, with dynamic and centrifugal forces disabled. This assumption will be utilized to simplify our analysis.

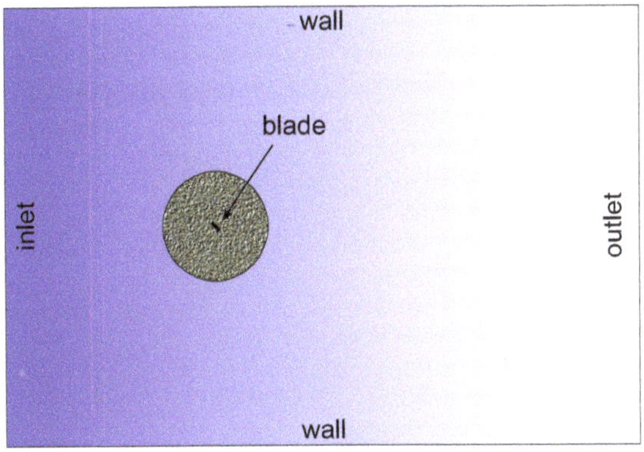

Figure 8. Geometry of CFD simulation domain $60\mathcal{D} \times 40\mathcal{D}$ (upstream $20\mathcal{D}$ and wake $40\mathcal{D}$), the near-field cylinder $10\mathcal{D}$ and the blade $1\mathcal{D}$ corresponding to the width of the beam b.

Figure 9 presents lift drag forces in the function of preset angle Θ and angular speed $\dot{\varphi}$. At zero angle of attack, no lift force was detected, and the drag forces were at their minimum compared to the entire chart. Moreover, in the steady-state flow of the considered velocities, oscillations did not occur. Increasing the angle of attack to 5° resulted in a significant increase in lift force with only a slight increase in drag force. Further increasing the angle of attack to 15° and 30° led to significant air resistance with only a minor rise in lift force. Karman vortices and the associated oscillations of forces in time histories appeared at an angle of attack of 45°. For this angle of attack, the values of lift and drag forces were nearly equal to each other. Subsequent changes in the angle of attack to 75° and 90° resulted in a significant increase in oscillations with increased drag and decreased lift forces. Furthermore, at a preset angle of 90°, the lift force oscillated around zero while drag reached its maximum values.

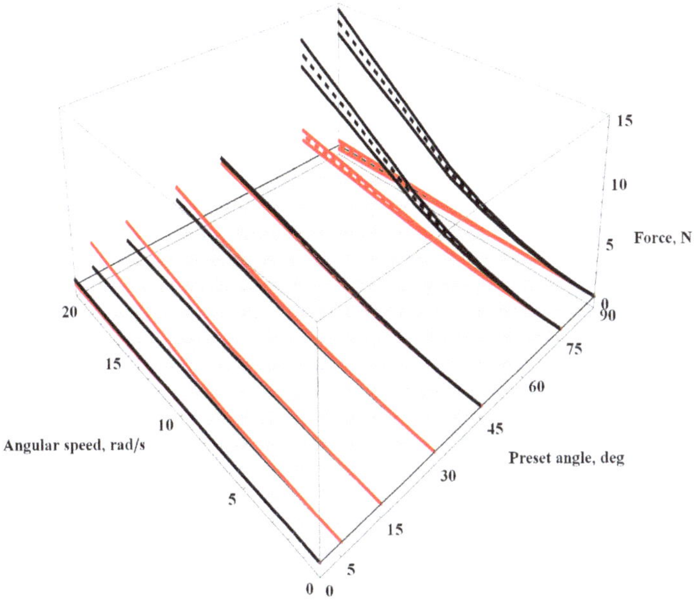

Figure 9. Aerodynamic lift (**red**) and drag (**black**) forces acting on the beam for the constant flow rate; see Equation (7).

The analogous chart depicting the values of the aerodynamic moment acting on the beam is presented in Figure 10. For high airflow velocities, negative values of the aerodynamic moment for angles of attack at 5°, 15°, 30°, and 45° draw attention. The remaining three angles of attack either exhibit zero moment values for 0° or symmetric oscillations around zero for 75° and 90°. It is worth noting that slight oscillations also occur at $\Theta = 45°$ and $v = 100$ m/s, but they diminish with increasing velocity.

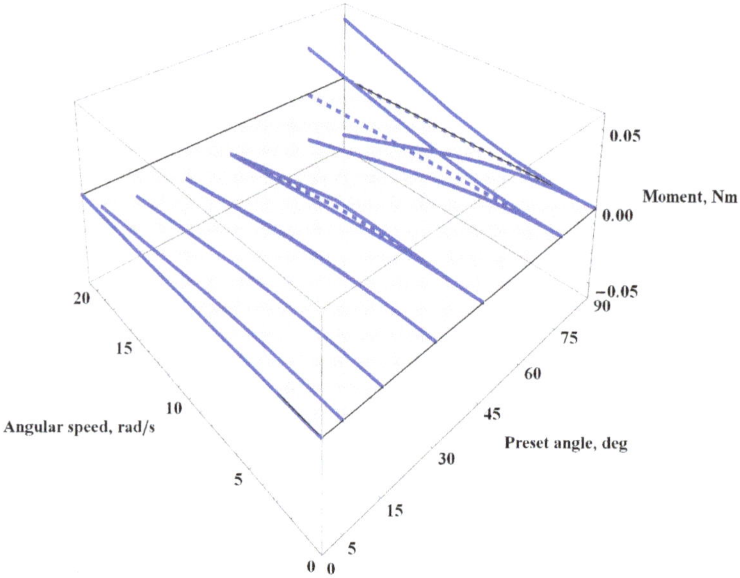

Figure 10. Aerodynamic momentum acting on the beam for the constant flow rate, see Equation (7).

The above-mentioned aerodynamic loads can induce quasi-static deformations of mechanical system or excite its vibrations near resonance frequencies. This provides the foundation for examining another crucial aspect; namely, the frequency windows that impact the sample, along with the measurement of their magnitudes. In essence, we are establishing the groundwork for a comprehensive analysis of how specific frequency ranges may affect the sample and the extent of their influence.

3.2. *Frequency Spectra*

The time histories were subjected to a Fast Fourier Transform (FFT) to determine the airflow frequencies. In many cases, the frequency–magnitude plots exhibited one or two peaks. In order to consolidate the results, Figure 11 depicts a bubble chart on the rotational speed vs. response frequency plane. The bubble sizes are normalized to the dominant value, corresponding to the highest indication, while the remaining values (if present) are proportionally smaller.

For rotational speeds below 5 rad/s, oscillations occur only at 30°, 45°, and 75°. Additionally, two harmonics are excited only for preset angle 30° and 45°. For angular speeds between 5 rad/s and 15 rad/s, a zero-degree angle of attack are inactive. However, for rotational speeds exceeding 20 rad/s, the first indication is observed at 150 rad/s, and the second at 300 rad/s.

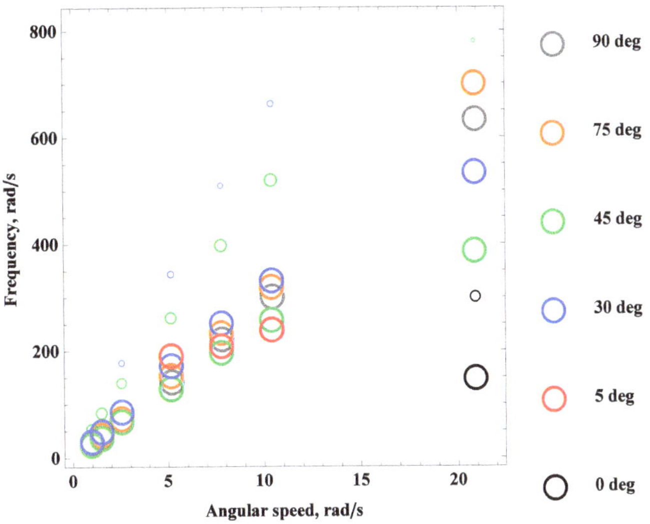

Figure 11. Frequency spectra of aerodynamic loads.

The arrangement of bubbles can be divided into two groups: the first group includes angles of attack of 30° and 45° with a nonlinear trend of decreasing frequency indication values, while the second group includes 75° and 90° degrees with a trend of increasing frequency with the rotational speed $\dot{\varphi}$. The angle of attack 5° has only three data points and appears to exhibit a linear trend. All time histories containing two distinct indications show that the lower harmonic has a greater value. However, it is essential to consider both frequencies to excite or avoid vibrations when their values coincide with the natural vibration frequencies, as depicted in the Campbell diagrams in Figures 5 and A1 as well as coupled vibration mode shapes presented in Figures 7 and A2–A8.

4. Final Remarks and Further Developments

The numerical tools presented in the paper for the FE method and CFD simulations depict the issues related to rotating composite laminates, in which, in addition to the specimen fabrication processes, rotational speed play an important role. After presenting the linear single modes of vibrations and their corresponding natural frequencies, the focus shifted towards linear modal interactions and their deformation field description using an analytical method, considering a combination of two bending directions, i and j; longitudinal motion k; and torsional l mode shapes. The mechanical system can be subjected to external loads arising from aerodynamic flow and centrifugal forces. Depending on whether we want to avoid vibrations or excite them, the natural frequencies must be either isolated from the excitation frequencies or targeted. Beyond the excitation frequency, the amplitude and the force/momentum distribution represent critical factors that have a direct impact on the efficiency of motion excitation. It is worth mentioning the possibility of indirectly exciting vibrations, for instance, by stimulating torsional modes and utilizing mechanical couplings to induce significant longitudinal motion. One should consider various types of internal resonances, external subharmonics, and superharmonics, which may arise from both linear and nonlinear mechanical couplings.

In the future research development of rotating composite structures, three main topics will be explored: (i) analytical modelling of vortex-induced vibrations, (ii) the utilization of an electromechanical system for energy harvesting from mechanical vibrations, and (iii) control of coupled vibrations via MFC patches. The first topic involves the expansion of an analytical model presented in the [17]. Besides the nonlinear beam model and aerodynamic flow, there is a plan to incorporate nonlinear Van der Pol equations based on the aerodynamic characteristics of the rectangular cross-section beam and its associated frequency spectra. The second research area includes experimental measurements on the prototype presented in [26]. Based on the vibration modes, it is possible to estimate the optimal location for a harvester that operates proportionally not to the maximum displacement amplitude, but to the maximum curvature. Furthermore, linearly coupled modes of vibrations appear to be a choice of higher efficiency. The final issue pertains to the vibration suppression during rotor operation, which involves avoiding resonant frequencies associated with aerodynamic flow or an vibration reduction active control by piezoelectric transducers.

5. Conclusions

The primary objective of this study was to perform experimental and numerical modal analysis of a composite cantilever beam. In the laboratory investigations with specialized vibration measurement equipment, detailed maps of beam deformations (mode shapes) and magnitude-frequency curves were executed. Additionally, in the absence of rotation, the study conclusively affirmed the absence of linear modal interactions in the composite beam. Through numerical simulations, we delved into the intricacies of Campbell diagrams, determining the natural frequencies and their corresponding modal shapes. Our findings were classified into two distinct categories: the classical single-mode behavior and the pioneering extension of linearly coupled modal analysis. The two fundamental preset angles of $\Theta = 0^\circ$ and $\Theta = 90^\circ$ did not have dynamic couplings within the linear range. However, when the preset angle was adjusted to $\Theta = 30^\circ$ and $\Theta = 60^\circ$, couplings between linear mode shapes in the plane emerged. Rotational speeds exceeding $\dot{\varphi} = 300$ rad/s and a preset angle of $\Theta = 45^\circ$ proved to be the most linearly coupled, with flexural–flexural and longitudinal–torsion modes strongly interfering with each other, respectively. The results for the preset angle were categorized based on the coupling type and presented graphically.

Notably, we introduced an analytical description of coupled mode shapes, encompassing various deformations such as bending, longitudinal deformations, and twisting. This contribution is a noteworthy advancement in understanding the behavior of rotating structures.

In addition to the modal analysis, our secondary objective was to assess the lift, drag forces, and moment characteristics of a rectangular profile in uniform flow. The preset angle variation between $\Theta = 0^\circ$ and $\Theta = 90^\circ$ demonstrated lift/drag force transmission, in which, for the preset angle of $\Theta = 45^\circ$, these forces were almost equal over the entire range of rotational speeds. Comprehensive insights into both the static and dynamic aerodynamic responses acting upon the beam within its operational frequency spectrum were provided. For preset angles 30° and 45°, the vortex-induced vibrations occurred at very small angular speed $\dot{\varphi} = 1$ rad/s, exhibiting two prominent harmonics.

This study represents a substantial step forward in the field of composite rotating beams, offering a deeper understanding of their dynamic characteristics and their interaction with aerodynamic forces. These findings hold significant promise for various engineering applications and contribute to the broader knowledge of dynamic systems.

Author Contributions: Conceptualization, L.K. and G.S.; methodology, L.K.; software, L.K. and G.S.; validation, Z.S.; formal analysis, L.K., Z.S. and G.S.; investigation, L.K.; resources, G.S.; data curation, L.K.; writing—original draft preparation, L.K. and Z.S.; writing—review and editing, L.K., Z.S. and G.S.; visualization, L.K. and Z.S.; supervision, L.K.; project administration, Z.S. and G.S.; funding acquisition, L.K. and G.S. All authors have read and agreed to the published version of the manuscript.

Funding: This research was funded in whole or in part by the National Science Centre, Poland 2021/41/B/ST8/03190. For the purpose of Open Access, the author has applied a CC-BY public copyright licence to any Author Accepted Manuscript (AAM) version arising from this submission. The first author would like to acknowledge the financial support of the National Centre for Research and Development under the project POWR.03.02.00-IP.08-00-DOK/16 within the framework of "The International PhD Course in Mechanics", provided by the Faculty of Mechanical Engineering at Lublin University of Technology, POWR.03.02.00-00-I017/16-00.

Institutional Review Board Statement: Not applicable.

Informed Consent Statement: Not applicable.

Data Availability Statement: The data presented in this study are available on request from the corresponding author.

Conflicts of Interest: The authors declare no conflict of interest.

Nomeclature

R	Radius of a hub, mm
$\dot{\varphi}$	Angular velocity of a hub, rad/s
L	Length of a beam, mm
b	Width of a beam, mm
h	Height of a beam, mm
Θ	Preset angle of a beam, °
ϱ	Effective density of a composite beam, kg/m^3
μ	Mass per unit length of a beam, kg/m
E	Effective Young modulus of a beam, GPa
G	Effective shear modulus of a beam, GPa
ν	Effective Poisson's ratio of a beam, -
i	Notation for out-of-plane bending deformation of a beam, -
j	Notation for in-plane bending deformation of a beam, -
k	Notation for longitudinal deformation of a beam, -
l	Notation for torsional deformation of a beam, -
$\Phi_i(x), \Phi_j(x), \Phi_k(x), \Phi_l(x)$	linear mode shape projections, m, m, m, rad
A_1–A_4	Normalized amplitudes of ith linear mode shape, m
B_1–B_4	Normalized amplitudes of jth linear mode shape, m
C_1–C_4	Normalized amplitudes of kth linear mode shape, m
D_1–D_4	Normalized amplitudes of lth linear mode shape, rad
$\lambda_1, \lambda_2, \lambda_3, \lambda_4$	Characteristic coefficients for ith, jth, kth and lth linear mode shape, respectively, 1/m, 1/m, 1/m, 1/rad
$\bar{\Phi}(t,x)$	Linearly coupled mode
ω_n	nth natural frequency of a beam, rad/s
$\Phi(i,j,k,l)$	Linearly coupled mode shape, m, m, m, rad
$\bar{A}, \bar{B}, \bar{C}, \bar{D}$	Weights of a linearly coupled mode shape, -
v	Air flow velocity, m/s

Appendix A

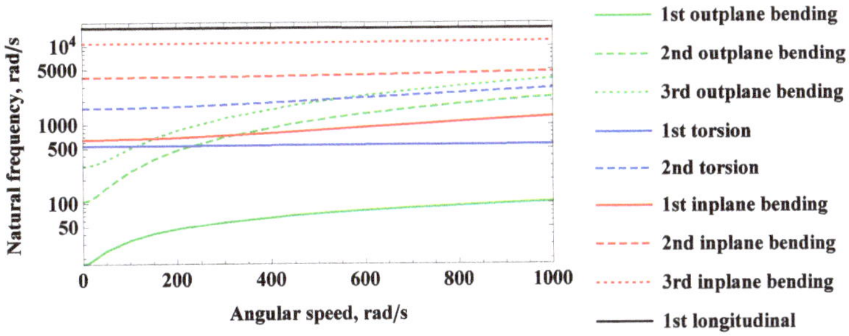

(a)

Figure A1. *Cont.*

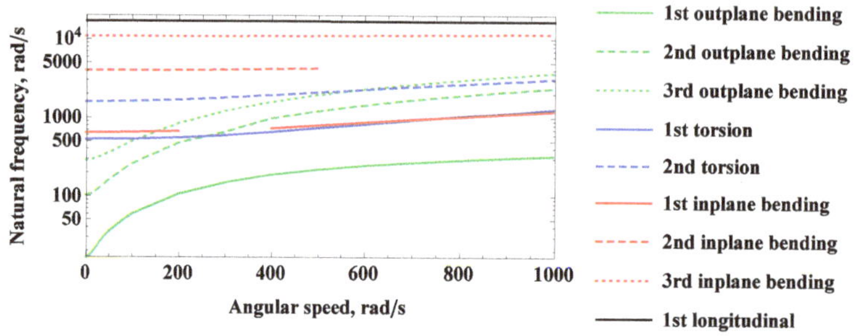

(**b**)

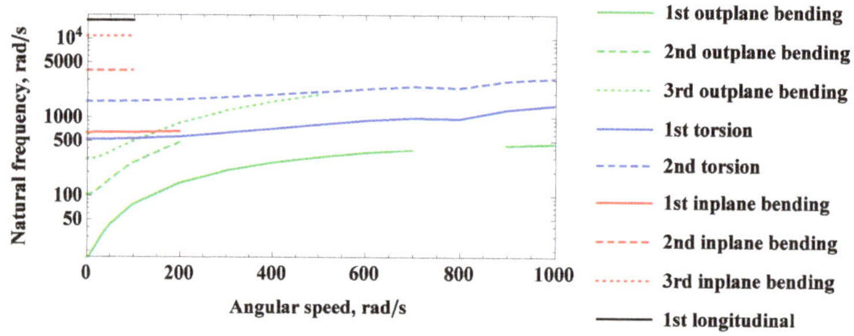

(**c**)

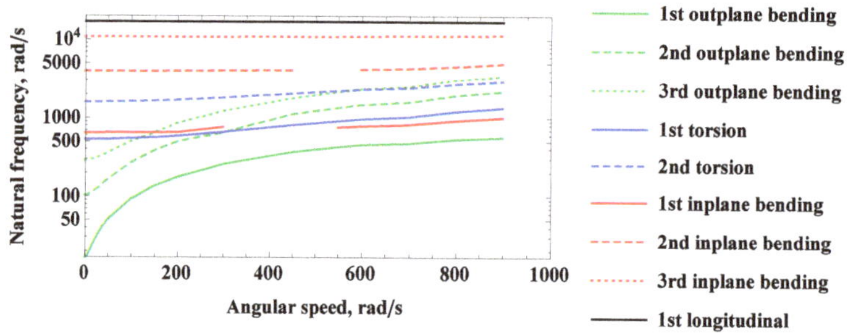

(**d**)

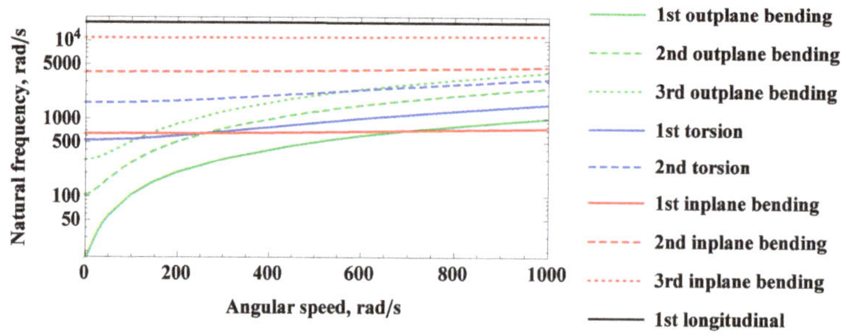

(**e**)

Figure A1. Campbell diagram of the rotating beam for predefined preset angle: (**a**) 0°, (**b**) 30°, (**c**) 45°, (**d**) 60°, (**e**) 90°.

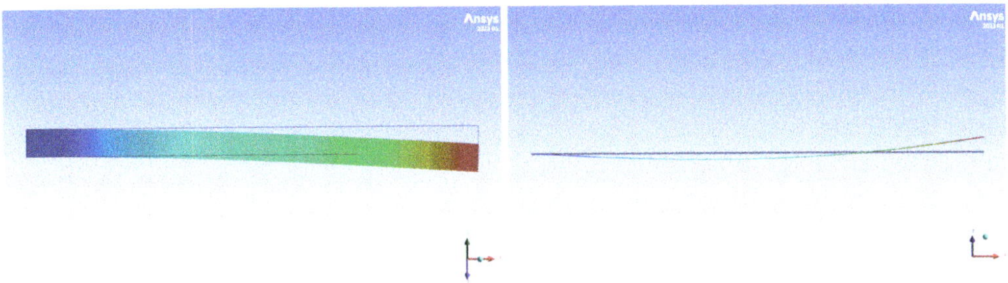

(**a**) $\Phi(2,1,0,0)$, $\omega_3 = 654.46$ rad/s

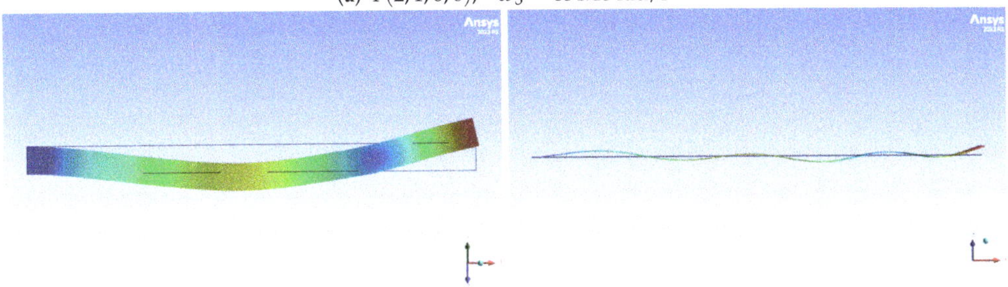

(**b**) $\Phi(7,2,0,0)$, $\omega_{12} = 4061.58$ rad/s

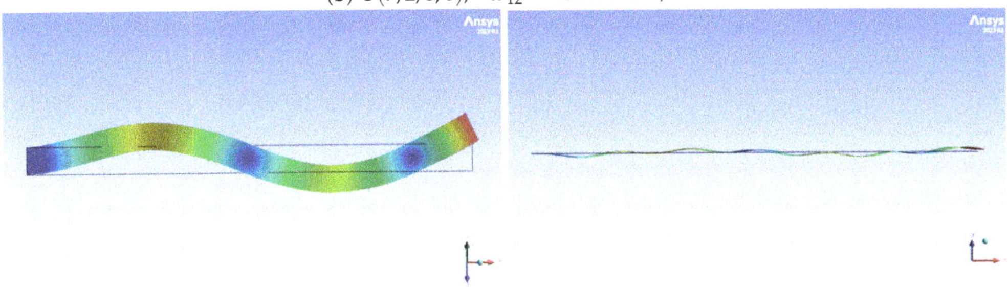

(**c**) $\Phi(14,3,0,0)$, $\omega_{25} = 10918.92$ rad/s

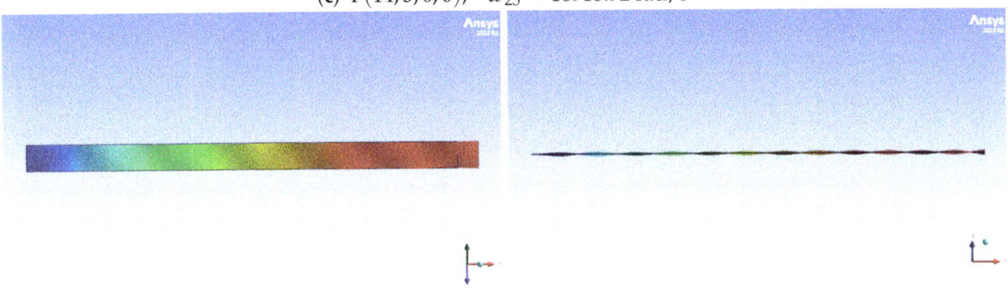

(**d**) $\Phi(0,0,1,13)$, $\omega_{34} = 16968.37$ rad/s

Figure A2. Linear mode couplings for preset angle $\Theta = 45^\circ$ and $\dot{\varphi} = 300$ rad/s.

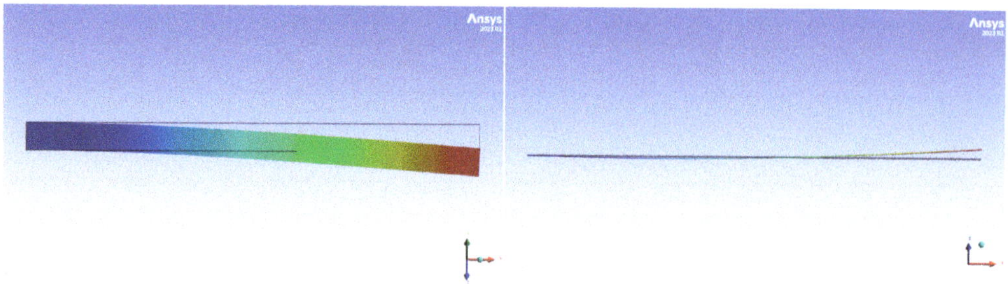

(**a**) $\Phi(2, 1, 0, 0), \quad \omega_2 = 712.89$ rad/s

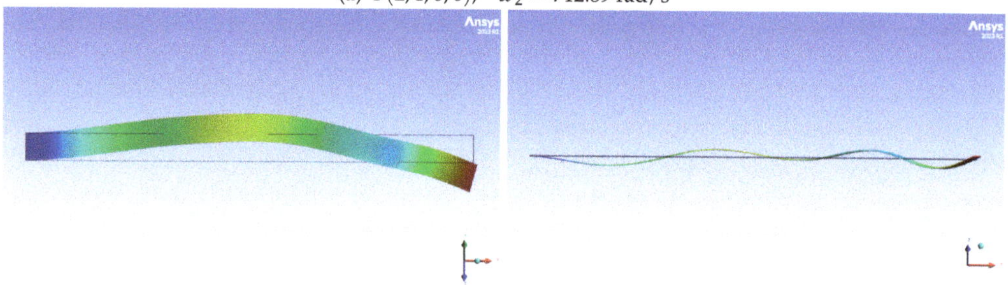

(**b**) $\Phi(6, 2, 0, 0), \quad \omega_{11} = 4143.64$ rad/s

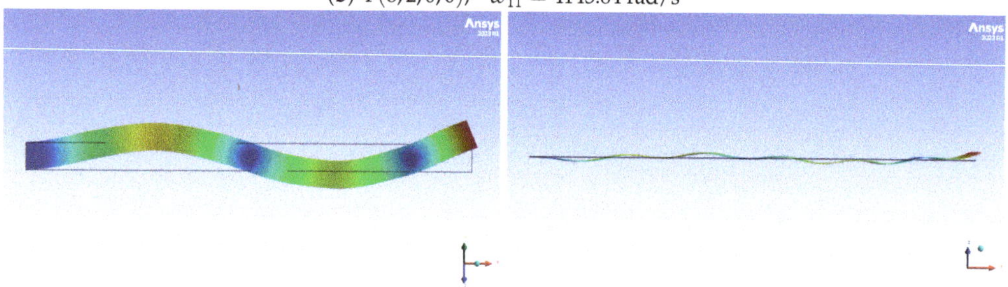

(**c**) $\Phi(12, 3, 0, 0), \quad \omega_{23} = 10986.78$ rad/s

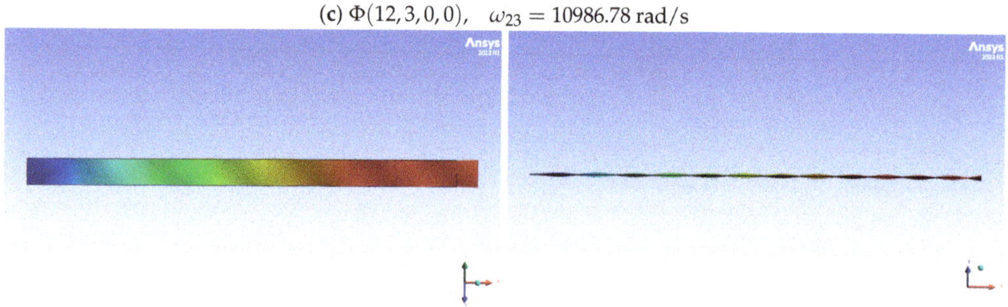

(**d**) $\Phi(0, 0, 1, 13), \quad \omega_{32} = 16959.57$ rad/s

Figure A3. Linear mode couplings for preset angle $\Theta = 45^\circ$ and $\dot{\varphi} = 400$ rad/s.

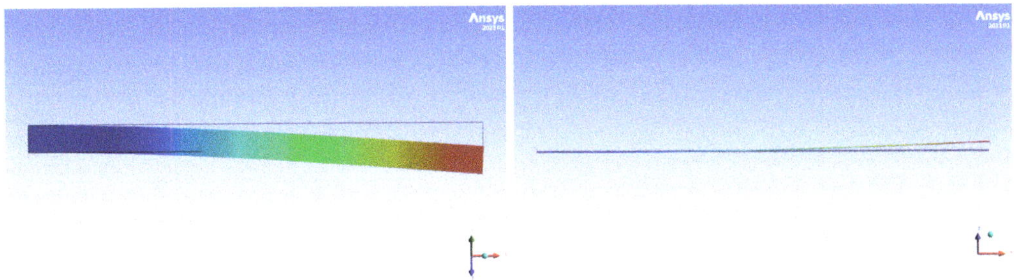

(**a**) $\Phi(1,1,0,0), \quad \omega_2 = 769.82$ rad/s

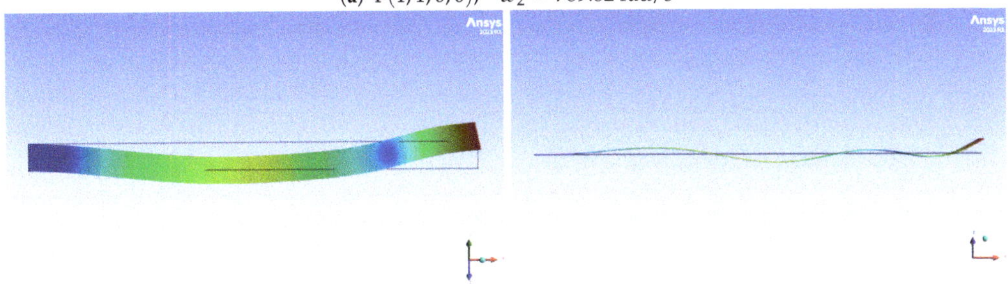

(**b**) $\Phi(5,2,0,0), \quad \omega_{10} = 4216.58$ rad/s

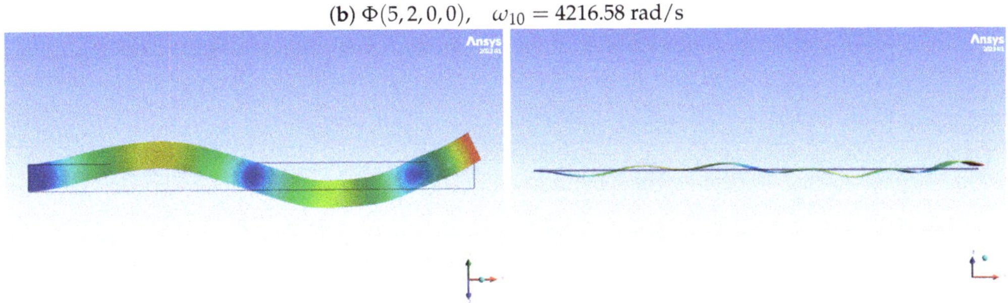

(**c**) $\Phi(11,3,0,0), \quad \omega_{22} = 11073.49$ rad/s

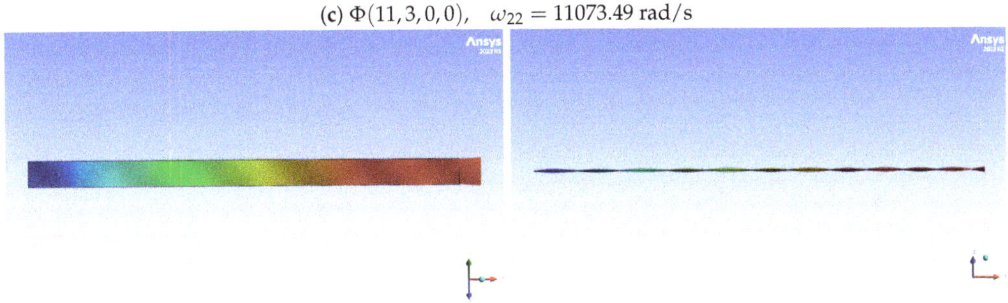

(**d**) $\Phi(0,0,1,12), \quad \omega_{30} = 16948.89$ rad/s

Figure A4. Linear mode couplings for preset angle $\Theta = 45^\circ$ and $\dot{\varphi} = 500$ rad/s.

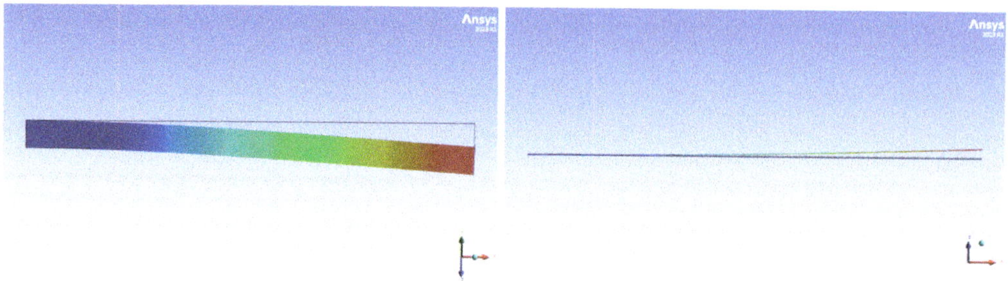

(**a**) $\Phi(1,1,0,0), \quad \omega_2 = 835.41$ rad/s

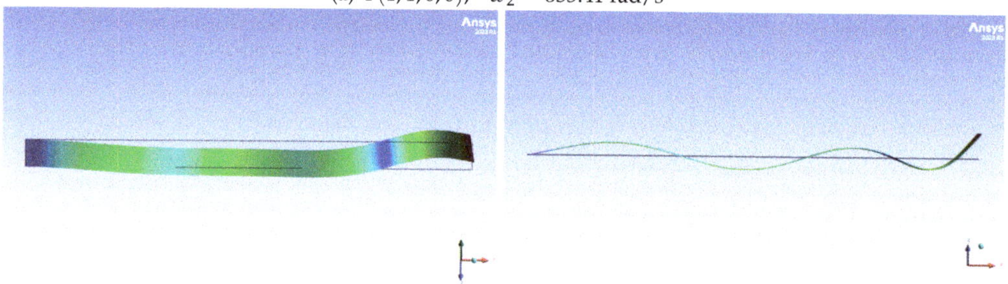

(**b**) $\Phi(5,2,0,0), \quad \omega_{10} = 4462.57$ rad/s

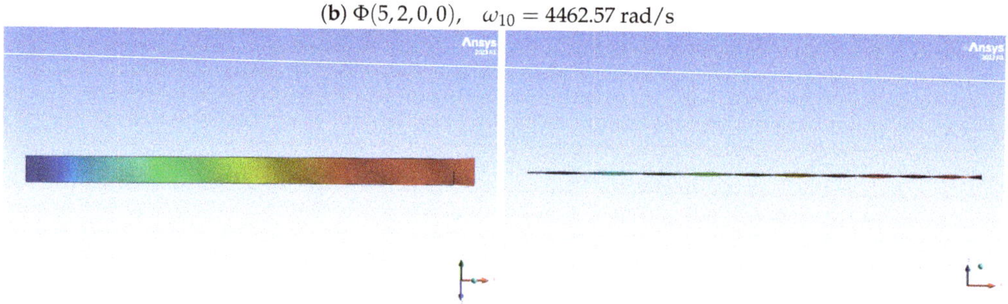

(**c**) $\Phi(0,0,1,11), \quad \omega_{29} = 16936.95$ rad/s

Figure A5. Linear mode couplings for preset angle $\Theta = 45^\circ$ and $\dot{\varphi} = 600$ rad/s.

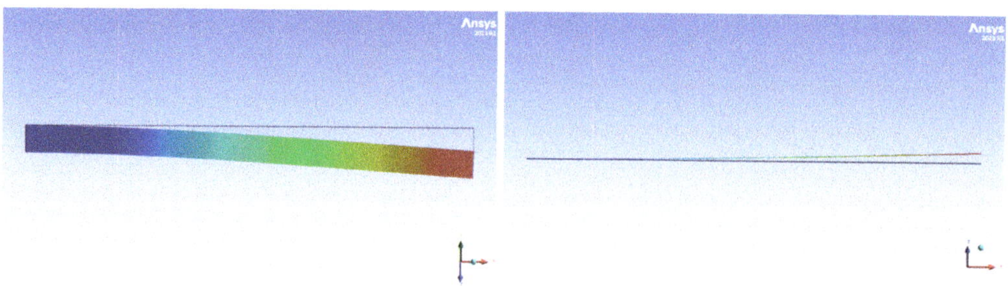

(**a**) $\Phi(1,1,0,0), \quad \omega_2 = 909.30$ rad/s

Figure A6. *Cont.*

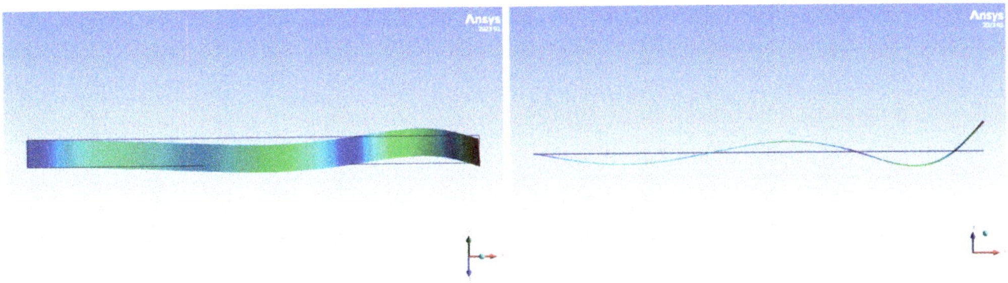

(**b**) $\Phi(4,2,0,0)$, $\omega_7 = 3745.28$ rad/s

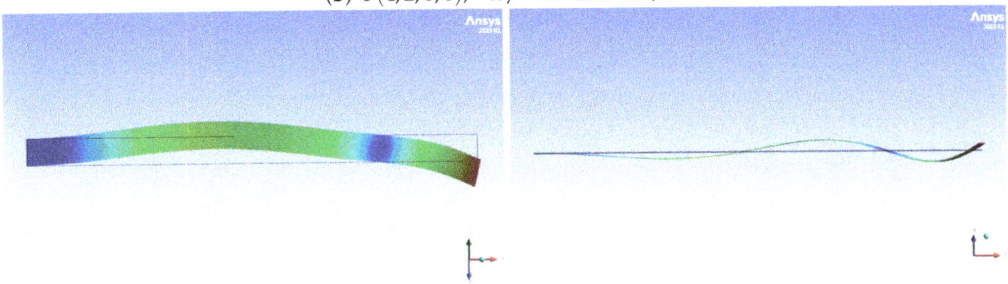

(**c**) $\Phi(4,2,0,0)*$, $\omega_9 = 4428.26$ rad/s

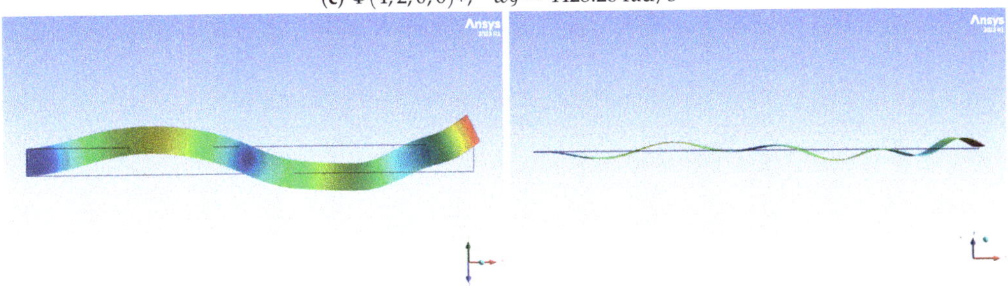

(**d**) $\Phi(9,3,0,0)$, $\omega_{19} = 11268.89$ rad/s

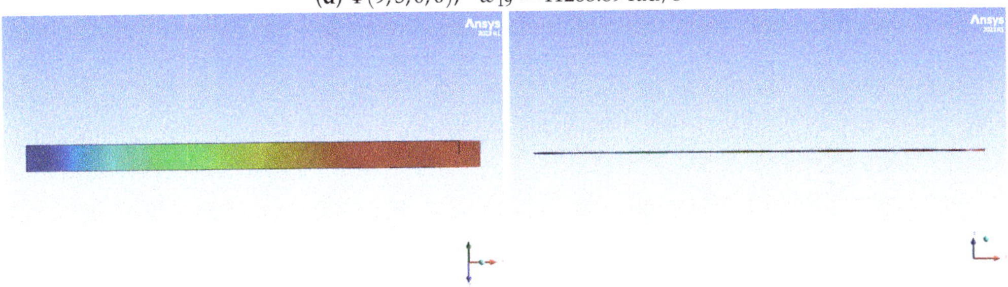

(**e**) $\Phi(0,0,1,9)$, $\omega_{27} = 16921.87$ rad/s

Figure A6. Linear mode couplings for preset angle $\Theta = 45°$ and $\dot{\varphi} = 700$ rad/s.

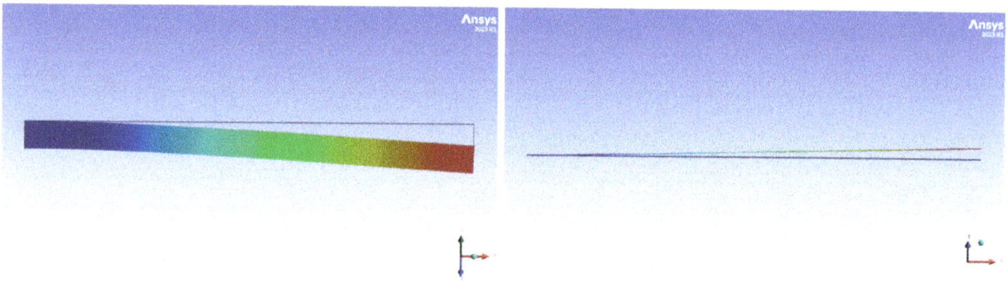

(**a**) $\Phi(1, 1, 0, 0)$, $\omega_2 = 1073.67$ rad/s

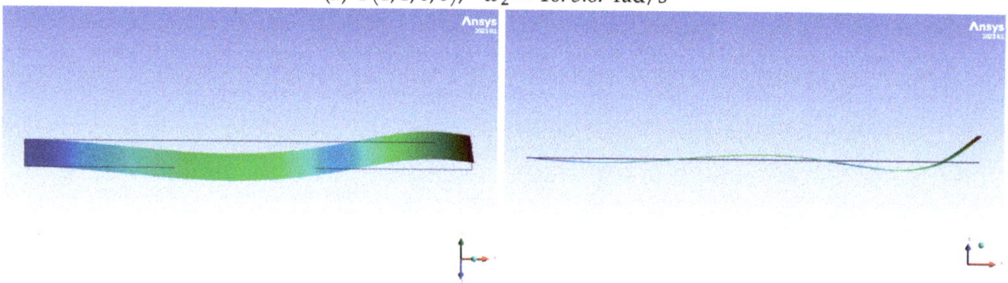

(**b**) $\Phi(4, 2, 0, 0)$, $\omega_7 = 4467.28$ rad/s

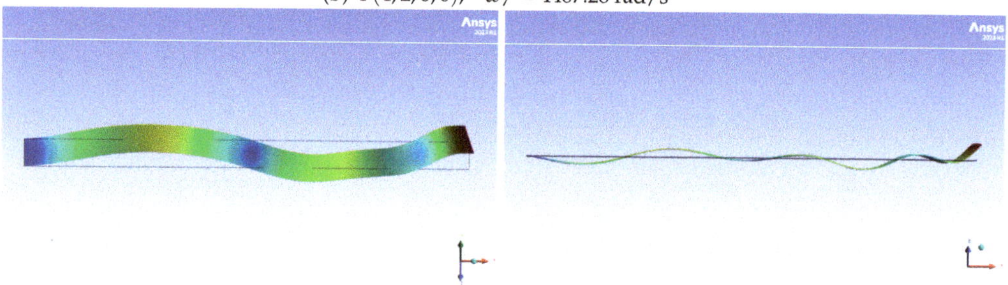

(**c**) $\Phi(8, 3, 0, 0)$, $\omega_{17} = 11555.41$ rad/s

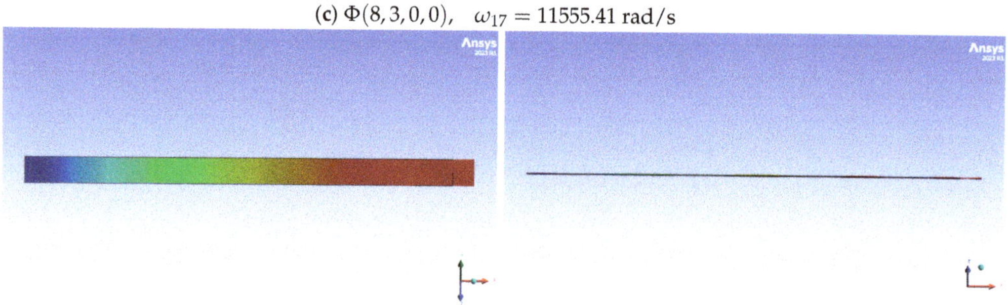

(**d**) $\Phi(0, 0, 1, 10)$, $\omega_{24} = 16884.80$ rad/s

Figure A7. Linear mode couplings for preset angle $\Theta = 45^\circ$ and $\dot{\varphi} = 900$ rad/s.

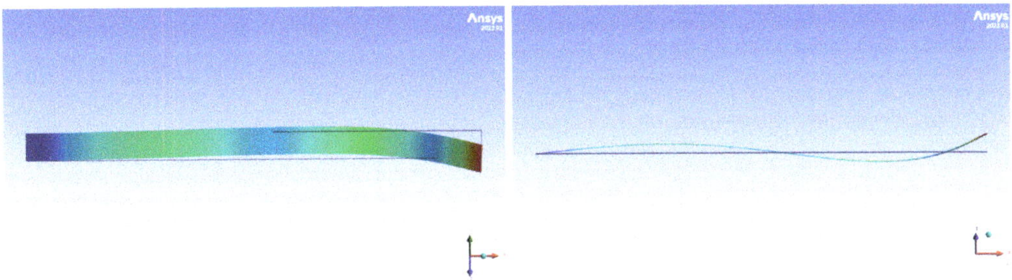

(**a**) $\Phi(3,2,0,0)$, $\omega_6 = 3745.47$ rad/s

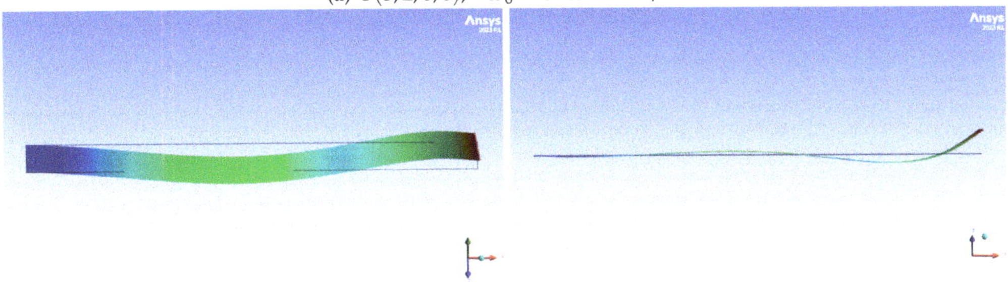

(**b**) $\Phi(4,2,0,0)$, $\omega_7 = 4693.92$ rad/s

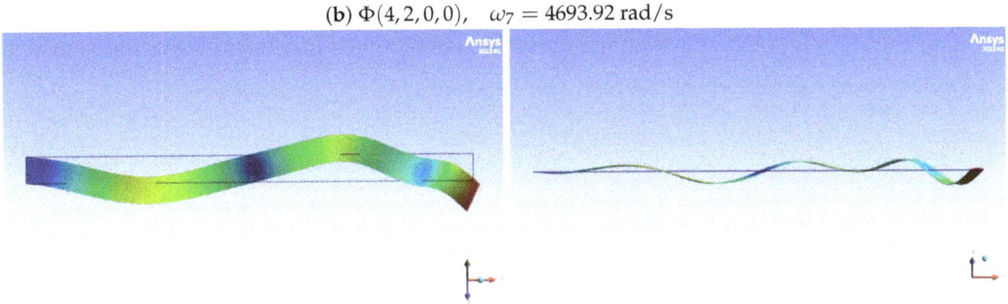

(**c**) $\Phi(8,3,0,0)$, $\omega_{16} = 11626.41$ rad/s

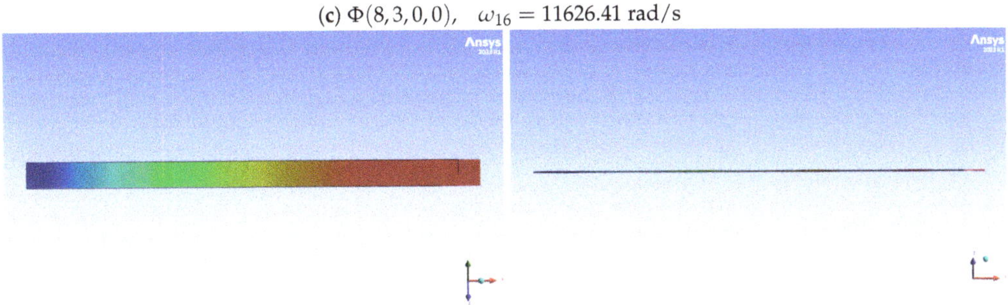

(**d**) $\Phi(0,0,1,8)$, $\omega_{22} = 16862.81$ rad/s

Figure A8. Linear mode couplings for preset angle $\Theta = 45°$ and $\dot{\varphi} = 1000$ rad/s.

References

1. Carrera, E.; Pagani, A.; Petrolo, M.; Zappino, E. Recent developments on refined theories for beams with applications. *Mech. Eng. Rev.* **2015**, *2*, 14-00298. [CrossRef]
2. Wang, D.; Chen, Y.; Wiercigroch, M.; Cao, Q. Bifurcation and dynamic response analysis of rotating blade excited by upstream vortices. *Appl. Math. Mech.* **2016**, *37*, 1251–1274. [CrossRef]

3. Lenci, S.; Clementi, F.; Rega, G. Comparing nonlinear free vibrations of Timoshenko beams with mechanical or geometric curvature definition. *Procedia IUTAM* **2017**, *20*, 34–41. [CrossRef]
4. Babilio, E.; Lenci, S. Consequences of different definitions of bending curvature on nonlinear dynamics of beams. *Procedia Eng.* **2017**, *199*, 1411–1416. [CrossRef]
5. Lacarbonara, W.; Arvin, H.; Bakhtiari-Nejad, F. A geometrically exact approach to the overall dynamics of elastic rotating blades—Part 1: linear modal properties. *Nonlinear Dyn.* **2012**, *70*, 659–675. [CrossRef]
6. Arvin, H.; Lacarbonara, W.; Bakhtiari-Nejad, F. A geometrically exact approach to the overall dynamics of elastic rotating blades—Part 2: flapping nonlinear normal modes. *Nonlinear Dyn.* **2012**, *70*, 2279–2301. [CrossRef]
7. Thomas, O.; Sénéchal, A.; Deü, J.F. Hardening/softening behavior and reduced order modeling of nonlinear vibrations of rotating cantilever beams. *Nonlinear Dyn.* **2016**, *86*, 1293–1318. [CrossRef]
8. Stoykov, S.; Ribeiro, P. Vibration analysis of rotating 3D beams by the p-version finite element method. *Finite Elem. Anal. Des.* **2013**, *65*, 76–88. [CrossRef]
9. Carrera, E.; Filippi, M.; Zappino, E. Free vibration analysis of rotating composite blades via Carrera Unified Formulation. *Compos. Struct.* **2013**, *106*, 317–325. [CrossRef]
10. Younesian, D.; Esmailzadeh, E. Non-linear vibration of variable speed rotating viscoelastic beams. *Nonlinear Dyn.* **2010**, *60*, 193–205. [CrossRef]
11. Georgiades, F.; Latalski, J.; Warminski, J. Equations of motion of rotating composite beam with a nonconstant rotation speed and an arbitrary preset angle. *Meccanica* **2014**, *49*, 1833–1858. [CrossRef]
12. Di Nino, S.; Luongo, A. Nonlinear dynamics of a base-isolated beam under turbulent wind flow. *Nonlinear Dyn.* **2022**, *107*, 1529–1544. [CrossRef]
13. Elmiligui, A.; Abdol-Hamid, K.S.; Massey, S.J.; Pao, S.P. Numerical study of flow past a circular cylinder using hybrid turbulence formulations. *J. Aircr.* **2010**, *47*, 434–440. [CrossRef]
14. Yao, M.H.; Chen, Y.P.; Zhang, W. Nonlinear vibrations of blade with varying rotating speed. *Nonlinear Dyn.* **2012**, *68*, 487–504. [CrossRef]
15. Eftekhari, M.; Owhadi, S. Nonlinear dynamics of the rotating beam with time-varying speed under aerodynamic loads. *Int. J. Dyn. Control* **2022**, *10*, 49–68. [CrossRef]
16. Warminski, J.; Kloda, L.; Latalski, J.; Mitura, A.; Kowalczuk, M. Nonlinear vibrations and time delay control of an extensible slowly rotating beam. *Nonlinear Dyn.* **2021**, *103*, 3255–3281. [CrossRef]
17. Kloda, L.; Warminski, J. Nonlinear longitudinal–bending–twisting vibrations of extensible slowly rotating beam with tip mass. *Int. J. Mech. Sci.* **2022**, *220*, 107153. [CrossRef]
18. Kumar, P. Modal analysis of viscoelastic three-dimensional rotating beam with generic tip mass. *Eur. J. Mech.-A/Solids* **2022**, *96*, 104734. [CrossRef]
19. Huang, J.; Wang, K.; Tang, J.; Xu, J.; Song, H. An experimental study of the centrifugal hardening effect on rotating cantilever beams. *Mech. Syst. Signal Process.* **2022**, *165*, 108291. [CrossRef]
20. Gawryluk, J.; Mitura, A.; Teter, A. Dynamic response of a composite beam rotating at constant speed caused by harmonic excitation with MFC actuator. *Compos. Struct.* **2019**, *210*, 657–662. [CrossRef]
21. Rafiee, M.; Nitzsche, F.; Labrosse, M. Dynamics, vibration and control of rotating composite beams and blades: a critical review. *Thin-Walled Struct.* **2017**, *119*, 795–819. [CrossRef]
22. Teter, A.; Gawryluk, J. Experimental modal analysis of a rotor with active composite blades. *Compos. Struct.* **2016**, *153*, 451–467. [CrossRef]
23. Mitura, A.; Gawryluk, J.; Teter, A. Numerical and experimental studies on the rotating rotor with three active composite blades. *Eksploat. Niezawodn. Maint. Reliab.* **2017**, *19*, 571–579. [CrossRef]
24. Warminski, J.; Latalski, J.; Szmit, Z. Vibration of a mistuned three-bladed rotor under regular and chaotic excitations. *J. Theor. Appl. Mech.* **2018**, *56*, 549–566. [CrossRef]
25. Szmit, Z. Vibration, synchronization and localization of three-bladed rotor: Theoretical and experimental studies. *Eur. Phys. J. Spec. Top.* **2021**, *230*, 3615–3625. [CrossRef]
26. Szmit, Z.; Kloda, L.; Kowalczuk, M.; Stachyra, G.; Warmiński, J. Experimental analysis of aerodynamic loads of three-bladed rotor. *Materials* **2022**, *15*, 3335. [CrossRef] [PubMed]
27. Vannucci, P.; Verchery, G. A new method for generating fully isotropic laminates. *Compos. Struct.* **2002**, *58*, 75–82. [CrossRef]
28. Warminski, J.; Kloda, L.; Lenci, S. Nonlinear vibrations of an extensional beam with tip mass in slewing motion. *Meccanica* **2020**, *55*, 2311–2335. [CrossRef]
29. Ehrhardt, D.A.; Hill, T.L.; Neild, S.A.; Cooper, J.E. Veering and nonlinear interactions of a clamped beam in bending and torsion. *J. Sound Vib.* **2018**, *416*, 1–16. [CrossRef]

materials

MDPI

Article

Delamination Behavior of CFRP Laminated Plates under the Combination of Tensile Preloading and Impact Loading

Kaiwei Lan [1], Haodong Wang [1] and Cunxian Wang [1,2,3,*]

1 School of Aeronautics, Northwestern Polytechnical University, Xi'an 710072, China; lankaiwei2023@163.com (K.L.); wanghaodong724@mail.nwpu.edu.cn (H.W.)
2 Institute of Extreme Mechanics, Northwestern Polytechnical University, Xi'an 710072, China
3 Shaanxi Key Laboratory of Impact Dynamics and Its Engineering Application (IDEA), Northwestern Polytechnical University, Xi'an 710072, China
* Correspondence: wangcunxian@nwpu.edu.cn

Abstract: When subjected to impact loading, aircraft composite structures are usually in a specific preloading condition (such as tension and compression). In this study, ballistic tests were conducted using a high-speed gas gun system to investigate the effect of biaxial in-plane tensile preload on the delamination of CFRP laminates during high-speed impact. These tests covered central and near-edge locations for both unloaded and preloaded targets, with the test speeds including 50 m/s, 70 m/s, and 90 m/s. The delamination areas, when impacting the center location under 1000 $\mu\varepsilon$, show a 14.2~36.7% decrease. However, the cases when impacting the near-edge location show no more than a 19.3% decrease, and even more delamination areas were observed. In addition, in order to enhance the understanding of experimental phenomena, numerical simulations were conducted using the ABAQUS/Explicit solver, combined with the user subroutine VUMAT with modified Hou criteria. The experimental and simulation results were in good agreement, and the maximum error was approximately 12.9%. The results showed that not only the preloading value but also the impact velocity have significant influences on the delamination behavior of preloaded CFRP laminated plates. Combining detailed discussions, the biaxial tensile preload enhanced the resistance to out-of-plane displacement and caused laminate interface stiffness degradation. By analyzing the influence of the preloading value and impact velocity on competing mechanisms between the stress-stiffening effect and interface stiffness degradation effect, the complex delamination behaviors of laminates under various preloading degrees and impact velocities at different impact locations were reasonably explained.

Keywords: biaxial tensile preload; delamination; CFRP laminated plate; competition mechanism; impact loading

1. Introduction

Carbon-fiber-reinforced polymer (CFRP) composites present many excellent mechanical properties compared to traditional metal materials, making them widely used in aircraft [1]. In the daily usage of civil aircraft, composites may produce damage when suffering various forms of impact loading, such as the impact of birds, gravel, falling tools, and hail. The predominant impact-induced damages primarily encompass fiber fracture and matrix crack, with the latter often leading to non-apparent delamination, invisible to the unaided eye. However, delamination significantly affects the integral stiffness and strength of composite structures, which may finally result in serious security threats to aircraft.

Over the years, investigations on the impact response of composites have attracted extensive attention from many scholars, and a number of good papers have been published. Due to the limited science and technology available in its early days, the original research mainly focused on ballistic tests and damage measurements. Lee et al. [2] and

Cheng et al. [3] carried out a series of impact tests on composites, and the results showed that the typical damage of composites mainly comprised crushing, fiber tensile fracture, and delamination. In addition, the ballistic limits of laminates were determined, and the effect of the projectile shape on perforation was also discussed. Yew and Kendrick [4] introduced ultrasonic C-scan and cross-section staining methods to detect internal damage in laminates under impact loads, revealing the anisotropic nature of composite impact damage. With the development of finite element (FE) technology, more and more scholars have attempted to investigate this topic by combining tests and simulations. Binienda and Zhang et al. [5,6] devised an array of experimental approaches and multi-scale numerical methodologies. These were employed to comprehensively examine the mechanical responses of triaxial braided composites, effectively establishing the impact threshold and delineating the various mechanisms governing damage patterns. Camanho et al. [7], as well as Sridharan and Pankow [8], posited that cohesive elements exhibit the capability to forecast the initiation and extension of delamination in composite materials. Rajaneesh et al. [9] formulated finite element models to explore the high-velocity impact responses of composite materials, introducing a physically grounded delayed damage model in their research. Furthermore, Xu et al. [10] conducted ballistic impact tests on CFRP plates under three different conditions: preloaded uniaxial tension, preloaded uniaxial compression, and no preload. They analyzed the energy absorption efficiency and failure modes and established a theoretical model based on energy absorption theory. Meanwhile, Qaderi, Ebrahimi, and Vingas [11] conducted a dynamic theoretical analysis of multi-layer composite beams reinforced with GPLs using the high-order shear deformation beam theory. They obtained the trends in the natural frequencies of the system as a function of the system parameters for different distribution patterns.

Nowadays, the majority of existing studies primarily concentrate on CFRP laminated plates without prior preloading. Notably, the reviews by Abrate [12] and the investigations by Reid and Zhou [13] predominantly emphasize the impact behavior of unloaded composite structures. Consequently, there exists a limited understanding of the impact properties of composites subjected to preloading conditions. However, composite structures on aircraft are usually in a specific loading condition (such as tension or compression) when subjected to impact loading [1]. For instance, during aircraft takeoff and landing maneuvers, the wing may undergo upward bending, potentially resulting in impact from debris on the runway. During this phase, the lower skin structure is subjected to significant tensile loads. Similarly, the upper skin of the wing can encounter impacts from hailstones, leading to substantial compressive loads on the upper skin structure. Additionally, engine blades may experience significant centrifugal dynamic loads and aerodynamic forces when impacted by flying birds. Hence, investigating the influence of preloads on the impact characteristics of composite laminates becomes imperative.

As an illustration, fuselage skins in aircraft service frequently encounter operational strains reaching up to 1500 με [11]. NASA held a seminar and put forward that the preloading effect was critical to the impact damage of composites. The LIBCOS and MAAXIMUS projects launched by EASA and Advisory Circular 20-107B of the FAA also stated that the actual tensile or compressive preloads should be considered when analyzing accidental impact damage. Williams et al. [14] pointed out that the preload's effect on composite structures' impact behavior is a potential problem in aircraft certification. However, most research focuses on the impact behavior of unloaded composites, and it is vital to study the impact responses of composites under specific load conditions, especially under preloading conditions.

At present, some studies have been devoted to the experimental and numerical evaluation of the impact behavior of preloaded composites. Both low-velocity and high-velocity impact tests on preloaded composites have been carried out by using drop-weight test instruments [14–16], impact pendulums [17], and gas gun devices [18,19]. Moreover, Mikkor et al. [20] and Pickett et al. [21] conducted explicit FE simulations on the impact process of unloaded and preloaded laminates using Pam-Crash. Heimbs et al. [1,22] simulated preloading and impact using a "layered shell" model and "implicit explicit coupling" in LS-DYNA and ABAQUS, respectively. It should be pointed out that there existed significant differences between these presented research studies. Garnier [19], Chiu [23], and Kelkar [24] proposed that tensile preload leads to an increase in the impact damage area for laminates. However, the works by Heimbs [22], Robb [25], García Castillo [26], Zhikharev [27], and Guillaud et al. [28] showed that tensile preload can reduce the delamination area. In addition, investigations by Mitrevski [29], Moallemzadeh [30], and Choi et al. [31] indicated that tensile preloading levels do not have an effect on impact damage under low-velocity impact loading. Wang et al. [32] found different delamination tendencies when impacting two different locations and that the interlaminar stresses induced a weakening effect on delamination. While these investigations present reasonable findings, further experimental, numerical, and theoretical research is needed to understand how preloads influence the impact response of composite laminates under different preloading levels and impact velocities.

In the present work, a series of ballistic impact tests were carried out on CFRP laminated plates with two different stacking sequences to understand this topic further. Three preloading levels (0 με, 500 με, and 1000 με) were realized using a specially designed preloading device. Moreover, three different impact velocities (50 m/s, 70 m/s, 90 m/s) and two typical impact locations (center location and near-edge location) were considered in the ballistic impact tests. Additional numerical simulations were also conducted in ABAQUS/Explicit to gain further insights into the effects of preloading levels and impact velocities on the delamination behavior of CFRP laminated plates under impact loading. Based on the comparison of experimental and numerical results, the influence of the preloading degree and impact velocity on the competing mechanisms between the stress-stiffening effect and interface stiffness degradation effect was analyzed.

2. Material Preparation and Experimental Procedure

2.1. Material Preparation, Preloading Fixture, and Test Conditions

The CFRP composite specimens tested in the ballistic impact tests were made of T700/epoxy resin M10R laminated plates. The laminated plates were prepared to consist of 16 plies with two types of stacking sequences ($[0/90]_8$ and $[0/90/+45/-45]_{2S}$), and each panel was molded in one shot to produce a nominal fiber volume fraction of 58%. All specimens were prepared with a size of (300 × 300) mm^2 by using water jet cutting to avoid initial damage as much as possible during processing. Figure 1 shows the preloading fixture assembly, which consists of eight steel clamping pieces, four thick steel frames, and an extensive steel support. Each side of the specimen was clamped by two clamping pieces, thus leaving a 210 × 210 mm^2 free impact zone. A set of eight strain gauges was affixed to the rear surface of each panel, serving a dual purpose: recording in-plane strain histories during impact and providing pre-strain control feedback prior to impact. With the support of these strain gauges, the specific pre-strain could be accurately applied by adjusting the bolts connecting the clamping pieces with the frames before the impact tests. In the present work, a series of ballistic impact tests were carried out under different preloads (0 με, 500 με, and 1000 με) and impact velocities (50 m/s, 70 m/s, and 90 m/s). In addition, two impact positions (point 1 and point 2) were predetermined for the ballistic tests in which point 1 was located in the center of the target and point 2 was 70 mm apart from point 1 in the horizontal direction, as shown in Figure 1. For each loading condition, at least three valid tests were performed. More than 120 pieces of CFRP laminated plates were consumed in the ballistic tests.

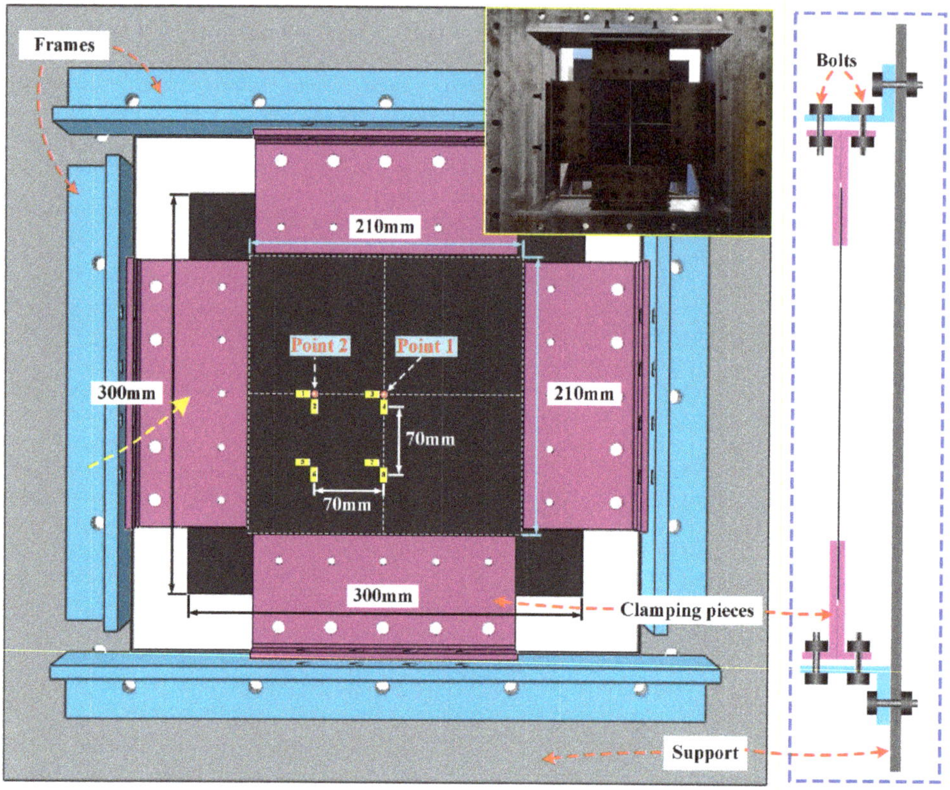

Figure 1. Configuration of preloading fixture and strain measurement scheme.

2.2. Ballistic Impact Test Devices

Figure 2 illustrates the one-stage compressed gas gun system used for the high-speed impact tests for preloaded CFRP laminated plates. The gas gun system comprised a pressure vessel with a volume of 0.2 m^3 and a 6.8 m long gun barrel with an inner diameter of 80 mm. As projectiles, steel spheres with a diameter of 10 mm and a corresponding mass of 4.05 g were chosen. Specially designed 3D-printed PLA shells were used to support the projectiles and were accelerated by compressive air in the gun barrel during the impact test, contributing to the control of the attitude and velocity of projectiles. Two high-speed cameras were positioned in front of the CFRP laminated plate for specific purposes: one camera was dedicated to recording the impact position, while the other was utilized to measure velocity. In addition to using high-speed cameras, the impact velocities of the projectiles were also measured using a laser velocimeter, which helped us to obtain reliable and accurate impact velocities. After the impact tests, the specimen's visible failures were evaluated first, and then the internal delamination damages were detected using a PAC ultrasonic C-scan.

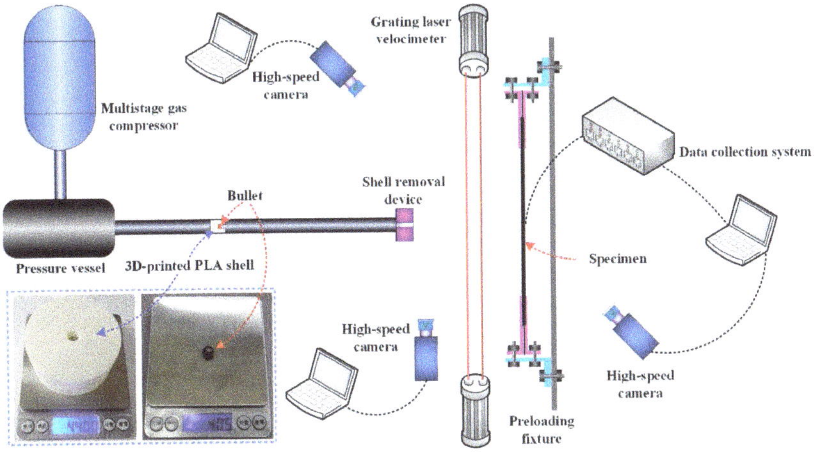

Figure 2. Schematic representation of a single-stage compressed gas gun system.

3. Experimental Results and Analysis

Figure 3a,b show the typical front-view high-speed photographs of point 1 and point 2 on the CFRP laminated plates impacted by steel sphere projectile under the speed of 50 m/s, respectively. Via the statistical analysis of all experimental data, the deviations between the actual impact positions and predetermined positions do not exceed 9.8 mm, and the deviations in the impact speeds do not exceed 4 m/s. The CFRP laminated plates present smooth circular dents in different degrees on the front side under impact velocities of 70 m/s and 90 m/s, while there is no visible dent in the case of 50 m/s. Furthermore, there was no observable damage detected on the rear side of any of the targets subjected to impact velocities of 50 m/s, 70 m/s, and 90 m/s. The delamination areas of the CFRP laminated plates subjected to various biaxial tensile preloads and impact speeds are summarized in Table 1 and Figure 4.

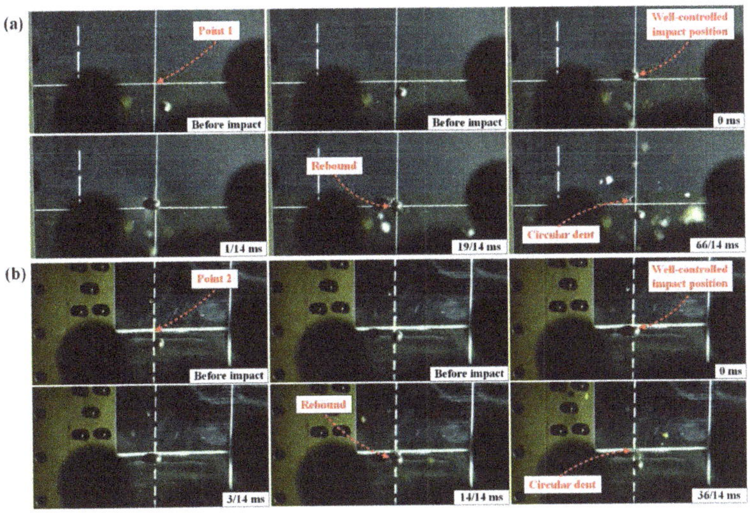

Figure 3. Typical high-speed photographs of CFRP laminated plates impacted by sphere steel projectile under the speed of 50 m/s; (**a**) impact on point 1 (center location); (**b**) impact on point 2 (near-edge location).

Table 1. Delamination areas from C-scan experiments.

Layups	Impact Location	Pre-Strain	Delamination Areas (mm^2)		
			V = 50 m/s	V = 70 m/s	V = 90 m/s
$[0/90]_8$	Point 1	0 με	161 ± 6.6	523 ± 23.2	998 ± 11.2
		500 με	144 ± 5.8	378 ± 26.5	828 ± 39.7
		1000 με	132 ± 3.4	331 ± 19.3	751 ± 50.4
	Point 2	0 με	158 ± 7.1	525 ± 19.7	979 ± 28.1
		500 με	184 ± 7.7	523 ± 12.3	859 ± 28.3
		1000 με	173 ± 6.2	464 ± 34.1	791 ± 42.9
$[0/90/+45/-45]_{2S}$	Point 1	0 με	134 ± 8.9	409 ± 11.8	826 ± 15.5
		500 με	118 ± 6.4	321 ± 19.1	689 ± 44.6
		1000 με	115 ± 11.2	309 ± 33.6	630 ± 32.2
	Point 2	0 με	131 ± 5.2	415 ± 6.2	837 ± 23.7
		500 με	164 ± 10.7	358 ± 38.7	737 ± 55.2
		1000 με	154 ± 12.3	337 ± 11.4	724 ± 18.8

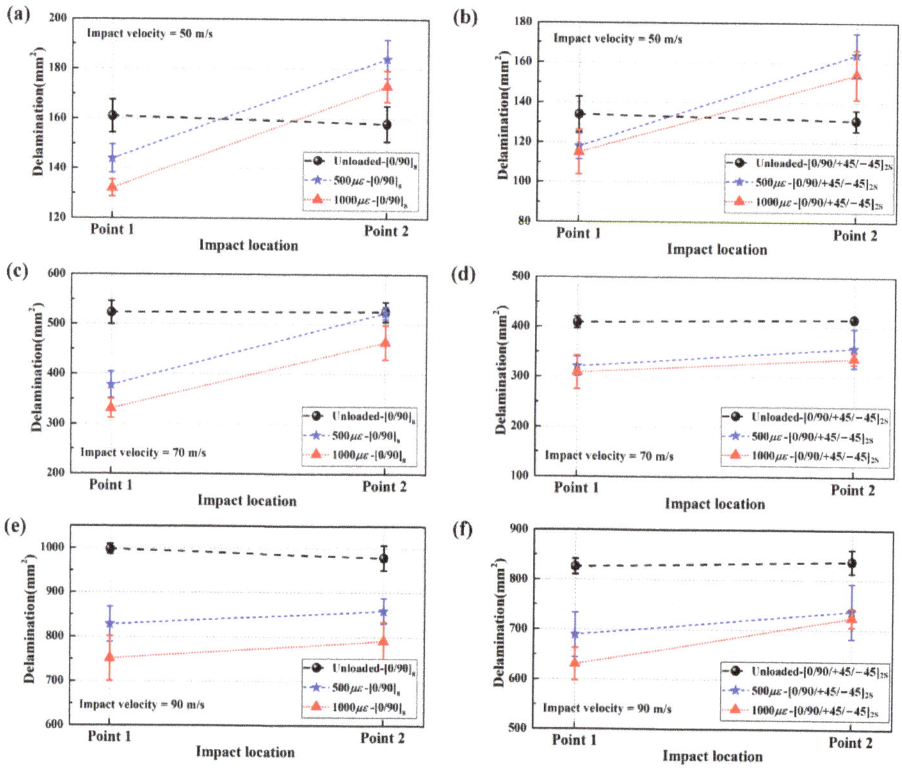

Figure 4. Delamination areas of CFRP laminated plates under various biaxial tensile preloads at impact speeds of 50 m/s, 70 m/s, and 90 m/s; (**a**,**b**) delamination areas of $[0/90]_8$ plates and $[0/90/+45/-45]_{2S}$ plates under 50 m/s; (**c**,**d**) delamination areas of $[0/90]_8$ plates and $[0/90/+45/-45]_{2S}$ plates under 70 m/s; (**e**,**f**) delamination areas of $[0/90]_8$ plates and $[0/90/+45/-45]_{2S}$ plates under 90 m/s.

As vividly shown in Table 1 and Figure 4, for the CFRP laminated plates subjected to the same impact loading, the delamination areas show a decreasing tendency as the biaxial in-plane tensile pre-strains increase when impacting the center location (point 1). Moreover, the typical in-plane strain histories during impact for the CFRP laminated plates under various biaxial tensile preloads are illustrated in Figure 5a–c, and the strain amplitudes under various biaxial tensile preloads at different impact velocities (50 m/s, 70 m/s, and 90 m/s) are summarized in Figure 5d–f, based on the in-plane strain histories. The strain amplitudes at all three impact velocities show a decreasing tendency with the increasing biaxial in-plane tensile pre-strains.

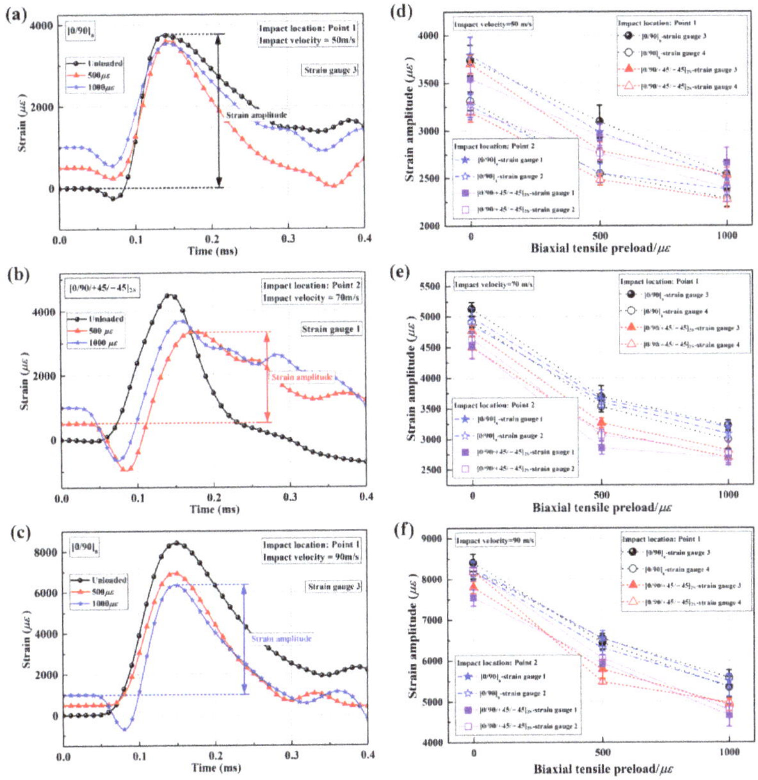

Figure 5. In-plane strain histories and strain amplitudes of CFRP laminated plates under various biaxial tensile preloads; (**a–c**) typical in-plane strain histories at impact speeds of 50 m/s, 70 m/s, and 90 m/s; (**d–f**) strain amplitudes calculated from in-plane strain histories at impact speeds of 50 m/s, 70 m/s, and 90 m/s.

Theoretically, when the projectile impacts the CFRP laminated plate, two kinds of stress waves (transverse wave and longitudinal wave) are generated at the impact point. These then propagate in the laminated plate along the in-plane direction. The longitudinal wave only induces longitudinal strain without changing the shape of the CFRP laminated plate. In contrast, the transverse wave only causes the shape of the CFRP laminated plate to change but does not produce longitudinal strain. Since the longitudinal wave propagates outward with a higher speed than the transverse wave [33], the target plate component passing through the transverse wavefront usually presents with a "V" shape, as shown in Figure 6a,b. Figure 6c,d illustrates the enhancement mechanism of the resistance to out-of-plane displacement for CFRP laminated plates when applying in-plane tensile preloads. As shown, the unloaded CFRP laminated plate produces resistance force $(2 \cdot F'_t \cdot \sin\theta')$ when suffering impact loading, and such resistance force is derived from the fiber tensile load. In comparison, the laminated plate provides additional resistance force $(2 \cdot F_p \sin\theta)$ when applying in-plane tensile preloads. It requires more significant impact loading to achieve the same bending degree for the preloaded CFRP laminated plate, proving that applying in-plane tensile preloads enhances the resistance to out-of-plane displacement. That is, when suffering the same impact loading, there will exist a reduction in the bending degree for preloaded targets in contrast to unloaded targets, as shown in Figure 6c,d. Correspondingly, the delamination propagation of CFRP laminated plates under impact loading is prevented indirectly when applying in-plane tensile preloads. Therefore, the biaxial in-plane tensile preloads are supposed to play a positive role in delamination resistance when impacting the center location of CFRP laminated plates, and such a positive effect is defined as the stress-stiffening effect.

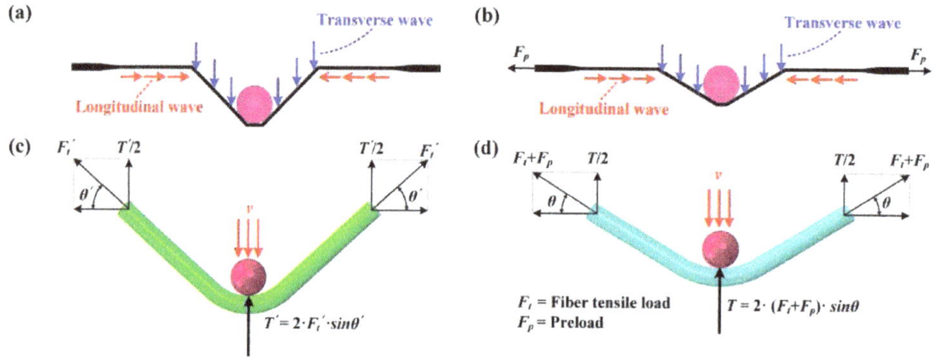

Figure 6. Enhancement mechanism of resistance to out-of-plane displacement for CFRP laminated plates when applying in-plane tensile preloads; (**a**,**b**) "V" shape caused by longitudinal wave and transverse wave with and without tensile preloads; (**c**,**d**) bending degree of plates with and without tensile preloads.

In addition, the delamination areas impacting the near-edge location (point 2) of preloaded CFRP laminated plates are significantly higher than those impacting the center location (point 1). For both $[0/90]_8$ and $[0/90/+45/-45]_{2S}$ laminated plates under 50 m/s impact loading, the preloaded targets even show more delamination than unloaded targets, which can also be observed in the previous work [32]. However, the strain amplitudes (seen in Figure 5d–f) still decrease as the biaxial in-plane tensile pre-strains increase, which is almost the same as when impacting the center location of the targets. Moreover, the energy absorption rate is introduced to further prove the existence of the stress-stiffening effect when impacting the near-edge location (point 2). During the impact process, most of the initial kinetic energy of the projectile is absorbed by the target plate in the form of local damage and global deformation, and a small part of the energy is transferred to the rebound kinetic energy of the projectile. Therefore, the energy of the projectile absorbed by the target can be assumed as

$$E = E_0 - E_r = \frac{1}{2} m \cdot ({v_0}^2 - {v_r}^2) \quad (1)$$

where E_0 and E_r denote the initial kinetic energy and rebound kinetic energy of the projectile, and m is the mass of the projectile. v_0 is the initial impact velocity of the projectile which can be obtained using the laser velocimeter. However, it is difficult to measure the rebound velocity of the projectile v_r. Figure 7 shows typical photographs of the impact process and rebound process of the projectiles taken using the high-speed camera. As shown, the relations between v_0 and v_r can be depicted as

$$\frac{v_0}{v_r} = \frac{N_1 + N_2 + N_3 + N_4}{n_1 + n_2 + n_3 + n_4} \quad (2)$$

where $N_i(i = 1,2,3,4)$ and $n_i(i = 1,2,3,4)$ represent the pixel numbers that are used for depicting the distances between the projectile positions at even intervals. Based on the measurements obtained from the high-speed camera and laser velocimeter, the initial impact velocity and rebound velocity of the projectile can be accurately measured, thus contributing to the reliable calculation of the energy of the projectile absorbed by the target. Thus, the energy of the projectile absorbed by the target E can be written as

$$E = \frac{1}{2}mv_0{}^2 \cdot \left[1 - \left(\frac{n_1 + n_2 + n_3 + n_4}{N_1 + N_2 + N_3 + N_4}\right)^2\right] \tag{3}$$

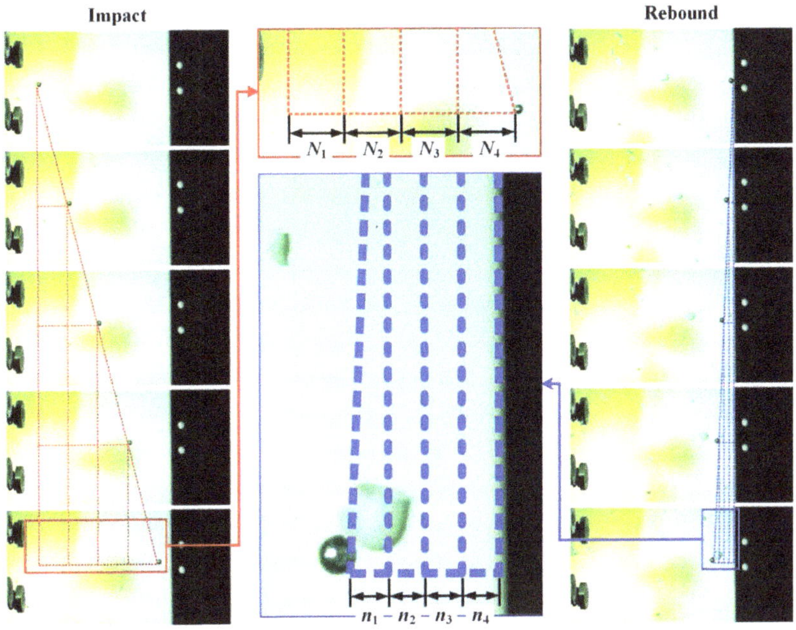

Figure 7. Typical photographs, taken with the high-speed camera, of the impact process and rebound process of the projectile.

The energy absorption rate α can be defined as

$$\alpha = \frac{E}{E_0} = 1 - \left(\frac{n_1 + n_2 + n_3 + n_4}{N_1 + N_2 + N_3 + N_4}\right)^2 \tag{4}$$

Figure 8 shows the energy absorption rates when impacting the center location (point 1) and near-edge location (point 2) of unloaded/preloaded CFRP laminated plates at impact velocities of 50 m/s, 70 m/s, and 90 m/s. As shown, for both $[0/90]_8$ and $[0/90/+45/-45]_{2S}$ laminated plates, the energy absorption rates still present decreasing tendencies as the biaxial in-plane tensile pre-strains increase when impacting the near-edge location, which is similar to the cases in which the center location is impacted. Therefore, the resistance to out-of-plane displacement for CFRP laminated plate is still enhanced by biaxial in-plane tensile preloads, and such stress-stiffening effect still has a positive effect on the delamination resistance when impacting the near-edge location (point 2) of CFRP laminated plates.

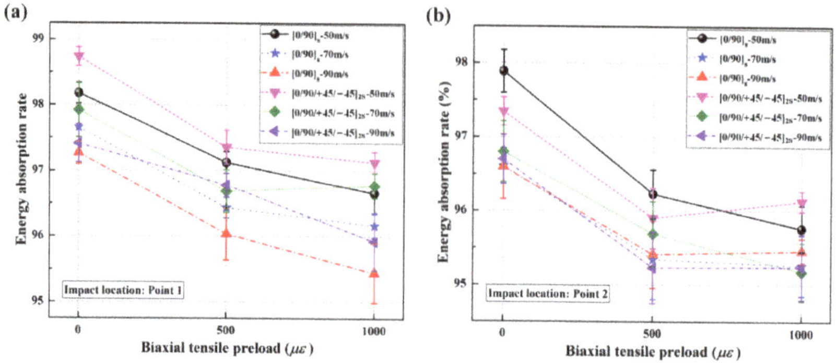

Figure 8. Energy absorption rates of CFRP laminated plates under various biaxial tensile preloads at impact speeds of 50 m/s, 70 m/s, and 90 m/s; (**a**) energy absorption rates when impacting point 1; (**b**) energy absorption rates when impacting point 2.

The interface stiffness degradation effect is supposed to have a negative effect on the delamination resistance of preloaded CFRP laminated plates. Generally, the laminated plate usually presents tensile-shear coupling or the mismatch of Poisson's ratio due to the different ply orientations, thus resulting in the generation of interlaminar normal stress and interlaminar shear stresses between each two adjacent layers when applying in-plane tensile preloads [31]. Such interlaminar stresses cause the degradation of the interface stiffness of CFRP laminated plates to different degrees. Therefore, the interfacial strengths of the laminated plates are weakened by the preloads induced by the interface stiffness degradation, which makes delamination even easier to generate and propagate in the laminated plates when subjected to impact loading. The previous work [32] proposed that competing mechanisms exist between the stress-stiffening effect and the interface stiffness degradation effect and influence the delamination behavior of preloaded CFRP laminated plates. Both the stress-stiffening effect and the interface stiffness degradation effect are due directly to the application of biaxial in-plane tensile preloads. Moreover, interface stiffness degradation is to a higher degree at the near-edge location than that at the center location, which could explain why the delamination areas when impacting the near-edge location (point 2), are significantly higher than those when impacting the center location (point 1).

It deserves to be noticed that the experimental results indicate that the preloading value and impact velocity have a significant influence on the competition between the stress-stiffening effect and the interface stiffness degradation effect. On the one hand, the decreased delamination areas also tend to increase as the preloading values increase when impacting the center location. On the other hand, the delamination areas of target plates under 1000 με biaxial tensile preloads are always fewer than those under 500 με biaxial tensile preloads, whether impacting the center location or the near-edge location. In the following sections, numerical modeling is carried out to gain further insights into the effect of the preloading value and impact velocity on the competition between the stress-stiffening effect and the interface stiffness degradation effect.

4. Numerical Modeling and Validation

4.1. Finite Element Modeling

The high-velocity impact tests conducted on both unloaded and preloaded CFRP laminated plates were simulated using the ABAQUS/Explicit solver. Figure 9 shows the mesh for the L × W = (300 × 300) mm^2 panel configuration with projectile. Given the widespread adoption of the 8-node hexahedral cohesive elements introduced by Hillerborg et al. [21] for predicting delamination in composite materials, a finite element model was constructed. This model comprised 16 three-dimensional solid plies, each with a thickness of 0.125 mm, and 15 cohesive plies, each with a thickness of 0.001 mm. It was employed to forecast the impact-induced damage in both unloaded and preloaded laminated plates. In detail, 242,064 three-dimensional elements with reduced integration were used to model the composite material, and 226,935 cohesive elements were applied to model the interface. An investigation into size dependency was conducted to identify an optimized mesh for impact analysis. This involved determining the most suitable mesh size through a comparison of numerical results and experimental data. The failure loads, overall deformation, and crack tip position, as predicted through the modification of the interfacial strength, exhibited a high degree of independence from mesh-related variations when the element sizes were maintained below 3 mm in Turon's work [34], so the element size was finally set as 2 mm. In addition, the contact behaviors between the projectile and the target were set using "General contact" combined with "surface contact" pairs. To perform an accurate contact analysis between the projectile and laminated plate, the projectile was set as a rigid body, and its mesh was refined to match the element size of the laminated plate. Additionally, in the simulations, a viscosity-based stabilization method was implemented to mitigate the occurrence of hourglass modes within the reduced integration elements.

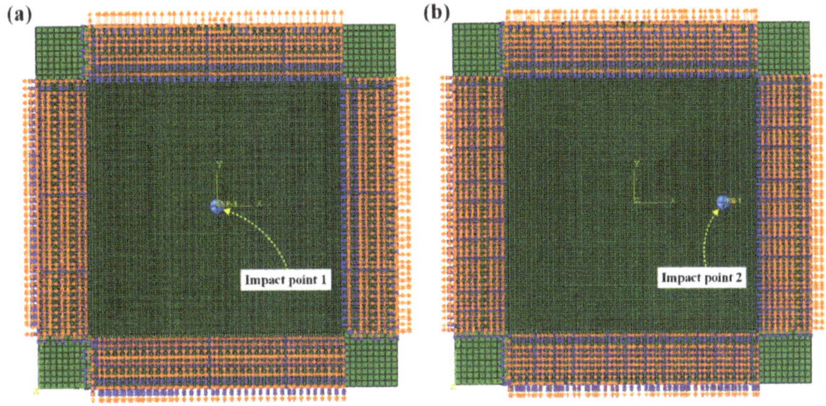

Figure 9. Mesh for L × W = (300 × 300) mm^2 laminated plate; (**a**) impact on point 1 (center location); (**b**) impact on point 2 (near-edge location).

4.2. Material Modeling

To anticipate the deformation and damage of CFRP laminated plates, whether unloaded or preloaded, when subjected to a combination of biaxial in-plane tensile preloads and impact loads, we employed a user subroutine VUMAT. This subroutine relies on an orthotropic material model and incorporates the revised Hou criterion [35]. This revised Hou criterion is a strain-based criterion and has been proven to satisfy the modeling of composite laminated plates under impact loading in previous work [36]. In the present work, the failure criterion is minorly modified as follows:

(1) Fiber failure:

$$e_f = \left(\frac{\varepsilon_{11}}{\varepsilon_{X_T}}\right)^2 + \left(\frac{\varepsilon_{12}}{\varepsilon_{S_f}^{12}}\right)^2 + \left(\frac{\varepsilon_{13}}{\varepsilon_{S_f}^{13}}\right)^2 \geq 1 \quad (5)$$

(2) Matrix cracking ($\sigma_{22} \geq 0$):

$$e_{mt} = \left(\frac{\varepsilon_{22}}{\varepsilon_{Y_t}}\right)^2 + \left(\frac{\varepsilon_{12}}{\varepsilon_{S_{12}}}\right)^2 + \left(\frac{\varepsilon_{23}}{\varepsilon_{S_{m23}}}\right)^2 \geq 1 \quad (6)$$

(3) Matrix crushing ($\sigma_{22} + \sigma_{33} < 0$):

$$e_{mc} = \frac{1}{4}\left(\frac{\varepsilon_{22}}{\varepsilon_{S_{12}}}\right)^2 + \frac{{Y_c}^2\varepsilon_{22}}{4{\varepsilon_{S_{12}}}^2\varepsilon_{Y_c}} - \frac{\varepsilon_{22}}{\varepsilon_{Y_c}} + \left(\frac{\varepsilon_{12}}{\varepsilon_{S_{12}}}\right)^2 \geq 1 \tag{7}$$

where

$$\varepsilon_{X_t} = \frac{X_t}{E_{11}}, \varepsilon_{Y_t} = \frac{Y_t}{E_{22}}, \varepsilon_{Y_c} = \frac{Y_c}{E_{22}}, \varepsilon_{S_{12}} = \frac{S_{12}}{G_{12}}, \varepsilon_{S_{m23}} = \frac{S_{m23}}{G_{23}}, \varepsilon_{S_f}^{12} = \frac{S_f}{G_{12}}, \varepsilon_{S_f}^{13} = \frac{S_f}{G_{13}} \tag{8}$$

e_f, e_{mt}, and e_{mc} denote the parameters for specific damage; ε_{11}, ε_{22}, and ε_{33} are the instant strains in the fiber, transverse, and through-thickness direction; and ε_{12}, ε_{13}, and ε_{23} are the instant shear strains. ε_{X_t}, ε_{Y_t}, ε_{Y_c}, $\varepsilon_{S_{12}}$, $\varepsilon_{S_{m23}}$, $\varepsilon_{S_f}^{12}$, and $\varepsilon_{S_f}^{13}$ represent the strain limits and shear strain limits. X_t and Y_t denote the tensile strength in the fiber and transverse direction, Y_c is the compressive strength in the fiber and transverse direction, S_{12} is the shear strength in the plane of the fiber and the transverse direction, S_{m23} is the shear strength for matrix cracking in the plane of transverse and through-thickness direction, and S_f is the shear strength involving fiber failure.

Moreover, a continuous damage evolution model that can characterize the stiffness variation for composites is established. Here, three failure factors, d_1, d_2, and d_3, are used to depict the degree of fiber failure, matrix cracking, and matrix crushing with a domain of [0, 1] as follows:

$$d_1 = 1 - e^{\left(-E_{11}(\varepsilon_{11}-\varepsilon_{X_t})^2(\sqrt{e_f}-1)L^c/G_f\right)} / \sqrt{e_f} \tag{9}$$

$$d_2 = 1 - e^{\left(-E_{22}(\varepsilon_{22}-\varepsilon_{Y_t})^2(\sqrt{e_{mt}}-1)L^c/G_m\right)} / \sqrt{e_{mt}} \tag{10}$$

$$d_3 = 1 - e^{\left(-E_{22}(\varepsilon_{22}-\varepsilon_{Y_c})^2(\sqrt{e_{mc}}-1)L^c/G_m\right)} / \sqrt{e_{mc}} \tag{11}$$

where L^c is the characteristic length, and G_f and G_m denote the fracture energy of the material along the longitudinal and transverse directions, respectively. Therefore, the variation in the stiffness of composites can be depicted as follows:

$$\begin{gathered} C_{11}^d = (1-d_1)C_{11}, C_{22}^d = (1-d_2)C_{22} \\ C_{12}^d = C_{21}^d = (1-d_1)(1-d_2)C_{12} \\ C_{13}^d = C_{31}^d = (1-d_1)(1-d_3)C_{13}, C_{23}^d = C_{32}^d = (1-d_2)(1-d_3)C_{23} \\ C_{44}^d = (1-d_1)(1-d_2)C_{44}, C_{55}^d = (1-d_1)(1-d_2)C_{55}, C_{66}^d = (1-d_1)C_{66} \end{gathered} \tag{12}$$

where C_{ij} and C_{ij}^d represent the stiffness coefficients before and after the onset of the particular damage, respectively.

The initiation of delamination in the applied cohesive elements is determined through the utilization of a mixed-mode secondary stress criterion. The progression of delamination is depicted by employing the coupling secondary critical energy release rate criterion. However, there are some difficulties in determining the interlaminar stiffness since the thickness of the cohesive layer is usually small in the numerical model. In this study, the interlaminar stiffness was obtained by referring to Daudeville et al.'s work [37] and Turon et al.'s work [34], expressed as follows:

$$\begin{aligned} K_{nn} &= \frac{\lambda E_{33}}{T} \\ K_{ss} &= \frac{2\lambda G_{13}}{T} \\ K_{tt} &= \frac{2\lambda G_{23}}{T} \end{aligned} \tag{13}$$

where T is the thickness of the single composite layer, and its value is 0.125 mm in this study. λ denotes an empirical coefficient. After applying the cohesive layers, the equivalent elastic modulus of the laminated plates along the thickness direction can be measured using

$$E_{eff} = \frac{E_{33}}{1 + \frac{1}{\lambda}} \tag{14}$$

where λ is taken as 50 to set the difference between E_{eff} and E_{33} as 5%. By referring to De Moura's work [38], the strengths of the cohesive layers are replaced by in-plane parameters, as follows:

$$\begin{aligned} \sigma_n &= Y_t \\ \sigma_s &= S_{12} \\ \sigma_t &= S_{m23} \end{aligned} \tag{15}$$

where σ_n is the normal strength of the cohesive layer, and σ_s and σ_t are the shear strengths of the cohesive layer. Table 2 presents the orthotropic material properties of the composite lamina and the properties of the cohesive elements governing the interface.

Table 2. Orthotropic material properties for composite lamina and cohesive element properties for interface.

Materials	Parameters	Values
Composite lamina	Density	ρ = 1510 kg/m^3
	Young's modulus	E_{11} = 151.8 GPa, E_{22} = 12 GPa $G_{12} = G_{13}$ = 3.3 GPa, G_{23} = 2.0 GPa
	Poisson's ratio	$\nu_{12} = \nu_{13}$ = 0.03, ν_{23} = 0.38
	Strength	X_t = 1872 MPa, Y_c = 150 MPa, Y_t = 34 MPa $S_{12} = S_{m23}$ = 100 MPa, S_f = 160 MPa
	Fracture energy	G_f = 92,000 J/m^2, G_m = 600 J/m^2
Interface	Density	ρ = 1000 kg/m^3
	Stiffness	K_{nn} = 4.8 × 10^6 N/mm^3, $K_{ss} = K_{tt}$ = 2.64 × 10^6 N/mm^3
	Strength	σ_n = 34 MPa, $\sigma_s = \sigma_t$ = 100 MPa
	Fracture energy	G_{IC} = 600 J/m^2, $G_{IIC} = G_{IIIC}$ =1200 J/m^2

4.3. Preloading Step and Impacting Step

In order to continuously simulate the impact process of the composite under preloading conditions, the restart analysis function was employed to realize a preloading step followed by an impacting step in sequence. Firstly, a dynamic explicit step was established to apply in-plane tensile preloads on the laminated plate. In this step, the displacement, defined by the smooth step amplitude curve, was applied to all nodes in the regions that connected with the loading blocks to avoid the influence of an inertia force caused by the discontinuous loading rate [37]. Figure 10 shows the in-plane strain distributions along the horizontal and vertical directions for $[0/90]_8$ laminated plates under 1000 $\mu\varepsilon$ biaxial tensile preloads. The strain distributions of the impact regions are almost smooth and uniform, indicating that the proposed method is feasible and efficient. After checking the value of the biaxial in-plane tensile pre-strains, the following impacting step was conducted based on the calculated result from the previous preloading step by employing restart analysis in the ABAQUS explicit solver.

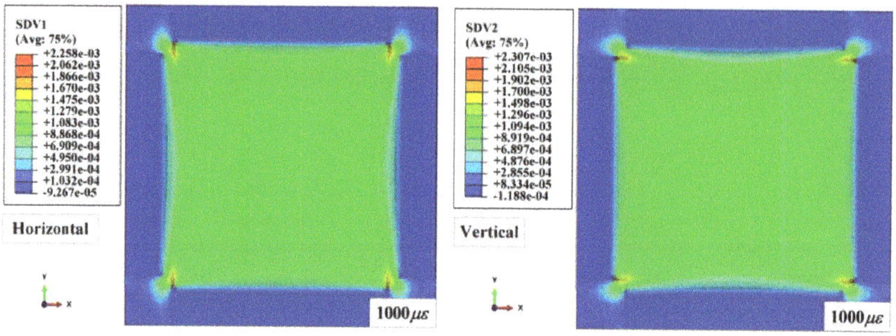

Figure 10. In-plane strain distributions along the horizontal and vertical direction of preloaded $[0/90]_8$ laminated plates.

4.4. Validation

To ensure the numerical model's accuracy, the strain histories of the experiments and numerical results were first compared. As shown in Figure 11, the levels and tendencies of the simulation strain histories were comparatively coincident with the experimental results. Moreover, the delamination areas of the targets between the measurements and computations were also compared to support the further validation of the numerical model, as shown in Figure 12 and Table 3. Since the projection of the delamination area could not be output by ABAQUS, a mini software processed by Python language was designed to calculate the total projected areas of delamination. The work logic of mini Python software is divided into the following steps: output the stiffness degradation field information of each cohesive element layer, compare the scalar stiffness degradation (SDEG) on the same vertical coordinate longitudinally, overwrite the SDEG value circularly, obtain the continuous stiffness degradation field via linear interpolation, and calculate the projected area for stiffness degradation SDEG $\geq$ 0.999. The numerical results also showed good agreement with the C-scan results, proving that the numerical simulation method applied in this study is appropriate for modeling the delamination behavior of preloaded CFRP laminated plates.

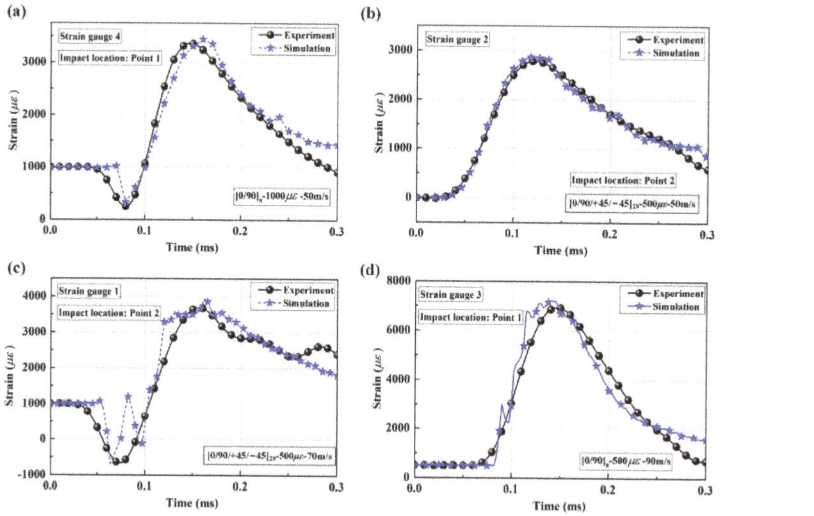

Figure 11. Comparison of typical strain histories between experiments and simulations; (**a**) strain gauge 4 on $[0/90]_8$ plate; (**b**) strain gauge 2 on $[0/90/+45/-45]_{2S}$ plate; (**c**) strain gauge 1 on $[0/90/+45/-45]_{2S}$ plate; (**d**) strain gauge 3 on $[0/90]_8$ plate.

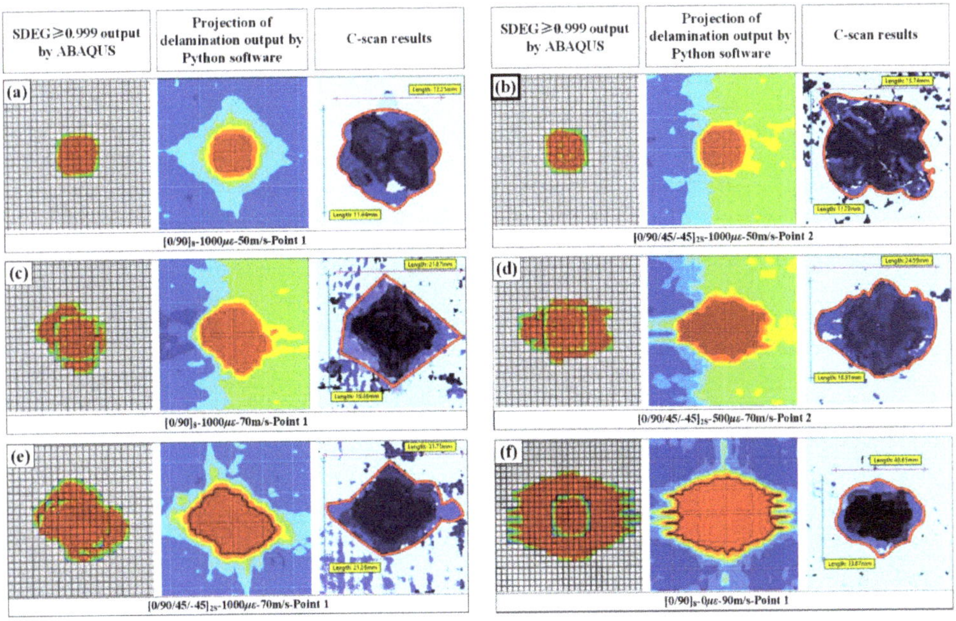

Figure 12. Comparison of typical delamination between experiments and simulations; (**a**) impact on point 1 of 1000 με preloaded $[0/90]_8$ plates at speeds of 50 m/s; (**b**) impact on point 2 of 1000 με preloaded $[0/90/+45/-45]_{2S}$ plates at speeds of 50 m/s; (**c**) impact on point 1 of 1000 με preloaded $[0/90]_8$ plates at speeds of 70 m/s; (**d**) impact on point 2 of 500 με preloaded $[0/90/+45/-45]_{2S}$ plates at speeds of 70 m/s; (**e**) impact on point 1 of 1000 με preloaded $[0/90/+45/-45]_{2S}$ plates at speeds of 70 m/s; (**f**) impact on point 1 of unloaded $[0/90]_8$ plates at speeds of 90 m/s.

Table 3. Comparison of delamination areas between experiments and simulations.

Layups Impact Location	Pre-Strain	Numerical Delamination Area (mm^2) and Errors					
		V = 50 m/s	Errors	V = 70 m/s	Errors	V = 90 m/s	Errors
$[0/90]_8$ Point1/Point 2	0 με	171	6.2%	513	−1.9%	970	−2.8%
	500 με	152	5.5%	364	−3.7%	758	−8.5%
	1000 με	135	2.3%	315	−5.1%	671	−10.7%
	0 με	171	8.2%	513	−2.3%	970	−0.9%
	500 με	193	4.9%	476	−9.0%	798	−7.1%
	1000 με	177	2.3%	428	−7.8%	755	−4.6%
$[0/90/+45/-45]_{2S}$ Point1/Point 2	0 με	142	6.0%	444	8.6%	849	2.8%
	500 με	125	5.9%	315	−1.9%	736	6.8%
	1000 με	113	−4.9%	269	−12.9%	613	−2.7%
	0 με	142	7.6%	444	7.0%	849	1.4%
	500 με	156	−4.9%	373	4.2%	769	4.3%
	1000 με	149	2.3%	300	−11.0%	662	−8.6%

5. Discussion

From the experimental and numerical results, it can be found that the preloading value and impact velocity have a significant influence on the delamination behavior of preloaded CFRP laminated plates. In this section, based on the verified numerical simulation method, the effects of the preloading degree and impact velocity on the delamination resistance will be discussed in detail by conducting a series of simulations.

5.1. Influence of Biaxial Tensile Pre-Strains Value

Figure 13 shows the numerical delamination areas of CFRP laminated plates under various biaxial tensile preloads when impacting different positions at velocities of 50 m/s, 70 m/s, and 90 m/s. As shown, the delamination areas gradually increase when moving the impact location from the center to the near-edge location. The $[0/90]_8$ and $[0/90/+45/-45]_{2S}$ targets under 500 µε and 1000 µε show more delamination than unloaded targets when impacting the near-edge location at 50 m/s. Figure 14 shows the stiffness degradation of $[0/90]_8$ CFRP laminated plates under various biaxial tensile preloads. Using the toggle global translucency function in ABAQUS/Explicit shows that the CFRP laminated plates present different degrees of interface stiffness degradation at different positions when applying various biaxial tensile preloads. The interface stiffness degradations at the near-edge location are significantly higher than those at the center location (almost no interface stiffness degradation), which results in the interfacial strengths of the laminated plates at the near-edge location being weaker than those at the center location. Thus, both the experimental and numerical results indicate that the preloaded targets generate more delamination at the near-edge location in comparison to the center location when suffering impact loading.

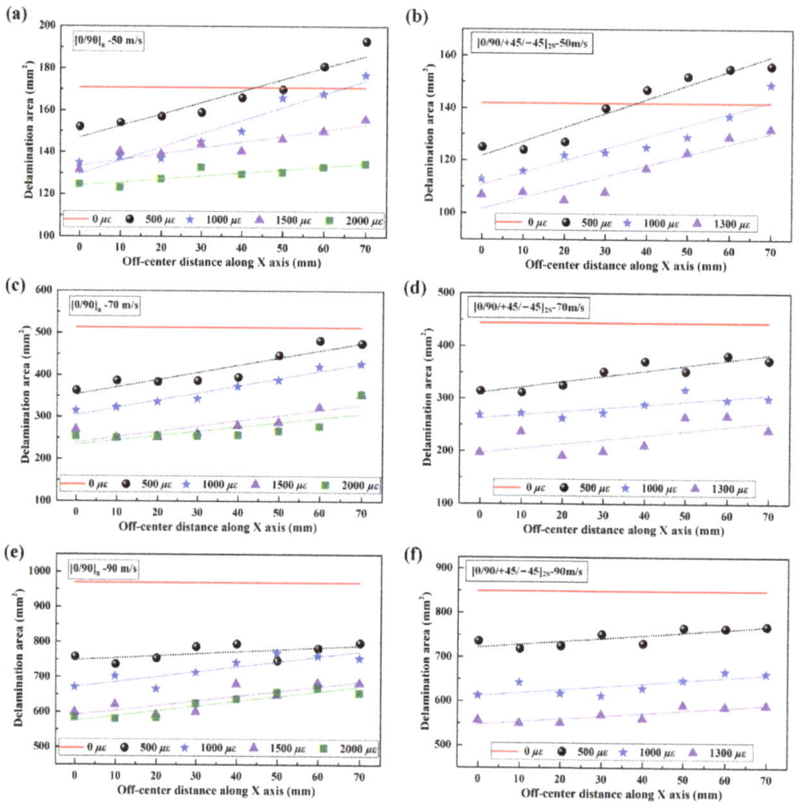

Figure 13. Numerical delamination areas of CFRP laminated plates under various biaxial tensile preloads when impacting different positions. (**a**,**b**) delamination areas of $[0/90]_8$ plates and $[0/90/+45/-45]_{2S}$ plates under 50 m/s; (**c**,**d**) delamination areas of $[0/90]_8$ plates and $[0/90/+45/-45]_{2S}$ plates under 70 m/s; (**e**,**f**) delamination areas of $[0/90]_8$ plates and $[0/90/+45/-45]_{2S}$ plates under 90 m/s.

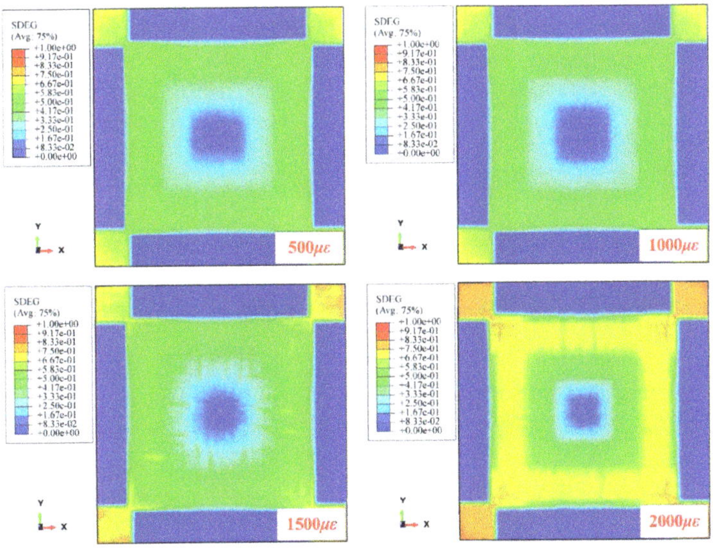

Figure 14. Stiffness degradation of $[0/90]_8$ CFRP laminated plates under various biaxial tensile preloads.

It should be noted that the delamination areas of target plates tend to decrease as the biaxial tensile pre-strains increase for all impact locations, from the center location (point 1, off-center distance is 0 mm) to the near-edge location (point 2, off-center distance is 70 mm). In particular, the preloaded targets present relatively fewer delamination areas at the near-edge location than unloaded targets when the preloading value is increased ($[0/90]_8$ from 500 με to 1500 με, $[0/90/+45/-45]_{2S}$ from 500 με to 1300 με). Since the delamination behavior of preloaded CFRP laminated plates is influenced by the competing mechanisms of the stress-stiffening effect and the interface stiffness degradation effect, it can be considered that the positive effect had on the delamination resistance by stress-stiffening gradually gains an advantage compared to the weakening effect had on the delamination resistance by interface stiffness degradation when the preloading value is increased. With the increase in the biaxial tensile pre-strains value, both the positive effect of stress-stiffening and the weakening effect of interface stiffness degradation are gradually enhanced. However, the enhancement degree of these two competitive factors is different. As can be seen in Figure 6, due to the linear elastic characteristic of composite material, the additional resistance force $2 \cdot F_p \sin\theta$ induced by the preload is also linear elastic. Thus, the stress-stiffening effect on the delamination resistance is almost linearly enhanced when the biaxial tensile preloads are increased linearly. Figure 15 illustrates the schematic diagram of the stiffness degradation for the interface layer based on the bilinear constitutive model. As shown, the degree of stiffness degradation for the interface gradually reduces when the pre-strain/deformation is increased linearly. Figure 16 also shows the SDEG output by Python software based on the numerical results. As shown, the stiffness degradation degrees at all five assigned locations tend to decrease as the biaxial tensile preloads increase linearly. Therefore, the stress-stiffening effect will gradually gain an advantage over the interface stiffness degradation effect when the biaxial tensile pre-strains value is increased.

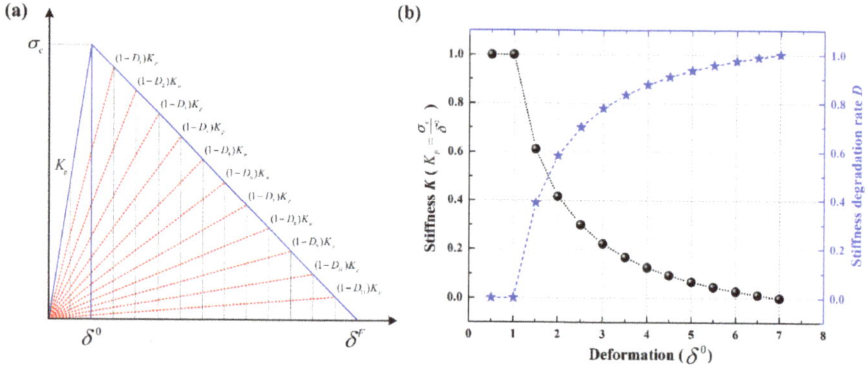

Figure 15. Schematic diagram of stiffness degradation for interface layer based on bilinear constitutive model; (**a**) bilinear constitutive model; (**b**) stiffness degradation for the interface layer.

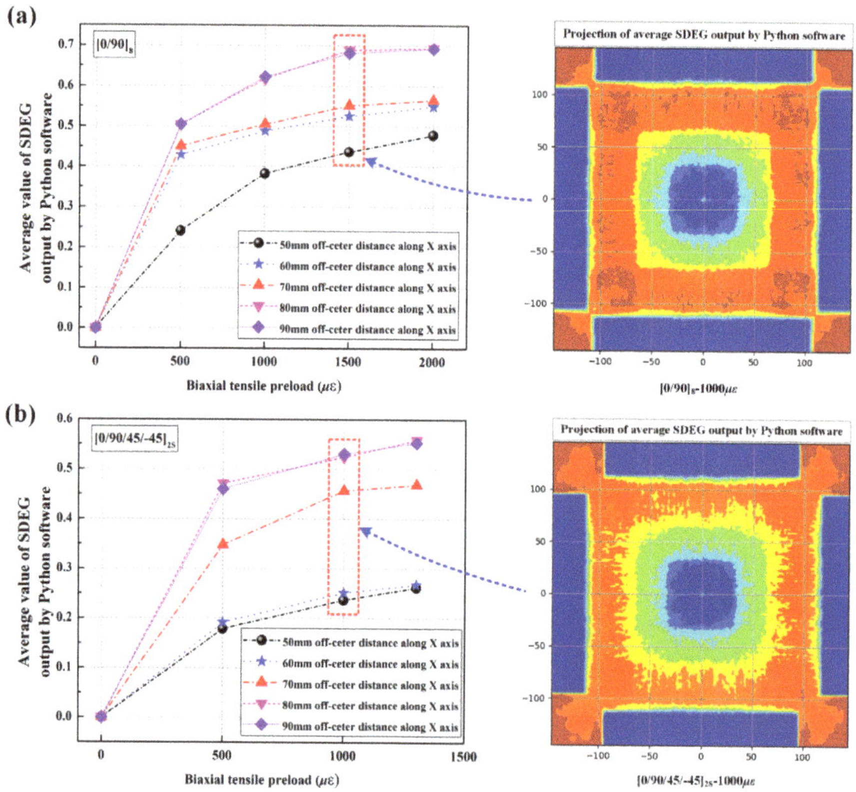

Figure 16. Stiffness degradations output by Python software based on simulations; (**a**) $[0/90]_8$ plates; (**b**) $[0/90/+45/-45]_{2S}$ plates.

5.2. Influence of Impact Velocity

Both experimental and numerical results showed that when impacting the near-edge location at impact velocities of 50 m/s, preloaded CFRP laminated plates present more delamination areas than unloaded targets. However, by increasing the impact velocity from 50 m/s to 70 m/s, the preloaded CFRP laminated plates show fewer delamination areas in comparison to unloaded targets, which can also be found in the case of impacting the near-edge location of CFRP laminated plates at the impact velocity of 90 m/s. Figure 17a shows the numerical delamination areas of $[0/90]_8$ plates under various biaxial tensile preloads (0 με, 500 με, and 1000 με) when impacting different positions at four different impact velocities (30 m/s, 50 m/s, 70 m/s, and 90 m/s). The preloaded targets present relatively fewer delamination areas at the near-edge location than unloaded targets when the impact velocity is increased to 70 m/s and 90 m/s, which stands in sharp contrast to the case at impact velocities of 30 m/s and 50 m/s. Assuming that the stiffening region of the preloaded CFRP laminated plate is the distribution range and that the delamination areas of the preloaded targets are less than those of unloaded targets, it can be seen that the stiffening regions constantly expand their scopes with the increase in the impact velocity. For example, the stiffening region of the 500 με preloaded target expands its scope from the 30 mm off-center distance to 43 mm off-center distance by increasing the impact velocity from 30 m/s to 50 m/s. In addition, when suffering the 70 m/s impact loading and 90 m/s impact loading, the delamination areas of the preloaded targets are less than those of the unloaded targets at all eight impact locations from the center location to the 70 mm off-center distance. This means that the stiffening regions for both the 500 με and 1000 με preloaded targets expand their scope to all impact locations.

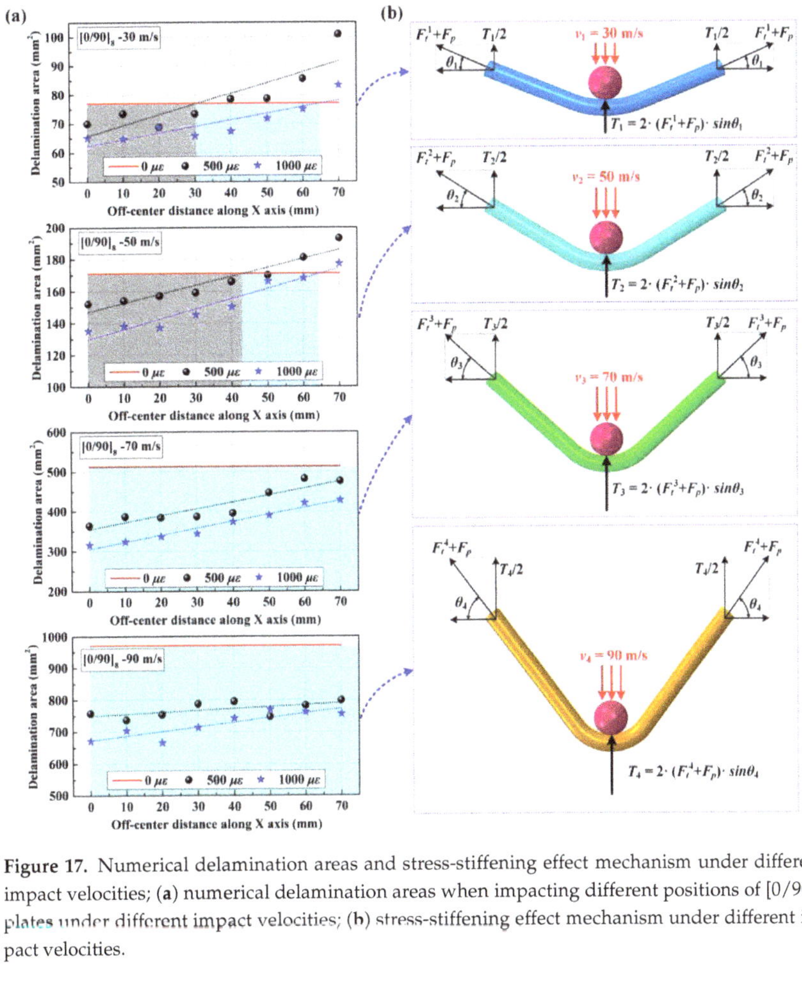

Figure 17. Numerical delamination areas and stress-stiffening effect mechanism under different impact velocities; (**a**) numerical delamination areas when impacting different positions of $[0/90]_8$ plates under different impact velocities; (**b**) stress-stiffening effect mechanism under different impact velocities.

The mechanism of the stress-stiffening effect at various impact velocities is depicted in Figure 17b. It is evident that the out-of-plane displacements of targets subjected to identical biaxial tensile pre-strains increase proportionally with the projectile's impact velocity, thus resulting in $\theta_1 < \theta_2 < \theta_3 < \theta_4$. Therefore, the additional resistance forces induced by biaxial tensile preloads also increase as the impact velocity increases, depicted as follows:

$$2 \cdot F_t \sin\theta_1 < 2 \cdot F_t \sin\theta_2 < 2 \cdot F_t \sin\theta_3 < 2 \cdot F_t \sin\theta_4 \quad (16)$$

Therefore, as the impact velocity increases, the positive effect of stress stiffening on the delamination resistance gradually enhances while the weakening effect of interface stiffness degradation on delamination resistance remains the same, indicating that the stress-stiffening effect gains an advantage over the interface stiffness degradation effect when competing to have an effect on delamination resistance.

6. Conclusions

An extensive investigation was conducted by combining ballistic impact tests, dynamic finite element simulations, and theoretical analysis to further understand the impact delamination behaviors of CFRP laminated plates under biaxial tensile preloads. The main conclusions were as follows:

(1) Both the experimental findings and simulations consistently illustrated that biaxial tensile preloading could bolster resistance to out-of-plane displacement, thereby exerting a beneficial influence on the delamination resistance of CFRP laminated plates (14.2~36.7% decrease in delamination areas under 1000 $\mu\varepsilon$). However, no more than a 19.3% decrease in delamination areas was observed when impacting the near-edge location, and the case under impact velocity of 50 m/s even showed increasing delamination areas. This phenomenon indicates that applying preloads was also supposed to induce the interface stiffness degradation effect, which was considered a negative effect on delamination resistance.

(2) The impact velocity can influence the competing mechanisms of the stress-stiffening effect and the interface stiffness degradation effect. Since the degree of interface stiffness degradation for CFRP laminated plates with specific stacking sequences is the same when applying the same biaxial tensile preloads, the influence of impact velocity on the competing mechanisms of the stress-stiffening effect and the interface stiffness degradation effect is mainly caused by the influence of impact velocity on the stress-stiffening effect.

(3) In near-edge location ballistic tests, it was noted that the preloaded CFRP laminated plates exhibited increased delamination compared to the unloaded target at an impact velocity of 50 m/s. However, at impact velocities of 70 m/s and 90 m/s, the preloaded CFRP laminate plates displayed reduced instances of delamination in comparison to the unloaded target. It can be concluded that the stress-stiffening effect becomes progressively more dominant than the interface stiffness degradation effect as the preloading value and impact velocity increase.

(4) A conceptual framework involving the competing mechanisms of the stress-stiffening effect and the interface stiffness degradation effect has been formulated to elucidate the impact of biaxial in-plane tensile preloads on delamination behavior, and such a competitive mechanism was found to be influenced by the preloading value and impact velocity. With an increase in the biaxial tensile pre-strains value or impact velocity, the stress-stiffening effect gradually gained an advantage over the interface stiffness degradation effect. Considering the effect of the preloading degree and impact velocity on the competitive mechanism, the complex delamination behaviors at different impact locations of CFRP laminated plates under various preloading degrees and impact velocities were reasonably explained.

Author Contributions: Conceptualization, C.W.; methodology, K.L. and C.W.; software, H.W.; validation, L.K and C.W.; formal analysis, K.L.; investigation, K.L. and H.W.; resources, C.W.; data curation, K.L. and H.W.; writing—original draft preparation, C.W.; writing—review and editing, C.W.; visualization, C.W.; supervision, C.W.; project administration, C.W.; funding acquisition, C.W. All authors have read and agreed to the published version of the manuscript.

Funding: This research was funded by [National Natural Science Foundation of China] grant number [No. 12002264].

Institutional Review Board Statement: Not applicable.

Informed Consent Statement: Not applicable.

Data Availability Statement: The data are available from the corresponding author on reasonable request.

Conflicts of Interest: The authors declare no conflict of interest.

References

1. Heimbs, S.; Heller, S.; Middendorf, P.; Hähnel, F.; Weiße, J. Low Velocity Impact on CFRP Plates with Compressive Preload: Test and Modelling. *Int. J. Impact Eng.* **2009**, *36*, 1182–1193. [CrossRef]
2. Lee, S.-W.R.; Sun, C.T. Dynamic Penetration of Graphite/Epoxy Laminates Impacted by a Blunt-Ended Projectile. *Compos. Sci. Technol.* **1993**, *49*, 369–380. [CrossRef]
3. Chen, J.K.; Medina, D.F. The Effects of Projectile Shape on Laminated Composite Perforation. *Compos. Sci. Technol.* **1998**, *58*, 1629–1639. [CrossRef]
4. Yew, C.H.; Kendrick, R.B. A Study of Damage in Composite Panels Produced by Hypervelocity Impact. *Int. J. Impact Eng.* **1987**, *5*, 729–738. [CrossRef]
5. Zhang, C.; Binienda, W.K.; Morscher, G.N.; Martin, R.E.; Kohlman, L.W. Experimental and FEM Study of Thermal Cycling Induced Microcracking in Carbon/Epoxy Triaxial Braided Composites. *Compos. Part A Appl. Sci. Manuf.* **2013**, *46*, 34–44. [CrossRef]
6. Zhang, C.; Li, N.; Wang, W.; Binienda, W.K.; Fang, H. Progressive Damage Simulation of Triaxially Braided Composite Using a 3D Meso-Scale Finite Element Model. *Compos. Struct.* **2015**, *125*, 104–116. [CrossRef]
7. Camanho, P.P.; Davila, C.G.; De Moura, M.F. Numerical Simulation of Mixed-Mode Progressive Delamination in Composite Materials. *J. Compos. Mater.* **2003**, *37*, 1415–1438. [CrossRef]
8. Sridharan, S.; Pankow, M. Performance Evaluation of Two Progressive Damage Models for Composite Laminates under Various Speed Impact Loading. *Int. J. Impact Eng.* **2020**, *143*, 103615. [CrossRef]
9. Rajaneesh, A.; Ponthot, J.P.; Bruyneel, M. High Velocity Impact Response of Composite Laminates Using Modified Meso-Scale Damage Models. *Int. J. Impact Eng.* **2021**, *147*, 103701. [CrossRef]
10. Xu, S.; Li, Y.; Zhou, S.; Jiang, X.; Xie, W.; Zhang, W. Ballistic Performance and Damage Analysis of CFRP Laminates under Uniaxial Pretension and Precompression. *Int. J. Impact Eng.* **2023**, *178*, 104620. [CrossRef]
11. Qaderi, S.; Ebrahimi, F.; Vinyas, M. Dynamic Analysis of Multi-Layered Composite Beams Reinforced with Graphene Platelets Resting on Two-Parameter Viscoelastic Foundation. *Eur. Phys. J. Plus* **2019**, *134*, 339. [CrossRef]
12. Abrate, S. Modeling of Impacts on Composite Structures. *Compos. Struct.* **2001**, *51*, 129–138. [CrossRef]
13. Reid, S.R.; Zhou, G. (Eds.) *Impact Behaviour of Fibre-Reinforced Composite Materials and Structures*; CRC Press: Boca Raton, FL, USA; Woodhead Pub: Cambridge, UK, 2000; ISBN 978-0-8493-0847-5.
14. Williams, J.G.; Anderson, M.S.; Rhodes, M.D.; Starnes, J.H.; Stroud, W.J. Recent Development in the Design, Testing and Impact-Damage Tolerance of Stiffened Composite Panels. In Proceedings of the 4th Conference on Fibrous Composites in Structural Design, San Diego, CA, USA, 14–17 November 1978. [CrossRef]
15. Whittingham, B.; Marshall, I.H.; Mitrevski, T.; Jones, R. The Response of Composite Structures with Pre-Stress Subject to Low Velocity Impact Damage. *Compos. Struct.* **2004**, *66*, 685–698. [CrossRef]
16. Butcher, B.R.; Fernback, P.J. Impact resistance of unidirectional cfrp under tensile stress: Further experimental variables. *Fibre Sci. Technol.* **1981**, *14*, 41–58. [CrossRef]
17. Nettles, A.; Daniel, V.; Branscomb, C. The Effects of Tensile Preloads on the Impact Response of Carbon/Epoxy Laminates. In *Composite Materials: Fatigue and Fracture: 7th Volume*; Bucinell, R., Ed.; ASTM International: West Conshohocken, PA, USA, 1998; pp. 249–262.
18. Zhang, X.; Davies, G.A.O.; Hitchings, D. Impact Damage with Compressive Preload and Post-Impact Compression of Carbon Composite Plates. *Int. J. Impact Eng.* **1999**, *22*, 485–509. [CrossRef]
19. Garnier, C.; Djilali, T.; Brault, R.; Mistou, S. Impact Resistance of Composite Materials under Biaxial Preloading. *KEM* **2011**, *482*, 39–48. [CrossRef]
20. Mikkor, K.M.; Thomson, R.S.; Herszberg, I.; Weller, T.; Mouritz, A.P. Finite Element Modelling of Impact on Preloaded Composite Panels. *Compos. Struct.* **2006**, *75*, 501–513. [CrossRef]

21. Pickett, A.K.; Fouinneteau, M.R.C.; Middendorf, P. Test and Modelling of Impact on Pre-Loaded Composite Panels. *Appl Compos Mater* **2009**, *16*, 225–244. [CrossRef]
22. Heimbs, S.; Bergmann, T.; Schueler, D.; Toso-Pentecôte, N. High Velocity Impact on Preloaded Composite Plates. *Compos. Struct.* **2014**, *111*, 158–168. [CrossRef]
23. Chiu, S.-T.; Liou, Y.-Y.; Chang, Y.-C.; Ong, C. Low Velocity Impact Behavior of Prestressed Composite Laminates. *Mater. Chem. Phys.* **1997**, *47*, 268–272. [CrossRef]
24. Kelkar, A.; Sankar, J.; Rajeev, K.; Aschenbrenner, R.; Schoeppner, G. Analysis of Tensile Preloaded Composites Subjected to Low-Velocity Impact Loads. In Proceedings of the 39th AIAA/ASME/ASCE/AHS/ASC Structures, Structural Dynamics, and Materials Conference and Exhibit, Long Beach, CA, USA, 20–23 April 1998. [CrossRef]
25. Robb, M.D.; Arnold, W.S.; Marshall, I.H. The Damage Tolerance of GRP Laminates under Biaxial Prestress. *Compos. Struct.* **1995**, *32*, 141–149. [CrossRef]
26. García-Castillo, S.K.; Sánchez-Sáez, S.; López-Puente, J.; Barbero, E.; Navarro, C. Impact Behaviour of Preloaded Glass/Polyester Woven Plates. *Compos. Sci. Technol.* **2009**, *69*, 711–717. [CrossRef]
27. Zhikharev, M.V.; Sapozhnikov, S.B.; Kudryavtsev, O.A.; Zhikharev, V.M. Effect of Tensile Preloading on the Ballistic Properties of GFRP. *Compos. Part B Eng.* **2019**, *168*, 524–531. [CrossRef]
28. Guillaud, N.; Froustey, C.; Dau, F.; Viot, P. Impact Response of Thick Composite Plates under Uniaxial Tensile Preloading. *Compos. Struct.* **2015**, *121*, 172–181. [CrossRef]
29. Mitrevski, T.; Marshall, I.H.; Thomson, R.S.; Jones, R. Low-Velocity Impacts on Preloaded GFRP Specimens with Various Impactor Shapes. *Compos. Struct.* **2006**, *76*, 209–217. [CrossRef]
30. Moallemzadeh, A.R.; Sabet, S.A.R.; Abedini, H. Preloaded Composite Panels under High Velocity Impact. *Int. J. Impact Eng.* **2018**, *114*, 153–159. [CrossRef]
31. Choi, I.-H. Low-Velocity Impact Analysis of Composite Laminates under Initial in-Plane Load. *Compos. Struct.* **2008**, *86*, 251–257. [CrossRef]
32. Wang, C.; Suo, T.; Hang, C.; Li, Y.; Xue, P.; Deng, Q. Influence of In-Plane Tensile Preloads on Impact Responses of Composite Laminated Plates. *Int. J. Mech. Sci.* **2019**, *161–162*, 105012. [CrossRef]
33. Liang, J.; Zhang, G.; Ba, Z.; Liang, J. Development of a 3D Fluid-Saturated Element for Dynamic Analysis of Two-Phase Media in ABAQUS Based on u-U Formed Equations. *Comput. Geotech.* **2021**, *139*, 104377. [CrossRef]
34. Turon, A.; Dávila, C.G.; Camanho, P.P.; Costa, J. An Engineering Solution for Mesh Size Effects in the Simulation of Delamination Using Cohesive Zone Models. *Eng. Fract. Mech.* **2007**, *74*, 1665–1682. [CrossRef]
35. Hou, J.P.; Petrinic, N.; Ruiz, C.; Hallett, S.R. Prediction of Impact Damage in Composite Plates. *Compos. Sci. Technol.* **2000**, *60*, 273–281. [CrossRef]
36. Wang, C.; Ren, T.; Miao, Y.; Suo, T.; Tang, Z.; Li, Y. High-Velocity Impact Response of CFRP Panels Reinforced with Stiffeners. *Compos. Struct.* **2020**, *246*, 112392. [CrossRef]
37. Daudeville, L.; Allix, O.; Ladeveze, P. Delamination analysis by damage mechanics: Some applications. *Compos. Eng.* **1995**, *5*, 17–24. [CrossRef]
38. De Moura, M.F.S.F.; Gonçalves, J.P.M. Modelling the Interaction between Matrix Cracking and Delamination in Carbon–Epoxy Laminates under Low Velocity Impact. *Compos. Sci. Technol.* **2004**, *64*, 1021–1027. [CrossRef]

materials

MDPI

Article

Dynamic Characteristic Analysis of a Toothed Electromagnetic Spring Based on the Improved Bouc—Wen Model

Xiaoyuan Zheng [1,2], Cheng Zhang [1,2], Yifang Lou [1,2], Guangming Xue [1,2,*] and Hongbai Bai [1,2,*]

1 College of Mechanical Engineering and Automation, Fuzhou University, Fuzhou 350116, China
2 Metal Rubber Engineering Research Center, Fuzhou University, Fuzhou 350116, China
* Correspondence: xueguangming@fzu.edu.cn (G.X.); bhbk11@sina.com (H.B.)

Abstract: Electromagnetic spring active isolators have attracted extensive attention in recent years. The standard Bouc–Wen model is widely used to describe hysteretic behavior but cannot accurately describe asymmetric behavior. The standard Bouc–Wen model is improved to better describe the dynamic characteristic of a toothed electromagnetic spring. The hysteresis model of toothed electromagnetic spring is established by adding mass, damping, and asymmetric correction terms with direction. Subsequently, the particle swarm optimization algorithm is used to identify the parameters of the established model, and the results are compared with those obtained from the experiment. The results show that the current has a significant impact on the dynamic curve. When the current increases from 0.5 A to 2.0 A, the electromagnetic force sharply increases from 49 N to 534 N. Under different excitations and currents, the residual points predicted by the model proposed in this work fall basically in the horizontal band region of −20–20 N (for an applied current of 1.0 A) and −40–80 N (for an application of 4.5 mm/s). Furthermore, the maximum relative error of the model is 12.75%. The R^2 of the model is higher than 0.98 and the highest value is 0.9993, proving the accuracy of the established model.

Keywords: electromagnetic spring; dynamic characteristic; Bouc–Wen model; particle swarm algorithm; experiment

1. Introduction

Vibration is a common physical phenomenon in engineering and technology, affecting the operation of the equipment and the mechanical power system while reducing its lifespan [1,2]. The traditional passive vibration isolation system has been widely used in vibration isolation due to its simple structure and low cost [3–5]. However, traditional vibration isolation systems cannot be adjusted in real time based on changes in external excitation frequency. With the high-precision requirements related to precision manufacturing and measurement and the demand for environmental noise reduction, traditional passive systems can hardly meet the increasingly strict vibration control requirements [6]. As an active vibration isolation system, electromagnetic springs have the characteristics of fast response, non-contact, and adjustable stiffness [7,8], indicating that it has broad application prospects, especially in the reduction in the vibration of marine engines and air compressors [9–11].

Researchers have studied many types of electromagnetic springs and their models. Batdorff et al. [12] proposed a method for calculating multiple-edge magnetic conductance and magnetic leakage in axisymmetric electromagnetic devices. Moreover, the authors established an analytical model of electromagnetic force using the equivalent magnetic circuit method, significantly improving the model's accuracy. Ertuğrul et al. [13] proposed a new segmented magnetic equivalent circuit method to analyze the magnetic force of the hybrid electromagnet system. In addition, the force characteristics and magnetic field distribution were studied and the results were compared with finite element analysis to verify the effectiveness of the method. Sun et al. [14] constructed a new type of electromagnetic negative stiffness spring using a coaxial permanent magnet ring magnet and rectangular cross-section coil. Furthermore, they established an analytical model based on the filament method to quantitatively study the factors affecting electromagnetic force and stiffness characteristics. Wu et al. [15] established the stiffness analytical model of negative stiffness array magnetic spring based on the magnetic charge model and validated the results via static experiments. Li et al. [16] proposed an improved current filament method, which uses the equivalent circuit principle and considers the skin effect to improve the model's accuracy by calculating the current of the electromagnetic forming coil and the electromagnetic force of the workpiece. Wang et al. [17] conducted in-depth research on the multi-gap permanent magnet-biased axial magnetic bearing, solved the magnetic leakage of the magnetic bearing using Laplace's equation and established an accurate analytical model.

The Bouc–Wen model is widely used to represent the properties of materials. The model has an efficient shape control flexibility and was proposed to describe highly asymmetric hysteresis [18–20]. In this work, based on the standard Bouc–Wen model, the hysteresis model of a toothed electromagnetic spring is established by introducing the mass, damping, and asymmetric correction terms, while the accuracy of the model is proven by experiments.

2. Structure and Principle of a Toothed Electromagnetic Spring

A toothed electromagnetic spring has an axisymmetric structure mainly composed of a motor, stator, coil, and air gap, as shown in Figure 1. The rotor and stator of the toothed electromagnetic spring are arranged with annular teeth of the same size with an air gap between each pair of teeth. The coil is usually wound around the actuator. When the current flows into the coil, the actuator generates an electromagnetic field and flows into the stator's tooth ring through the air gap. Then, the field flows back into the rotor through the air gap, forming a closed loop.

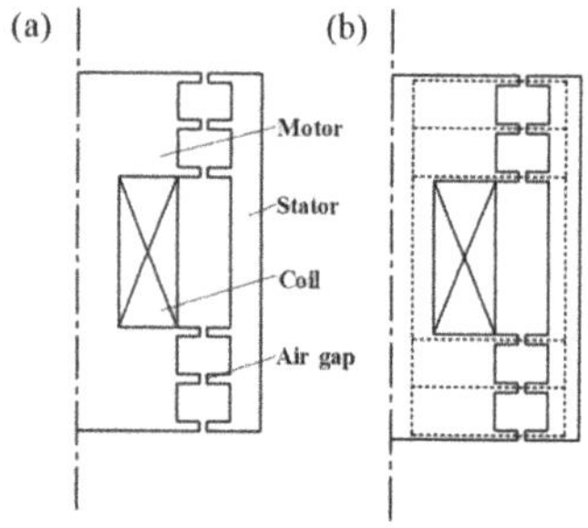

Figure 1. Structure and magnetic circuit of the electromagnetic spring: (**a**) schematic diagram of the toothed electromagnetic spring structure and (**b**) schematic diagram of the magnetic field.

The rotor and the stator of the toothed electromagnetic spring are usually processed with high-permeability materials. At the same time, the air permeability is much lower than that of the rotor and stator. Therefore, the magnetic circuit of the electromagnetic spring is consistent with the direction along the dotted line in Figure 1b, and there is rarely magnetic leakage. Figure 2 shows the relationship between the axial displacement of the rotor (x) and the electromagnetic force generated by the toothed electromagnetic spring (F). When the axial displacement of the rotor is zero, the electromagnetic spring is in equilibrium. Consequently, the magnetic teeth of each rotor and stator correspond to each other. The electromagnetic force generated between the magnetic teeth only exists in the radial direction, and the resultant force is zero. When the rotor produces axial displacement, there is a displacement deviation between the rotor's magnetic teeth and the stator's. At this time, an electromagnetic force is generated between the magnetic teeth, which is approximately proportional to the axial displacement of the rotor within a certain range. Therefore, the electromagnetic force generated by the electromagnetic spring can be adjusted by modifying the coil current, ultimately adjusting the stiffness characteristics of the electromagnetic spring.

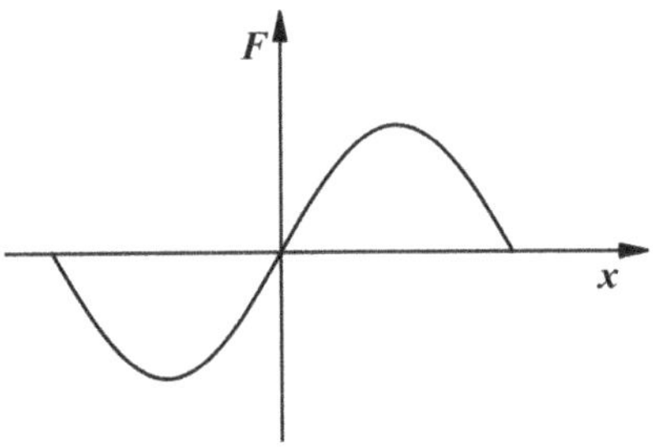

Figure 2. Electromagnetic force characteristics of a toothed electromagnetic spring.

3. Establishing the Model

3.1. Standard Bouc–Wen Model

The standard Bouc–Wen model was first proposed by Bouc [21] to characterize the hysteretic characteristics of materials. Then, the model was further extended by Wen [22]. The standard Bouc–Wen model has been widely used to describe hysteretic nonlinear models, such as piezoelectric actuators and magnetorheological dampers [23,24].

The standard Bouc–Wen model can be decomposed into linear yield and nonlinear hysteresis springs connected in parallel, as illustrated in Figure 3. The hysteresis characteristics of this model can be described through a first-order differential equation, as shown in Equation (1) [25]:

$$\begin{cases} F(t) = F_k + F_n = \alpha \frac{F_y}{u_y} x(t) + (1-\alpha) F_y Z \\ \frac{dZ}{dt} = v\{A - [\gamma + \beta \mathrm{sgn}(\dot{x}Z)]|Z|^n\} \end{cases} \quad (1)$$

where F represents the restoring force of the system, F_k represents the linear spring-restoring force, F_n represents the nonlinear hysteretic-restoring force, F_y represents the yield force, u_y represents the yield displacement, x represents the relative displacement, Z represents the dimensionless hysteretic variable, $sgn()$ represents the symbolic function, $\dot{x}$ represents the relative velocity, and A, γ, and β are the shape parameters of the standard Bouc–Wen model.

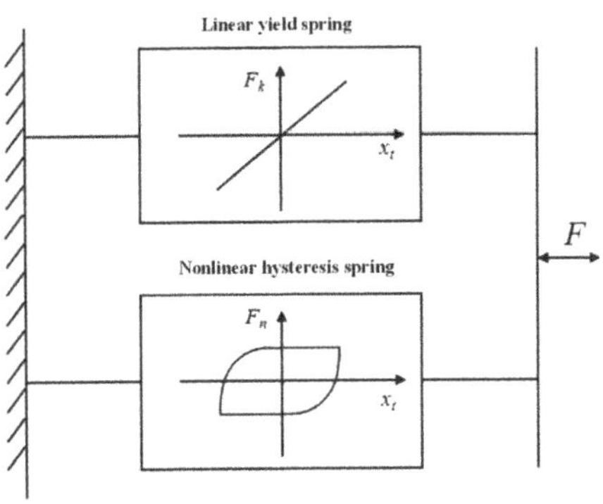

Figure 3. The standard Bouc–Wen model.

However, the standard Bouc–Wen model has many parameters, resulting in complex mathematical expressions. Therefore, Vincenzo [26] derived a standard model, which is more concise and conducive to studying parameter identification.

The standard Bouc–Wen model can be expressed as follows:

$$\frac{dz}{dt} = \rho\left[v - \sigma|\dot{x}||z|^{n-1}z + (\sigma - 1)\dot{x}|z|^{n}\right] \tag{2}$$

where ρ and σ are the shape parameters of the Bouc–Wen model, with $\rho = \frac{A}{u_y \varphi} > 0$ and $\sigma = \frac{\beta}{\beta+\gamma} > 0.5$; $\varphi = \sqrt[n]{\frac{A}{\beta+\gamma}}$, $n > 1$, and z is a non-observable dimensionless hysteretic variable, with $z \in [-1, 1]$.

This section aims to explore the impact of these parameters on the hysteresis curve of the system by adjusting the shape parameters of the standard Bouc–Wen model. The performance curve of Equation (2) is shown in Figure 4 by changing the parameters ρ, σ, and n. Figure 4a shows the hysteresis curve under different values of parameter ρ. It can be seen that ρ mainly changes the hysteresis characteristic of the hysteresis curve at the end of the forward and reverse motion. The higher the ρ, the more obvious the hysteresis phenomenon. Similar to parameter ρ, a higher value of σ increases the hysteresis area, while the influence is not so obvious when $\sigma > 8$. The parameter n mainly influences the bending shape of the hysteresis curve. With a smaller value of n, the bending radius of the curve increases, and the curve transition becomes smoother.

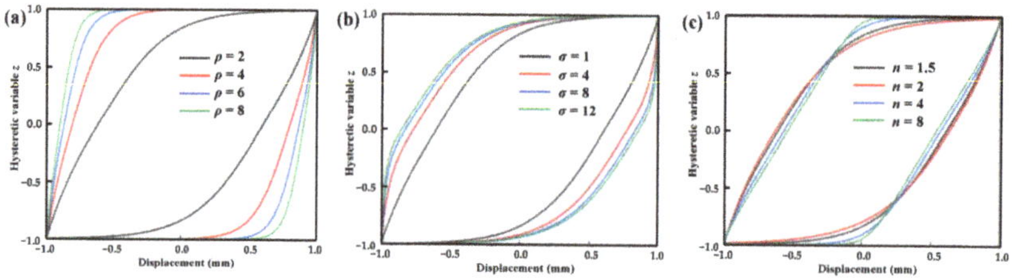

Figure 4. Effect of standard parameters on the hysteresis curve: (**a**) effect of parameter ρ on the hysteresis curve ($\sigma = 1$, $n = 2$), (**b**) effect of parameter σ on the hysteresis curve ($\rho = 2$, $n = 2$), and (**c**) effect of parameter n on the hysteresis curve ($\sigma = 1$, $\rho = 2$).

3.2. Improved Bouc–Wen Model

The hysteresis curve produces different magnitudes when the electromagnetic spring actuator moves forward and backward; therefore, the model exhibits asymmetry. Thus, the standard Bouc–Wen models cannot accurately describe its asymmetric hysteresis characteristics. Hence, it is necessary to improve the Bouc–Wen model.

The Bouc–Wen model shape control function is segmented based on the movement direction of the toothed electromagnetic spring to produce differences in the hysteresis curves of different movement directions, as shown in Equation (3):

$$\rho = |\lambda_1 \text{sgn}(v) + \lambda_2| \tag{3}$$

The dynamic forces of electromagnetic springs are divided into linear and hysteresis parts due to the nonlinearity of both static and dynamic forces of electromagnetic springs while considering the addition of mass and damping terms. Thus, Equation (1) can be written as Equation (4):

$$\begin{cases} F = k_1 z + k_2 x + cx + m\ddot{x} \\ \frac{dz}{dt} = |\lambda_1 \text{sgn}(\dot{x}) + \lambda_2|\left[v - \sigma|\dot{x}||z|^{n-1}z + (\sigma - 1)\dot{x}|z|^{n}\right] \end{cases} \tag{4}$$

where k_1 is the nonlinear stiffness of the electromagnetic spring; k_2 is the linear stiffness; λ_1, λ_2, σ, and n are the shape parameters of the improved Bouc–Wen model; m is the mass of the mover; and c is the system damping.

The experimental data at a current of 1.5 A were taken to study the effect of changes in parameters λ_1 and λ_2 on the dynamic characteristics of the toothed electromagnetic spring, and the results are shown in Figure 5.

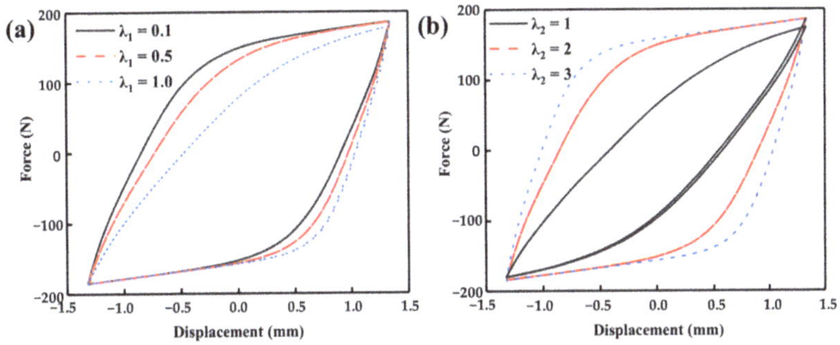

Figure 5. Hysteresis curves for different parameters: (**a**) λ_1 and (**b**) λ_2.

The experimental results of changing the parameter λ_1 ($\lambda_2 = 2$) and changing the parameter λ_2 ($\lambda_1 = 0.5$) are shown in Figure 5a,b, respectively. Figure 5 shows that, the larger λ_1 is, the more pronounced the hysteresis curve is in asymmetry. As λ_2 increases, the hysteresis phenomenon of the entire model becomes more pronounced, but the degree of asymmetry decreases accordingly.

3.3. Parameter Identification Method

After the dynamic output force and displacement data of the toothed electromagnetic spring were experimentally measured, the dynamic output force and displacement were simulated by the Bouc–Wen model, and the parameters were identified by the algorithm. The commonly used parameter identification algorithms for hysteresis models include the least square method, artificial neural networks, and particle swarm optimization. The least squares method requires multiple iterations to solve the nonlinear models, and the calculation is relatively complex. The artificial neural network method requires a large amount of training data, and the grid parameters must also be adjusted multiple times. The particle swarm optimization algorithm can perform a global search in parameter space and adaptively adjust the search direction and range, making it widely used in parameter recognition. Therefore, the particle swarm optimization algorithm was used to identify the hysteresis model parameters of the toothed electromagnetic spring, and the flowchart is shown in Figure 6.

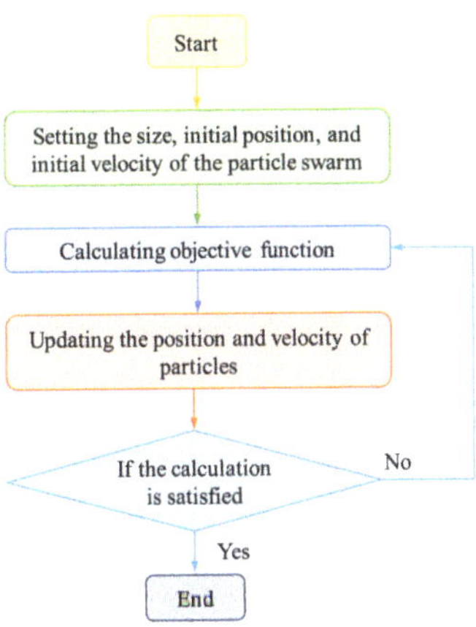

Figure 6. Flowchart of the particle swarm optimization algorithm.

According to Equation (4), the model has six unknown parameters. Due to differences in the Bouc–Wen model of the toothed electromagnetic spring, the model does not have a true solution, and it cannot be determined whether the solution is optimal. Therefore, the velocity and position of particles were updated according to Equation (5):

$$\begin{cases} x_i^{k+1} = x_i^k + x_i^{k+1} \\ w = w_{\max} - \frac{(w_{\max} - w_{\min})k}{k_{\max}} \\ v_i^{k+1} = wv_i^k + c_1 r_1 (p_i - x_i^k) + c_2 r_2 (p_g - x_i^k) \end{cases} \quad (5)$$

where $p_{\rm i}$ is the best position for oneself, p_g is the best position for the entire population, w is the inertia weight, c_1 and c_2 are learning factors, r_1 and r_2 are random numbers, $v_{\rm i}$ is particle velocity, $x_{\rm i}$ is the particle position, k is the current iteration number, $k_{\max}$ is the maximum iteration number, $v_{\max}$ is the inertia weight at the beginning of the iteration, and $v_{\min}$ is the weight at the end of the iteration.

A fitness function needs to be used to evaluate the advantages and disadvantages of the solution. In this work, the root-mean-square error between the test value of the electromagnetic spring output force and the calculation result was taken as the fitness function, as shown in Equation (6).

$$f_t = \sqrt{\frac{\sum\limits_{i=1}^{N_t} (F - F_t)^2}{N_t}} \quad (6)$$

where N_t is the total number of samples; and F_t and F are the measured and calculated values of the output force of the toothed electromagnetic spring, respectively.

4. Experimental Validation

4.1. Experimental System and Results

The test flowchart and platform of the toothed electromagnetic spring are shown in Figure 7. The DC power supply outputs different currents (or electric cylinder produce different moving speeds), so the electromagnetic spring receives current (or speed) pulses and produces electromagnetic force and displacement. Then, the force sensor and the displacement sensor collect the force and displacement signals, respectively, and send them to the computer for signal processing. During the test, the displacement of the actuator of the electromagnetic spring was adjusted to 1.3 mm, and a constant current was applied to the coil. The electric cylinder was controlled to move at a constant speed until the actuator moved to −1.3 mm. Then, the electric cylinder was controlled to move in the opposite direction at the same speed until the displacement of the actuator reached 1.3 mm. The main technical specifications for the experimental system are illustrated in Table 1. Additionally, a 1 mm polyester imide enameled wire was selected as the solenoid coil in this experiment, with a safe current of 2.75 A and 390 turns.

Table 1. The main technical specifications for the experimental system.

Equipment	Model	Parameters	Manufacture
Force sensor	AR-DN23	Range: 0–5 kN Accuracy: 0.015%F.S	Ailixun, Chian
Displacement sensor	ML33-12.5-A	Range: 0–12.5 mm Accuracy: 0.1 %F.S	Miran, China
Servo electric	ECMA-C200807SS	Output: 3000 rpm	Delta, China
Power supply	DC-3010D	Range: 0–10 A, 0–30 V	Yihua, China

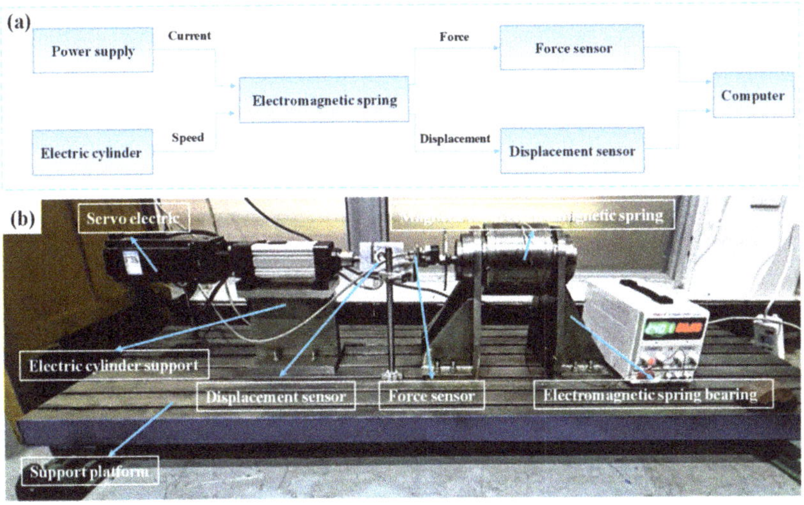

Figure 7. The toothed electromagnetic spring of (**a**) the test flow chart and (**b**) test platform.

Figure 8 shows a significant hysteresis phenomenon between the output force and displacement of the toothed electromagnetic spring. According to Figure 8a, the force–displacement curve exhibits asymmetry when the electromagnetic spring actuator moves in the forward and opposite directions. In addition, under the same current, when the velocity of the mover changes, a significant change in its hysteresis phenomenon is also observed. With the decrease in the moving speed of the actuator, the output force–displacement hysteresis of the electromagnetic spring decreases significantly, and the parallel part of the hysteresis curve with the static curve increases. When the velocity of the actuator is small enough, it can be assumed that the dynamic electromagnetic force characteristics of the electromagnetic spring are the same as those of the static electromagnetic force characteristics. Figure 8b shows the output force–displacement characteristics of the toothed electromagnetic spring under different currents at a rotor speed of 4.5 mm/s.

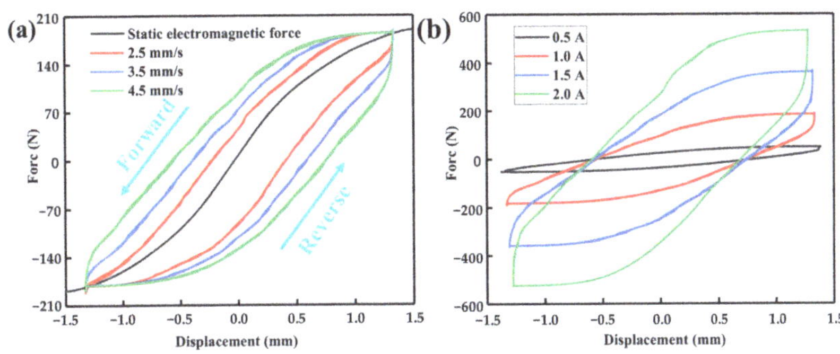

Figure 8. (**a**) Dynamic output force–displacement curve at the current of 1.0 A and (**b**) output force–displacement curve at different currents at the speed of 4.5 mm/s.

According to Figure 8b, the hysteresis of the electromagnetic spring gradually increases with the current, i.e., the output force under the same displacement gradually increases. This phenomenon indicates that the electromagnetic spring's stiffness increases with the current. The increase in the hysteresis also indicates that the nonlinear degree of electromagnetic spring increases gradually during dynamic works. Therefore, in practical applications, it is necessary to conduct reasonable current control according to specific requirements to provide the required output characteristics.

4.2. Model Verification

The parameters in the Bouc–Wen model change due to different hysteresis characteristics. In this work, the output force test data generated under different current and speed excitation conditions were selected as F_t. The particle swarm optimization was used to identify the parameters of the improved Bouc–Wen model, in which $w_{max} = 0.9$, $w_{min} = 0.4$, $c_{1int} = 2$, $c_{2int} = 2$, $c_{1fin} = 4$, and $c_{2fin} = 4.5$ were considered. Additionally, the number of population particles was set to 100. The maximum number of iterations was 200, and the range of values for each parameter of the Bouc–Wen model is shown in Table 2.

Table 2. Range of the model parameter values of the Bouc–Wen.

Parameters	Value
σ	(0.5, 100]
n	(1, 20]
λ_1	[0, 10]
λ_2	(0, 10]
k_1	(0, 1000]
k_2	(0, 1000]
c	$(-\infty, +\infty)$

Table 3 shows the parameter identification results of the improved Bouc–Wen model using particle swarm optimization. The hysteresis loop of the restoring force of the toothed electromagnetic spring can be reconstructed according to the identification results. A comparison between the estimated curves fitted by the proposed model and experimentally measured results was performed to verify the accuracy of the established model, as illustrated in Figure 9, which shows that the experimental curve is highly consistent with the predicted curve. Therefore, the improved Bouc–Wen model established in this work can accurately describe toothed electromagnetic springs' dynamic output force–displacement characteristics.

Table 3. Identification results of the particle swarm optimization algorithm.

Current (A)	Speed (mm/s)	σ	n	λ_1	λ_2	k_1 (N)	k_2 (N/mm)	c (N·s/m)
1.0	2.5	0.25	8.13	0.15	1.19	130.69	42.34	589.90
1.0	3.5	1.91	9.98	0.15	1.19	140.59	29.96	611.74
1.0	4.5	18.74	7.03	0.12	1.04	158.37	20.35	646.39
1.5	4.5	21.19	6.59	0.12	1.03	308.78	39.93	654.99
2.0	4.5	20.53	5.42	0.06	0.97	541.70	62.31	665.09

It is necessary to analyze the model error additionally. A residual analysis, which indicates the difference between the measured and predicted results, was adopted to evaluate the prediction accuracy. The results of the residual analysis are shown in Figure 10. It can be seen in Figure 10 that the residual points predicted by the model proposed in this work and the actual experimental results fell basically in the horizontal band region of –20 N–20 N (for an applied current of 1.0 A) and –40 N–80 N (for an application of 4.5 mm/s). This indicates that the constitutive model developed in this work can effectively reflect the dynamic output force–displacement characteristics.

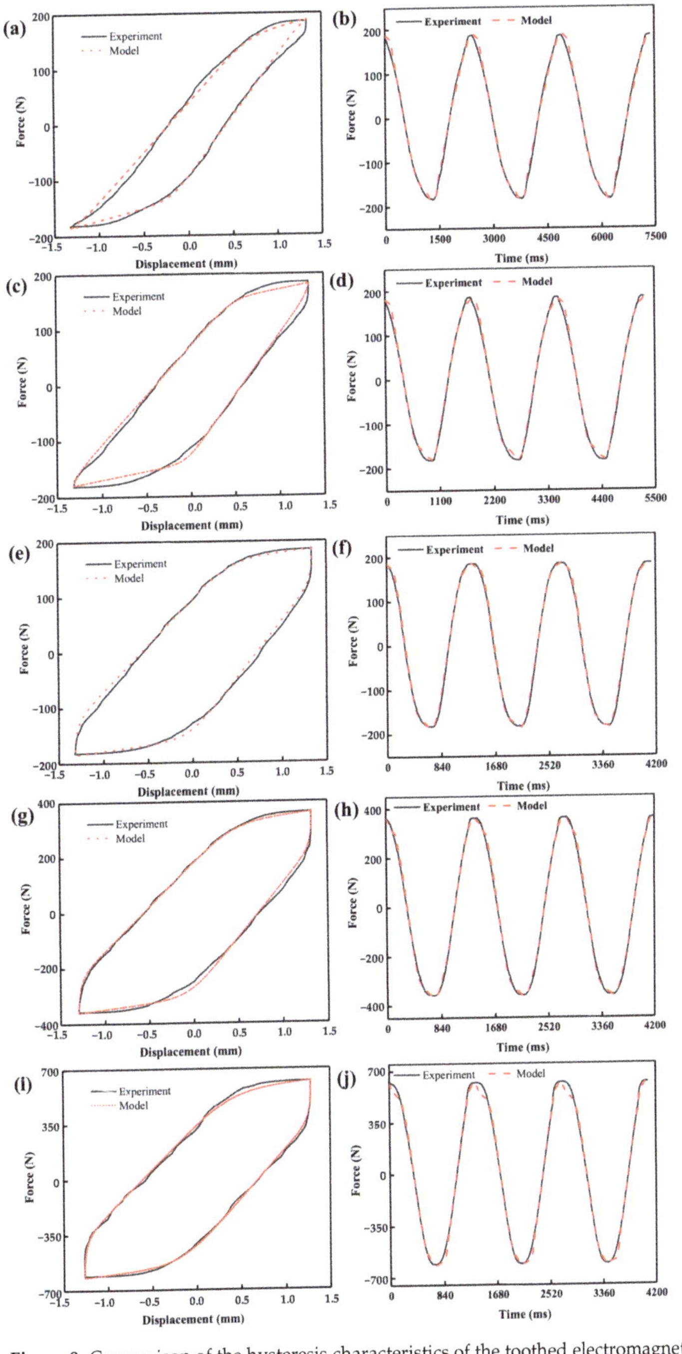

Figure 9. Comparison of the hysteresis characteristics of the toothed electromagnetic spring: (**a**,**b**) for the applied current of 1.0 A and speed of 2.5 mm/s, (**c**,**d**) for the applied current of 1.0 A and speed of 3.5 mm/s, (**e**,**f**) for the applied current of 1.0 A and speed of 4.5 mm/s, (**g**,**h**) for the applied current of 1.5 A and speed of 4.5 mm/s, and (**i**,**j**) for the applied current of 2.0 A and speed of 4.5 mm/s.

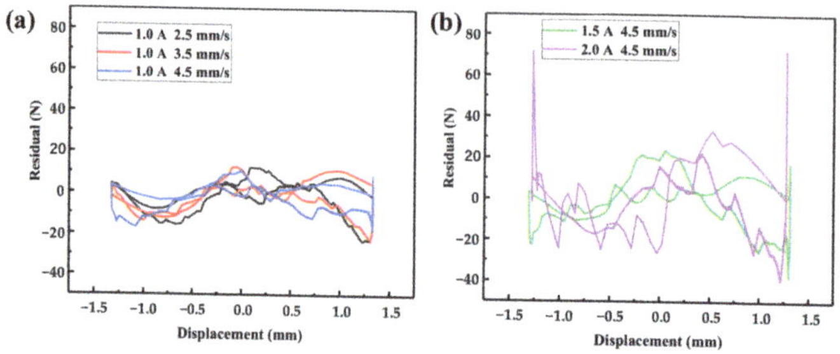

Figure 10. Residual analysis of (**a**) for the applied current of 1.0 A and (**b**) for the application of 4.5 mm/s.

To provide a more quantitative explanation, the maximum relative error of the improved model was defined as follows:

$$\delta_{MAX} = \frac{\Delta F_{MAX}}{F_{MAX}} \times 100\% \tag{7}$$

where F_{MAX} is the maximum output force measured in the experiment and ΔF_{MAX} is the maximum output force error of the model compared to the experiment. Meanwhile, R^2 was defined to test the goodness of fit of the improved model, as shown in Equation (7):

$$R^2 = 1 - \frac{\sum_{i=1}^{n} (\hat{F}_i - \overline{F})^2}{\sum_{i=1}^{n} (F_i - \overline{F})^2} \tag{8}$$

where F_i is the dynamic output force measured in the experiment, $\overline{F}$ is the average value of the dynamic output force, and $\hat{F}_i$ is the dynamic output force obtained through model simulation fitting. The correlation index R^2 was used to describe the fitting degree of the model; the closer R^2 is to 1, the better the prediction accuracy of the improved model, and the stronger the linear correlation between the actual variables and the predicted variables [27].

Table 4 illustrates the error results under different excitation currents and speeds. It can be seen that the maximum relative error of the standard Bouc–Wen model was 12.75%. Generally, a Bouc–Wen model with a maximum relative error of less than 15% is considered to have a high accuracy [28]. In addition, the correlation index R^2 of the Bouc–Wen model was higher than 0.98 under different working conditions. The highest value was 0.9993, indicating that the model used in this work has a good fitting accuracy and verifies the accuracy of the model.

Table 4. Error analysis of the Bouc–Wen model.

Current (A)	Speed (mm/s)	ΔF_{MAX} (N)	F_{MAX}	δ_{MAX}	R^2
1.0	2.5	23.65	185.49	12.75%	0.9991
1.0	3.5	23.46	185.49	12.68%	0.9959
1.0	4.5	18.04	185.49	9.68%	0.9971
1.5	4.5	37.84	363.49	10.41%	0.9993
2.0	4.5	72.65	621.47	11.69%	0.9868

5. Conclusions

In this work, the standard Bouc–Wen model was improved, and the hysteresis model of a toothed electromagnetic spring was established. Then, the particle swarm optimization algorithm was used to identify the model's parameters. The main conclusions are as follows:

(1) The influence of various parameters of the standard Bou–Wen model was studied. The results show that the higher the parameter ρ, the more serious the hysteresis phenomenon, and the greater the coincidence range between the hysteresis curve and the upper and lower boundary of the hysteresis variable. The higher the σ, the larger the coincidence range between the hysteresis curve and the boundary. The higher the parameter n, the smaller the bending radius of the hysteresis curve.
(2) The electromagnetic spring exhibits a hysteresis phenomenon under different currents and velocities. The hysteresis phenomenon becomes more pronounced as the speed increases. The current has a significant impact on the dynamic curve. When the current increases from 0.5 A to 2.0 A, the electromagnetic force sharply increases from 49 N to 534 N.
(3) An asymmetric correction term was proposed based on the asymmetry of the dynamic characteristics to replace the parameter ρ and improve the standard Bou–Wen model. Compared with the experimental results, the residual points predicted by the model proposed in this work fell basically in the horizontal band region of −20 N–20 N (for an applied current of 1.0 A) and −40 N–80 N (for an application of 4.5 mm/s) and the maximum relative error of the model was 12.75%. The R^2 of the model was greater than 0.98 and the highest value was 0.9993, indicating that the model has a good fitting accuracy.

Author Contributions: X.Z.: Methodology, Formal analysis, Data curation, Investigation, Writing—original draft. C.Z.: Conceptualization, Methodology, Supervision, Writing—review and editing. Y.L.: Conceptualization, Methodology, Supervision, Writing—review and editing. G.X.: Conceptualization, Methodology, Supervision, Writing—review and editing. H.B.: Supervision, Writing—review and editing. All authors have read and agreed to the published version of the manuscript.

Funding: This work was supported by the key technology project of Inner Mongolia (No. 82321160) and Young and Middle-aged Teachers Education and Research Project (Science and Technology) of Fujian Province (No. JAT220016).

Institutional Review Board Statement: Not applicable.

Informed Consent Statement: Not applicable.

Data Availability Statement: The data presented in this study are available upon request from the corresponding author. The data are not publicly available due to privacy.

Conflicts of Interest: The authors declare no conflict of interest.

References

1. Liu, Y.; Yun, F.; Wei, H.; Hui, W. Modeling and observer-based vibration control of a flexible spacecraft with external disturbances. *IEEE Trans. Ind. Electron.* **2019**, *66*, 8648–8658. [CrossRef]
2. Li, J.Y.; Jing, X.J.; Li, Z.C.; Huang, X.L. Fuzzy adaptive control for nonlinear suspension systems based on a bioinspired reference model with deliberately designed nonlinear damping. *IEEE Trans. Ind. Electron.* **2019**, *66*, 8713–8723. [CrossRef]
3. Zheng, X.Y.; Wu, Y.W.; Zheng, C.; Bai, H.B.; Liu, R. Quasistatic and low-velocity impact properties of entangled metallic wire material–silicone rubber composite. *J. Mater. Res. Technol.* **2023**, *23*, 143–152. [CrossRef]
4. Wu, M.K.; Wu, J.L.; Chen, J.X.; Gao, R.Q.; Chen, X.D.; Li, X.Q.; Zeng, L.Z.; Jiang, W. Analysis and experiment of a novel compact magnetic spring with high linear negative stiffness. *Mech. Syst. Signal Process.* **2023**, *198*, 110387. [CrossRef]
5. Zheng, X.Y.; Ren, Z.Y.; Bai, H.B.; Wu, Z.B.; Guo, Y.S. Mechanical behavior of entangled metallic wire materials-polyurethane interpenetrating composites. *Def. Technol.* **2023**, *20*, 120–136. [CrossRef]
6. Pu, H.Y.; Yuan, S.J.; Peng, Y.; Meng, K.; Zhao, J.L.; Xie, R.Q.; Huang, Y.N.; Sun, Y. Multi-layer electromagnetic spring with tunable negative stiffness for semi-active vibration isolation. *Mech. Syst. Signal Process.* **2019**, *121*, 942–960. [CrossRef]
7. Yu, S.Q.; Wang, A.B.; Zhao, L.P. Characteristics test study on electromagnetic actuator. *Mach. Des. Manuf.* **2011**, *1*, 136–138.
8. Zhang, B. Research on Hybrid Isolator Design Technique. Master's Thesis, Harbin Engineering University, Harbin, China, 2011.
9. Zhang, B.; Wang, X.; Wu, H.M.; Yang, B.T. Optimization design of electromagnetic vibration absorber for ship vibration suppression. *Noise Vibr. Control* **2020**, *40*, 213–218.
10. Ai, W.; Jin, L.A.; Chi, W. The status and prospect of ship vibration isolation technology. *Mar. Technol.* **2015**, *1*, 4–8.
11. Ma, J.G.; Shuai, C.G.; Li, Y. Anti-shock design for active-passive vibration isolation using electromagnetic actuator and air spring. *J. Ship. Mech.* **2021**, *25*, 120–127.

12. Batdorff, M.A.; Lumkes, J.H. High-fidelity magnetic equivalent circuit model for an axisymmetric electromagnetic actuator. *IEEE Trans. Magn.* **2009**, *45*, 3064–3072. [CrossRef]
13. Erturul, H.F.; Erkan, K.; Hüseyin, V. Detailed magnetic forcesnalysis of A4-Pole hybrid electromagnet by magnetic equivalent circuit method. *J. Magn.* **2018**, *23*, 499–508. [CrossRef]
14. Sun, Y.; Meng, K.; Yuan, S.J.; Zhao, J.L.; Xie, R.Q.; Yang, Y.; Luo, L.; Peng, Y.; Xie, S.R.; Pu, H.Y. Modeling electromagnetic force and axial-stiffness for an electromagnetic negative-stiffness spring toward vibration isolation. *IEEE Trans. Magn.* **2019**, *55*, 18470603. [CrossRef]
15. Wu, J.L.; Zeng, L.Z.; Han, B.; Zhou, Y.F.; Luo, X.; Li, X.Q.; Chen, X.D.; Jiang, W. Analysis and design of a novel arrayed magnetic spring with high negative stiffness for low-frequency vibration isolation. *Int. J. Mech. Sci.* **2021**, *216*, 106980. [CrossRef]
16. Li, Z.H.; Cao, Q.L.; Lai, Z.P.; Han, X.T. Application of current filament method on the calculation of current and force in electromagnetic forming. *Trans. China Electron. Soc.* **2018**, *33*, 4181–4190.
17. Wang, K.; Wang, D.; Wu, L.T.; Su, Z.Z.; Zhang, X.B.; Lu, X.Q. Optimal design of permanent magnet biased axial magnetic bearing with multiple air gaps based on analytical model. *Mot. Control Appl.* **2016**, *43*, 1–7.
18. Angelo, A.; Alaggio, R.; Köhler, J.; Fragiacomo, M. Extension of generalized Bouc–Wen hysteresis modeling of wood joints and structural systems. *J. Eng. Mech.* **2020**, *146*, 04020001.
19. Birhan, A.N.; You, W.; Lee, J.H.; Lee, K. Parameter identification of Bouc–Wen model for magnetorheological (MR) fluid damper by a novel genetic algorithm. *Adv. Mech. Eng.* **2020**, *12*, 2213–2224.
20. Nguyen, X.B.; Toshihiko, K.; Truong, H.T. Adaptive parameter identification of Bouc–Wen hysteresis model for a vibration system using magnetorheological elastomer. *Int. J. Mech. Sci.* **2022**, *213*, 106848. [CrossRef]
21. Bouc, R. A Mathematical model for hysteresis. *Acta Acust. Acust.* **1971**, *24*, 603–608.
22. Wen, Y.K.; Asce, M. Method for random vibration of hysteretic systems. *J. Eng. Mech. Div.* **1976**, *102*, 0002106. [CrossRef]
23. Alem, S.F.; Izadi, I.; Sheikholeslam, F.; Mohsen, E. Piezoelectric actuators with uncertainty: Observer-based hysteresis compensation and joint stability analysis. *IEEE Trans. Control Syst. Technol.* **2020**, *28*, 1997–2004. [CrossRef]
24. Li, R.; Gou, X.; Zhou, M.J.; Yang, P.A.; Li, P.H.; Shou, M.J.; Liao, C.R.; Wang, X.J. Modeling and validation of multifield coupled self-sensing characteristics of magnetorheological elastomer for vibration isolators. *Mater. Des.* **2022**, *217*, 110636. [CrossRef]
25. Ma, F.; Zhang, H.; Bockstedte, A.; Foliente, G.C.; Paevere, P. Parameter analysis of the differential model of hysteresis. *J. Appl. Mech.* **2004**, *71*, 342–349. [CrossRef]
26. Charalampakis, A.E.; Koumousis, V.K. Identification of Bouc–Wen hysteretic systems by a hybrid evolutionary algorithm. *J. Sound Vib.* **2008**, *314*, 571–585. [CrossRef]
27. Xin, X.; Ruan, S.X.; Bai, H.B.; Chen, X.C. An enhanced constitutive model for the nonlinear mechanical behavior of the elastic-porous metal rubber. *Mech. Mater.* **2020**, *148*, 103447.
28. Chen, S.X.; Xu, L.Y.; Zhang, S.; Zhao, S.X.; Liu, K. Parameter identification of the Bouc–Wen model for the magnetorheological damper using fireworks algorithm. *J. Mech. Sci. Technol.* **2022**, *36*, 2213–2224. [CrossRef]

materials

MDPI

Article

Investigation on Vibration Characteristics of Thin-Walled Steel Structures under Shock Waves

Zehao Li [1], Wenlong Xu [1,*], Cheng Wang [2,*], Xin Liu [2] and Yuanxiang Sun [2]

1 Institute of Advanced Technology, Shandong University, Jinan 250061, China
2 State Key Laboratory of Explosion Science and Technology, Beijing Institute of Technology, Beijing 100081, China
* Correspondence: xuwenlong@sdu.edu.cn (W.X.); wangcheng@bit.edu.cn (C.W.)

Abstract: Thin-walled steel structures, prized for their lightweight properties, material efficiency, and excellent mechanical characteristics, find wide-ranging applications in ships, aircraft, and vehicles. Given their typical role in various types of equipment, it is crucial to investigate the response of thin-walled structures to shock waves for the design and development of innovative equipment. In this study, a shock tube was employed to generate shock waves, and a rectangular steel plate with dimensions of 2400.0 mm × 1200.0 mm × 4.0 mm (length × width × thickness) was designed for conducting research on transient shock vibration. The steel plate was mounted on an adjustable bracket capable of moving vertically. Accelerometers were installed on the transverse and longitudinal symmetric axes of the steel plate. Transient shock loading was achieved at nine discrete positions on a steel plate by adjusting the horizontal position of the shock tube and the vertical position of the adjustable bracket. For each test, vibration data of eight different test positions were obtained. The wavelet transform (WT) and the improved ensemble empirical mode decomposition (EEMD) methods were introduced to perform a time-frequency analysis on the vibration of the steel plate. The results indicated that the EEMD method effectively alleviated the modal aliasing in the vibration response decomposition of thin-walled structures, as well as the incompletely continuous frequency domain issue in WT. Moreover, the duration of vibration at different frequencies and the variation of amplitude size with time under various shock conditions were determined for thin-walled structures. These findings offer valuable insights for the design and development of vehicles with enhanced resistance to shock wave loading.

Keywords: thin-walled structures; shock waves; shock tube; structural shock dynamic

1. Introduction

Thin-walled steel structures are commonly utilized in various applications such as ships, aircraft, and vehicles due to their lightweight nature, material efficiency, and favorable mechanical properties [1,2]. These structures are often subjected to transient shock loads during operation in challenging environments. As a result, the dynamic response characteristics of thin-walled structures under transient shock loads have received extensive attention from researchers and engineers in related fields [3,4].

The investigation of the dynamic response of thin-walled structures has predominantly focused on two distinct modes of failure. Specifically, the response of these structures to shock loads is generally characterized by either large plastic deformation [5–8] or tensile tearing [9,10]. For the study of large plastic deformation of thin-walled structures, Wang et al. [11] conducted experimental and numerical simulation studies on the response of free thin steel plates under intense loads and proposed a theoretical model to describe the deformation of free metal thin plates under powerful loading. They found that when the loading parameters are determined, the deformation velocity and deflection of the plate are only related to the width-to-thickness ratio of the plate. Xu et al. [12] presents experimental and numerical investigations on the response of thin aluminum plates to shock loading, reveals the relationship between the deformation region of counterintuitive behavior (CIB) and loading and geometric parameters of the structure, and derives a relationship between normalized duration and charge mass to predict the occurrence of CIB. Kaufmann et al. [13] introduced the Virtual Fields Method (VFM) to reconstruct surface pressures on thin steel plates by measuring full-field deformation of plate dynamics. A shock wave similar to an explosion is generated by a shock tube, demonstrating the crucial role of VFM modeling in predicting large plastic deformations of structures. Curry et al. [14] investigated the impact of different charge backing types on the plastic deformation of thin-walled steel plates using a combination of experimental and computational methods. The findings suggest that metal-backed charges increase impulse transfer by 3–5 times compared to air-backed charges. While the permanent deflection was greater with metal backing, the degree of increase was less pronounced compared to that of the impulse transfer. Kim et al. [15] studied the influence of the relative position of explosives and thin steel plates on the plate damage and found that tearing damage can be reduced by optimizing the inclination angle of the thin wall. Yao et al. [16] conducted a large-scale experimental study on the damage characteristics and dynamic response of thin-walled multi-steel box structures under restrained high-intensity loads and found that strengthening the corner of the box can prevent tearing of the thin-walled structure. McDonald et al. [17] examined the response of four high-strength thin-walled structures to localized blast loading. The findings indicate that higher strength steels and tailored microstructures provide enhanced rupture resistance. A new non-dimensional impulse correction parameter is introduced for assessing the impact of charge stand-off on deformation and rupture performance.

As the level of intelligence and informationization of vehicles increases, electronic and mechanical equipment becomes increasingly important in the carriers [18]. In the face of enemy attacks, even if the structure remains intact, strong vibrations can damage electronic and mechanical devices, thus affecting weapon and equipment effectiveness [19]. Research on the vibration characteristics of thin-walled structures mainly focuses on numerical simulations [20–22] and theoretical analysis [23–25].

In terms of numerical simulations, Park et al. [26] conduct a numerical investigation to evaluate the effectiveness of thin-walled panels in attenuating vibrational damage induced by explosive events. Their findings reveal that the implementation of blast-resistant panels significantly mitigates the propagation of acceleration, with the most favorable outcomes achieved through the utilization of thicker panels and lower explosion loads. In a separate study, Wu et al. [27] inspect the vibrational response of subterranean thin-walled structures subjected to surface blast loads by employing numerical simulations. The researchers observe an increase in peak velocity as the proximity to the explosion source decreases, with a predominance of vertical vibrations. Moreover, the study introduces a predictive model for damage assessment and delineates critical thresholds associated with distinct damage levels. Wu et al. [28] use numerical simulations to propose a data-driven approach for designing distributed Dynamic Vibration Absorbers (DVAs) to mitigate vibrations in thin-walled structures with tight modal spacing. Leveraging Singular Value Decomposition (SVD) on structural response data, without needing excitation or structural mode info, optimal DVA placement and parameters are determined. This method surpasses traditional techniques, showcasing its robustness and effectiveness on a simply supported square plate and a fairing, achieving broad-band vibration suppression utilizing only structural response data.

For theoretical studies, Pandey et al. [29] investigates the transient vibroacoustic response of functionally graded sandwich plates with varying thickness ratios and material gradation. A parametric study is conducted to investigate the influence of volume fraction index and thickness ratio on the transient vibroacoustic response. Vieira et al. [30] introduced a new high-order beam model for thin-walled structure response analysis, which considers the three-dimensional displacement characteristics of the thin-walled structure and the in-plane bending characteristics of the section. This model effectively analyzes local and global buckling phenomena of thin-walled structures under high-order modes. Xu et al. [31] found that the improved beam theory based on the Carrera Unified Formulation has higher reliability and accuracy in predicting the modal behavior of the structure than the classical beam theory. The complex interactions between shock waves and structures lead to substantial analytical challenges, making it difficult for theoretical analysis to effectively address such complexities [32,33]. With these factors in mind, the importance of experimental research must be highlighted, as it offers valuable insights and enables accurate validation and refinement of theoretical results [34,35].

In this work, an experimental device comprising a shock tube system and a steel plate fixed onto an adjustable support bracket is proposed. The device enables the application of transient shock loads at various positions along a thin-walled structure, facilitating the measurement of its vibration response under shock loading conditions. By conducting transient loading at nine different positions on the structure and using accelerometers at key positions, the transient shock vibration response characteristics were obtained. To address the frequency domain discontinuity problem inherent to traditional wavelet transformation methods, this study proposes a novel empirical mode decomposition method grounded in piecewise cubic Hermite interpolation (PCHIP), an approach that ensures monotonicity and circumvents the "over envelope" and "under envelope" fitting issues commonly associated with conventional interpolation methods, thereby enhancing the analysis of the vibration response of thin-walled structures under varying shock positions.

2. Experimental Design and Conditions

2.1. Experimental Setups

In this study, transient loads were generated using a shock tube system, as illustrated in Figure 1a. The shock tube, with an inner diameter of 90.0 mm, was partitioned into high-pressure and low-pressure sections, separated by a 0.5 mm-thick aluminum diaphragm with a 0.3 mm cross-shaped scratch. High-pressure nitrogen served as the driving gas in the high-pressure section, while the low-pressure section was initially connected to the atmosphere. When the pressure of the high-pressure section reaches about 1300.0 kPa and the low-pressure section maintains atmospheric pressure, the diaphragm ruptured rapidly along the scratch, creating a shock wave in the low-pressure section. A pressure sensor (Kistler 211B4) was installed at the shock tube opening to measure the pressure generated by the shock wave from the tube. In order to investigate the shock response of thin-walled structures, an adjustable bracket was designed and installed. The bracket featured an adjustable track that allowed for easy modification of the shock wave position by adjusting the plate's vertical and horizontal positions. A steel plate (measuring 2400.0 mm × 1200.0 mm × 4.0 mm in length, width, and thickness) was mounted onto the bracket. This steel plate is made from Q235 steel, with a density of 7.85 g/cm^3, a minimum yield strength of 235 MPa, and a Young's modulus of 210 GPa.

The plate's response to shock wave loading was analyzed using eight Kistler 8776B100A accelerometers, which were symmetrically positioned along the horizontal and vertical axes of the steel plate. Acceleration data was obtained by processing the voltage signals captured by these sensors with a Kistler TraNET 408DP data acquisition device. Figure 1b shows the arrangement of the accelerometers, with six (RM1-RM6) placed on the plate's horizontal axis and two (CM1, CM2) on the thin plate's horizontal axes. The first transverse sensor was placed 1200.0 mm to the left of the steel plate's center point, with subsequent sensors positioned every 200.0 mm to the right. Sensors CM1 and CM2, located on the vertical axis, were situated 200.0 mm and 500.0 mm above the center point, respectively. This setup was used to determine the structure's vibration characteristics. All nine shock positions were located on the right side of the thin plate. Shock positions M1, M2, and M3 were positioned on the horizontal axis of symmetry of the thin plate, lying 200.0 mm, 500.0 mm, and 800.0 mm to the right of the plate's center, respectively. Above these three shock loading points, shock positions H1, H2, and H3 were found at a height of 200.0 mm on the thin plate's horizontal axis of symmetry. Below these points, loading positions L1, L2, and L3 were situated at a height of 100.0 mm.

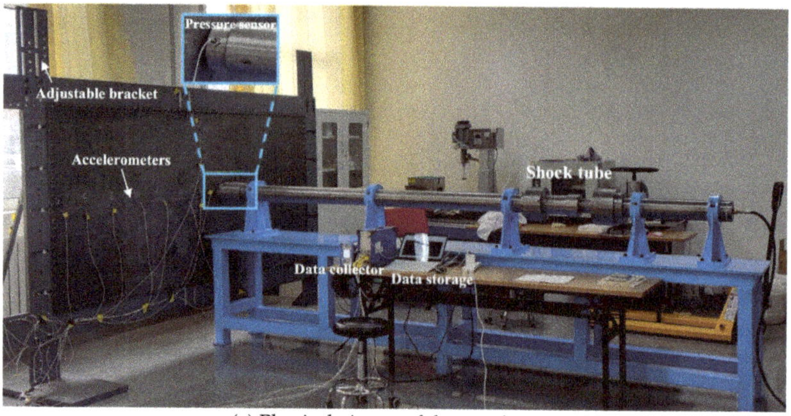

(**a**) Physical picture of the test device

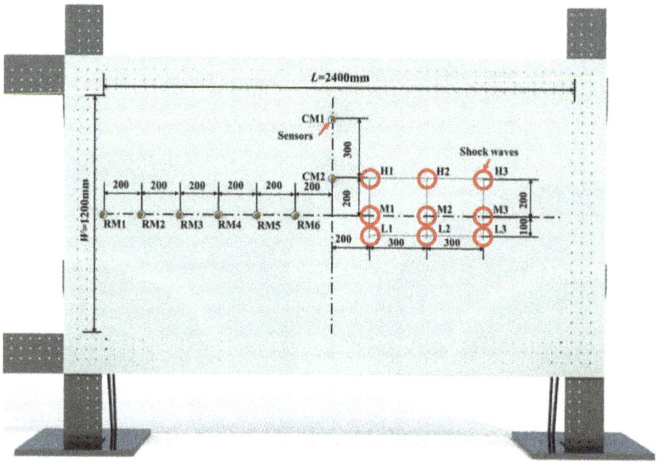

(**b**) Picture of shock conditions (The red circle in the picture represents the loading position of the shock wave)

Figure 1. Experimental schematic.

2.2. Repetition Verification

To ensure the reproducibility of shock wave loading, a repeatability verification experiment was conducted using a shock tube apparatus. Figure 2 depicts the pressure–time history curves obtained by a pressure sensor at the shock tube nozzle in two separate experiments. The initial peak pressures of the shock waves in the two tests were 129 kPa and 128 kPa, respectively, exhibiting a deviation of 0.7%. These results indicate that the shock wave transient load system used in the experiments demonstrates high consistency.

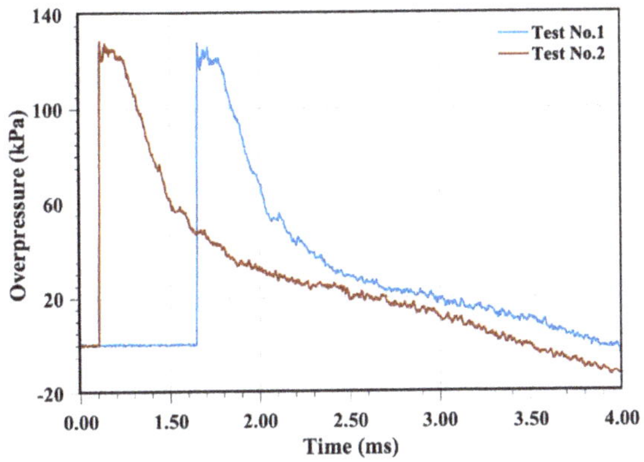

Figure 2. Independent experimental verification.

3. Experimental Data Analysis Methods

3.1. *Wavelet Transform Analysis Method*

Wavelet Transform (WT) can show the detailed information of signals in time-frequency domain. The wavelet transform $(W_\psi f)(a,b)$ of a signal $f(t)$ is defined as:

$$(W_\psi f)(a,b) = \langle f, \psi_{a,b} \rangle = |a|^{-1/2} \int_{-\infty}^{+\infty} f(t)\psi^* \left(\frac{t-b}{a} \right) \mathrm{dt} \tag{1}$$

where $a, b \in R$, and $\alpha \neq 0$ is called the scaling factor, and b is called the translation factor. $\psi\left(\frac{t-b}{a}\right)$ represents a series of basis wavelets determined by a and b, and $\psi^*\left(\frac{t-b}{a}\right)$ represents the complex conjugate of the basis wavelet. The role of WT is to transform one-dimensional impulse response signal data into a two-dimensional matrix, where each row represents the wavelet coefficients at different decomposition scales, and each column represents the different time of the impulse response signal data. Since the WT is discontinuous in the frequency domain, it is not feasible to achieve a fully continuous wavelet transform of signal [36]. To accurately capture the frequency domain characteristics of the structure, we have developed a novel method, namely, the improved ensemble empirical mode decomposition method.

3.2. *Improved EEMD-HHT Analysis Method*

3.2.1. EEMD-HHT Analysis Method

Hilbert–Huang transform (HHT) transform is a time-frequency localization analysis method with strong adaptability, which is suitable for processing and analyzing non-stationary signals. HHT transform consists of EMD decomposition and Hilbert transform, but EMD decomposition has the problem of mode aliasing, that is, the components of different frequency bands in the signal cannot be effectively separated. Wu et al. [37] proposed EEMD decomposition, which can prevent the diffusion of low-frequency modal components by adding white noise and alleviate modal aliasing. After the signal is decomposed by EEMD, the combination of multiple IMF components can be observed, and the analytical signal can be observed by Hilbert transform of the IMF component $c(t)$ of the original signal. After EEMD decomposition of the signal, a group of IMF components can be obtained. Hilbert transform is performed on each order of IMF components to construct the Analytic signal $z(t)$:

$$z(t) = c(t) + jH[c(t)] = a(t)e^{j\Phi(t)} \tag{2}$$

$H[c(t)]$ is the IMF component after the Hilbert transformation, and j is an imaginary unit. The instantaneous amplitude $a(t)$ corresponding to the analytical signal can be obtained:

$$a(t) = \sqrt{c^2(t) + H^2[c(t)]} \tag{3}$$

3.2.2. Improved Decomposition Method

In the ensemble empirical mode decomposition (EEMD) method, accurate envelope fitting is essential for signal decomposition. However, traditional cubic spline interpolation methods for fitting extreme points often lead to "over envelope" and "under envelope" issues, which can significantly impact the accuracy of decomposition. In this work, an EEMD method based on piecewise cubic Hermite interpolation (Pchip interpolation) was employed, which effectively resolves the fitting issues encountered in traditional methods. To decompose the signal, we used a low-pass filter to isolate the target frequency band and applied a set empirical mode decomposition based on cubic Hermite interpolation, Figure 3 shows the pchip-EEMD transformation process of real signals.

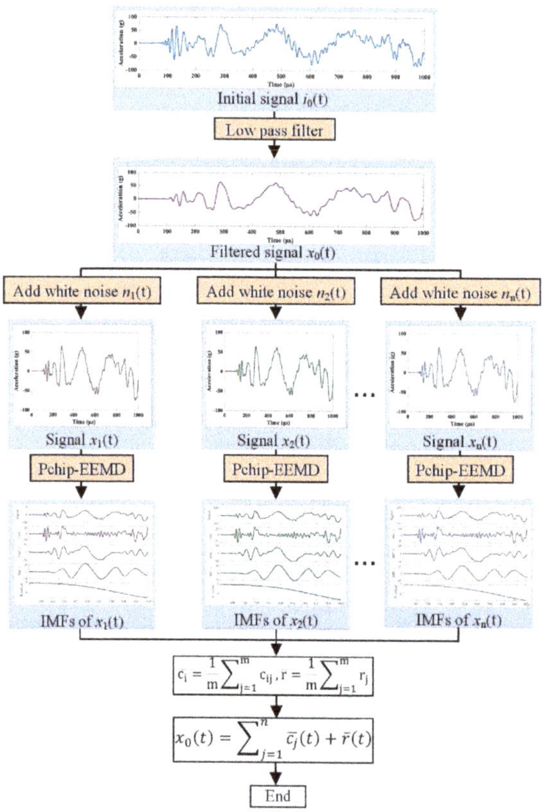

Figure 3. Illustration of the EEMD decomposition principle.

We then compared the decomposition results obtained using this approach with those obtained from EMD decomposition under identical parameters. As illustrated in Figure 4, the IMF components derived from decomposition show a relatively concentrated center frequency in each order, effectively reducing mode aliasing phenomenon.

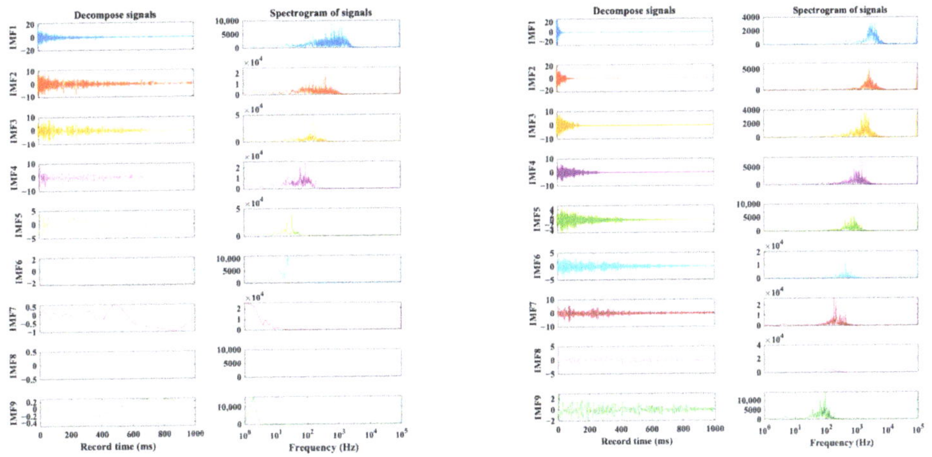

(**a**) EMD decomposition effect

(**b**) Improve the decomposition effect of EEMD

Figure 4. IMF and its spectrum. (Different colored curves represent different IMF components).

To validate the effectiveness of the Pchip-EEMD decomposition, Figure 5 provides a comparison between the improved EEMD decomposition results and the power spectral density estimates of different IMF components obtained via the original EMD decomposition. As illustrated in the figure, the frequency bands associated with each IMF component in the improved decomposition method are less prone to aliasing. Conversely, the EMD decomposition results reveal that most of the frequency bands in the signal are aliased in the IMF1 component. This comparison demonstrates that the Pchip-EEMD method effectively mitigates the mode aliasing phenomenon observed in the original decomposition method.

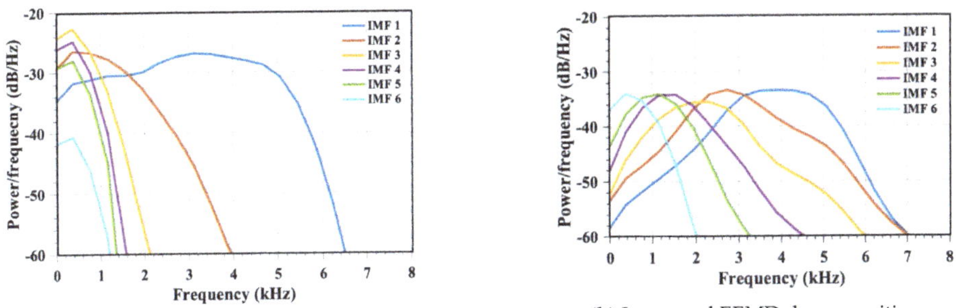

(**a**) EMD decomposition

(**b**) Improved EEMD decomposition

Figure 5. Spectral density estimation of two decomposition methods.

4. Experimental Results and Discussion

4.1. Vibration Characteristics of Steel Plates Suffering Higher Region's Shock

4.1.1. Shock Position of H1

Figure 6 illustrates the wavelet coefficient diagrams of vibration signals captured from different measuring points under the H1 shock condition, which is close to the vertical axis of symmetry of the steel plate. The diagrams of Figure 6a–f correspond to six transverse sensors RM1-RM6, while Figure 6g,h correspond to two longitudinal sensors CM1 and CM2. At the beginning of the loading process, the structure exhibited high-frequency vibration with a relatively high amplitude, which gradually transformed over time. In this context, f_L represents the frequency with the longest vibration duration, and T_L denotes the duration of that frequency. Measuring point RM1, which is close to the left boundary, showed no lower frequency vibration below 1000 Hz, and the high-frequency vibration duration was significantly shorter than that of other measuring points. The duration of higher frequency vibration was generally shorter than that of lower-frequency vibration, with an inverse correlation between the frequency and duration. At RM1, f_L = 2670 Hz, while for RM2-RM6, f_L was concentrated between 100–250 Hz with a T_L = 500 ms. For CM1, f_L = 546 Hz, and the vibration duration below 50 Hz decreased with a decrease in frequency. At CM2, f_L = 117 Hz, with a T_L = 470 ms.

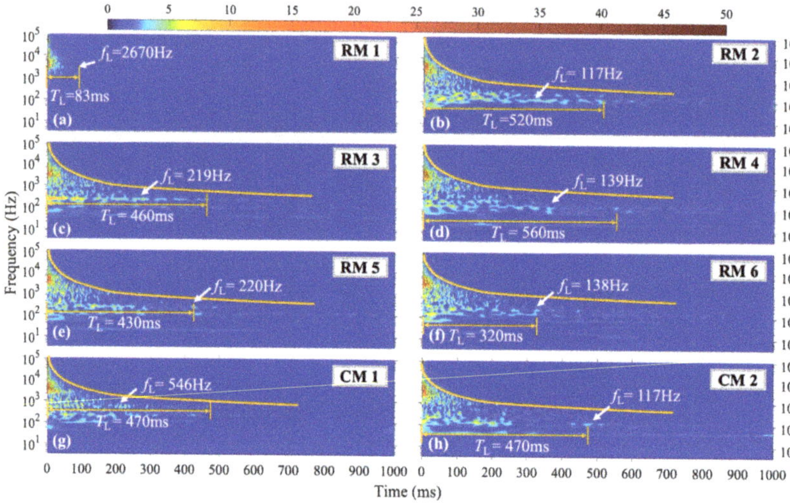

Figure 6. WT coefficient diagram under H1 condition.

4.1.2. Shock Position of H2

Figure 7 depicts the acquired data under H2 shock conditions, The vibration characteristics at each measurement point broadly resemble those under the H1 shock condition. For a given measuring point, as the vibration dissipates, the orange curve in the figure shows that high-frequency vibrations tend to have significantly shorter durations compared to low-frequency vibrations. At measurement point RM1 near the left boundary, f_L = 2606 Hz, which is the highest frequency with the longest vibration duration among all measurement points within the plate. At measurement point CM1 near the upper boundary of the plate, T_L = 360 ms and f_L = 325 Hz. The vibration at measurement points RM2, RM4, RM5, and RM6 within the plate exhibits the same f_L as under the H1 shock condition, while at measurement points RM3 and CM2, T_L occurs at f_L = 99 Hz, which is lower than that of the H1 shock condition (as depicted in Figure 7c,i).

4.1.3. Shock Position of H3

For the H3 shock condition, located to the right of the H1 and H2 conditions, the wavelet coefficient diagram in Figure 8 reveals that the vibration signals at each measuring point have shorter durations across all frequency ranges compared to the H1 and H2 conditions. When the loading is near the edge, the overall amplitude of the plate vibration is smaller, leading to shorter durations of vibration. As shown in Figure 8c, the low-frequency vibration duration and amplitude at measuring point RM3, which is equidistant from the vertical symmetry axis as the loading position, are significantly enhanced compared to the H1 and H2 conditions. At measuring point RM1, f_L = 1228 Hz, with a T_L = 64 ms, as shown in Figure 8a. The vibration duration at RM1 is consistently shorter than those of other measuring points across all shock conditions (H1, H2, and H3).

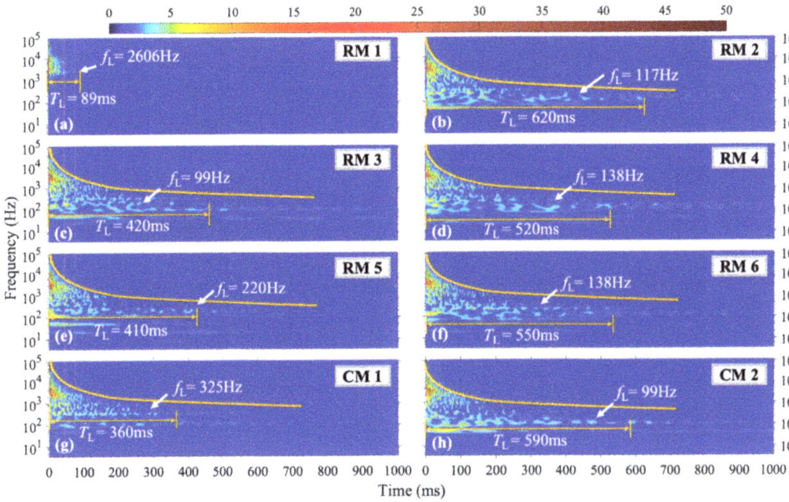

Figure 7. WT coefficient diagram under H2 condition.

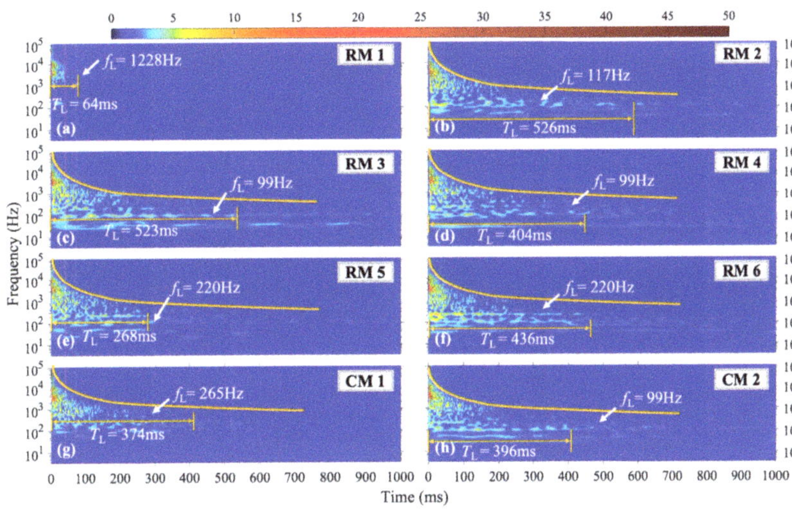

Figure 8. WT coefficient diagram under H3 condition.

4.2. Vibration Characteristics of Steel Plates Suffering Middle Region's Shock

4.2.1. Shock Position of M1

Under the M1 shock condition, the vibration characteristics of the structure change when the shock is located on the horizontal axis of symmetry of the steel plate. Figure 9 presents the wavelet coefficient diagram of the vibration signals at each measuring point. Due to the loading shock position being close to the center of the structure, the amplitude and duration of the vibration signals at each point are higher, as shown in Figure 9b,c,e,f. The duration of the f_L at measuring points RM2, RM3, RM5, and RM6 is significantly longer than that of the H1–H3 conditions, where the loading position is above the axis of symmetry. Specifically, at measuring point RM2, the T_L = 790 ms of the f_L = 117 Hz vibration signal is 32% higher than the average duration under H1–H3 conditions with the same frequency. Measuring point RM3 exhibits a f_L = 202 Hz vibration signal with a

T_L = 550 ms, while measuring point RM4 shows a relatively small f_L = 117 Hz vibration frequency with a T_L = 305 ms compared to other conditions. At measuring point RM5, the T_L = 800 ms of the f_L = 187 Hz vibration signal is significantly longer than that of other shock conditions. Moreover, as shown in Figure 9f, the 100–300 Hz vibration at measuring point RM6 has a higher initial amplitude due to its proximity to the loading position, and the f_L = 109 Hz vibration signal at this point has a T_L = 830 ms. As shown in Figure 9h, the vibration signal with the longest duration (T_L = 310 ms) at measuring point CM1, which is near the upper boundary of the plate, has a frequency of f_L = 378 Hz and is significantly higher than that of other measuring points under this shock condition.

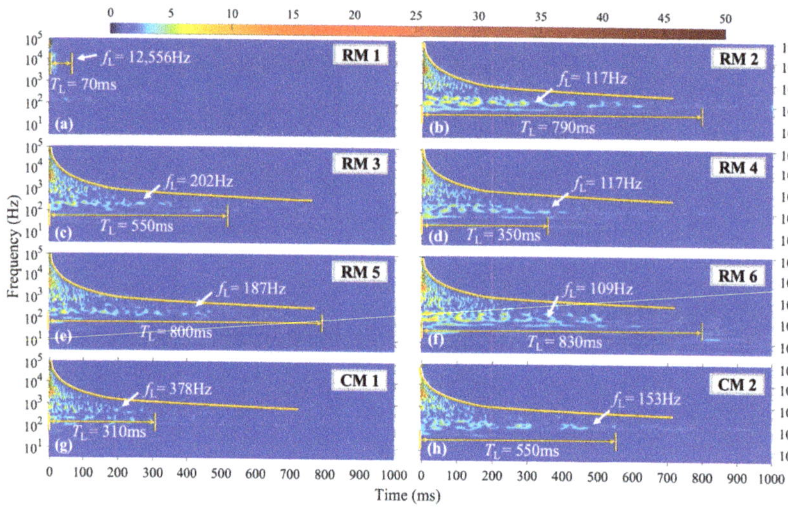

Figure 9. WT coefficient diagram under M1 condition.

4.2.2. Shock Position of M2

In the M2 shock condition, for RM1, f_L = 12,053 Hz, similar to the M1 condition. The vibration at f_L = 138 Hz in measuring points RM2, RM4, and RM6 has the longest duration and higher amplitude compared to other frequencies under this condition. For RM5, f_L = 85 Hz, with a T_L of 660 ms as shown in Figure 10e. At CM1, which is closer to the upper side boundary, f_L = 566 Hz, as shown in Figure 10g, and this frequency is significantly higher than other measuring points except RM1. At CM2, T_L = 470 ms and f_L = 69 Hz, which is lower than the frequency with the longest duration under other conditions.

4.2.3. Shock Position of M3

The vibration duration of frequencies above 250 Hz is prolonged at each measuring point under the M3 shock condition when the shock position is near the right boundary and located on the horizontal symmetry axis of the steel plate. The curve in Figure 11 represents the vibration duration of different frequencies at each measuring point under the M3 shock condition. Compared with the previous 100–250 Hz frequency band, the f_L at measuring points RM2, RM4, RM5, and RM6 has increased to 500 Hz or above under the M3 shock condition. Under these three shock conditions with the shock position located on the horizontal symmetry axis, the f_L at measuring point RM1 is consistently above 12,000 Hz. For the M3 shock conditions, the duration of vibrations above 250 Hz in the structure significantly increases, and the amplitude is also enhanced. Taking the third-layer IMF component with a center frequency of around 1000 Hz at the RM6 measurement point as an example, the average instantaneous amplitude of vibrations in the M3 condition is 0.93 g, which is significantly higher than that in the M2 condition at 0.81 g.

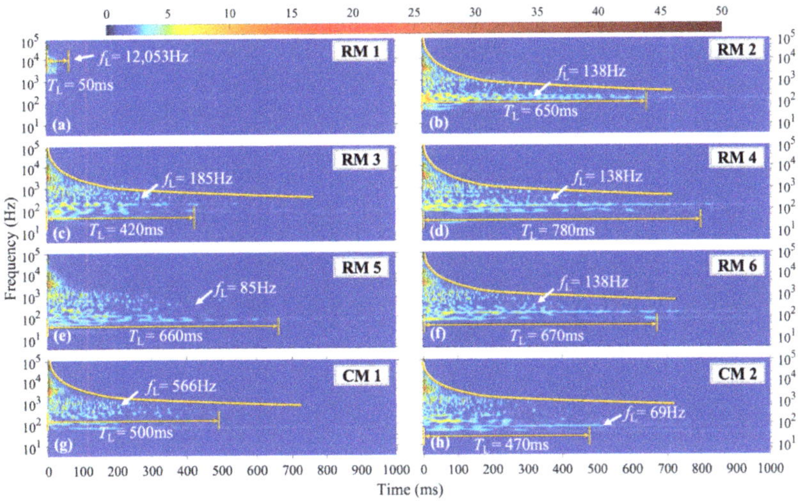

Figure 10. WT coefficient diagram under M2 condition.

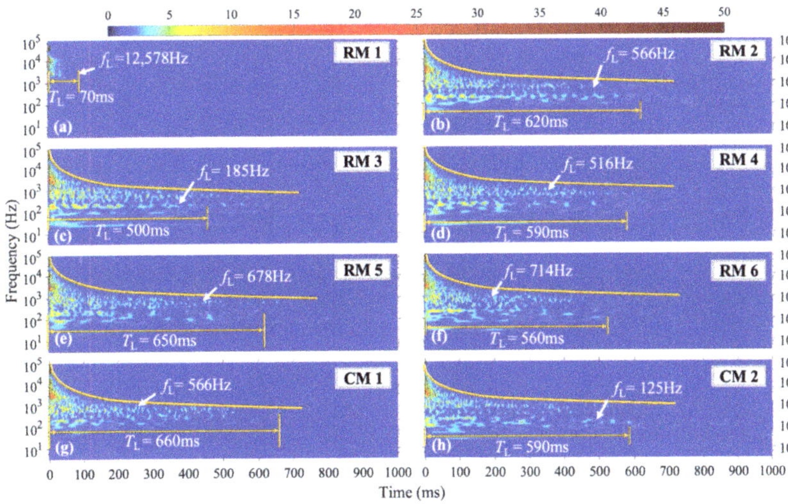

Figure 11. WT coefficient diagram under M3 condition.

4.3. Vibration Characteristics of Steel Plates Suffering Lower Region's Shock

4.3.1. Shock Position of L1

The vibration duration and frequency distribution of the structure under the L1 shock condition, as shown in Figure 12, are similar to those under the H1–H3 and M1, M2 shock conditions. However, one distinct difference is that the vibration energy at each measurement point stays focused within specific frequency ranges for extended durations. Measurement points RM2, RM4, and RM6 exhibit significant vibration at f_L = 110 Hz and 138 Hz for a T_L of over 800 ms, while measurement points RM3 and RM5 show concentrated vibration at f_L = 186 Hz. Moreover, measurement point CM1 displays pronounced vibration at f_L = 546 Hz.

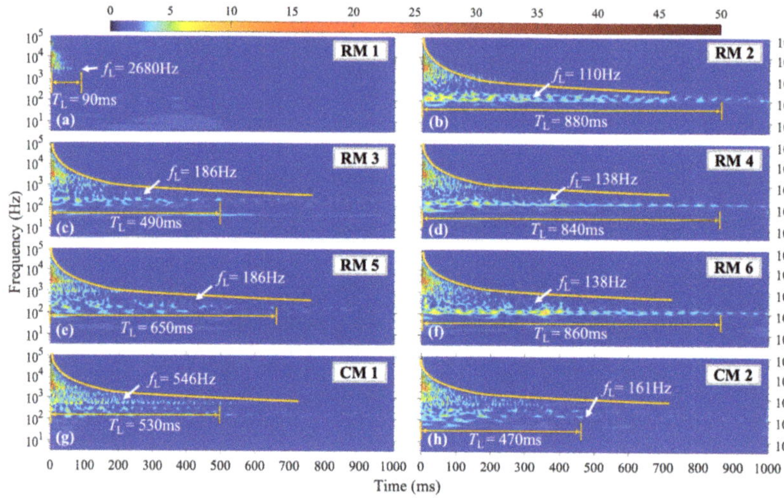

Figure 12. WT coefficient diagram under L1 condition.

4.3.2. Shock Position of L2

Figure 13 demonstrates that in the L2 condition, where the shock position is to the right of the vertical symmetry axis, the vibration patterns observed at each measurement point are similar to those in the L1 condition. However, in comparison to the L1 condition, the amplitudes of the vibrations at each measurement point are generally reduced. The vibration energy is concentrated in a few fixed frequencies, and in some measurement points, the T_L is longer than that observed in the L1 condition. At measurement points RM4 and RM6, the vibrations predominantly occur at f_L = 138 Hz, with a T_L of approximately 900 ms. At measurement points RM2, RM3, and CM2, the vibration energy is concentrated at f_L = 215 Hz, with a T_L of around 800 ms. At measurement point RM3, which is close to the vertical symmetry axis, the amplitude of the vibrations around 200 Hz abnormally increases, and the T_L is significantly longer than that observed in the L1 condition. Finally, at measurement point CM1, which is close to the upper boundary, the vibrations are concentrated at f_L = 638 Hz, with a slightly longer T_L than that observed in the L1 condition.

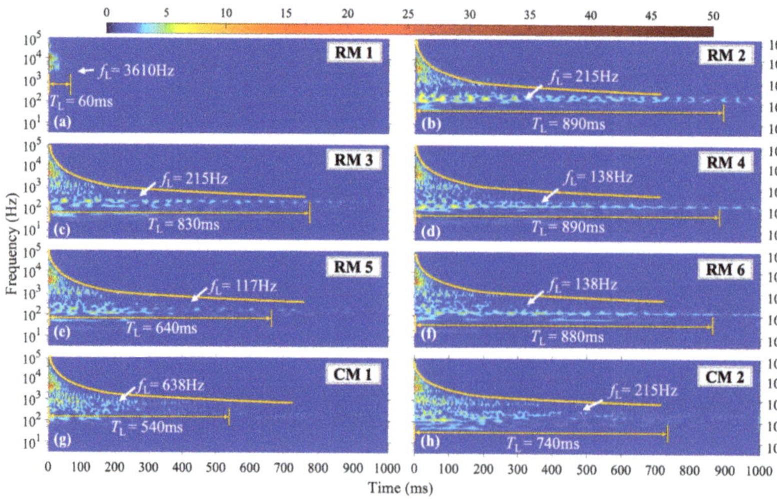

Figure 13. WT coefficient diagram under L2 condition.

4.3.3. Shock Position of L3

In the case of the L3 shock position, which is situated closest to the lower right corner of the steel plate, as depicted in Figure 14, the amplitude of each frequency at various measurement points throughout the structure has exhibited an overall reduction, along with a shortened T_L compared to the L1 and L2 shock conditions. Specifically, the maximum instantaneous amplitude of the fifth-level intrinsic mode function (IMF) at measurement point RM6 under L2 and L3 shock conditions has been reduced by 28% and 47%, respectively, in comparison to the L1 condition. The frequencies that exhibit the longest T_L at measurement points RM2, RM3, RM5, and RM6 are all f_L = 186 Hz, with the duration spanning from 445 ms to 892 ms. For the measurement points RM3 and RM4, which are situated nearer to the vertical symmetric position from the loading position, the amplitude below 200 Hz has been higher and the T_L has been longer, particularly for measurement point RM3.

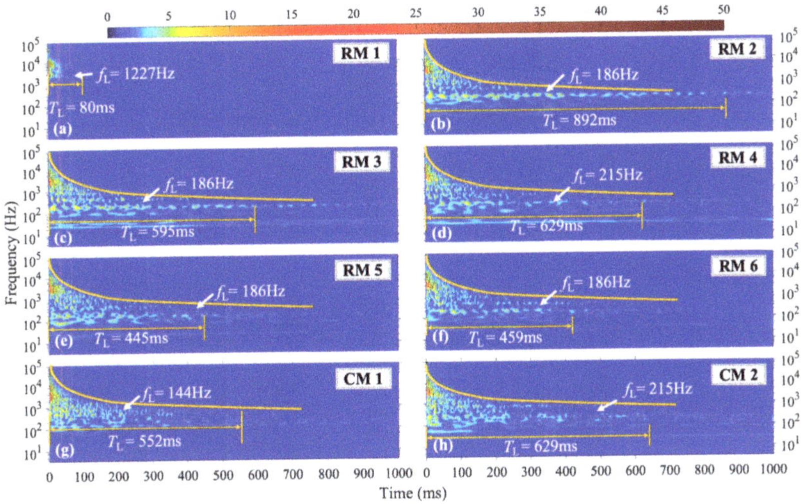

Figure 14. WT coefficient diagram under L3 condition.

4.4. Effect of Shock Position on Vibration Characteristics of Steel Plates

Figure 15 depicts the vibration characteristics observed at measuring point RM1. In the figure, the bar chart represents the f_L at measuring point RM1 under different loading conditions: H-f_L (above horizontal symmetry axis), M-f_L (on horizontal symmetry axis), L-f_L (below horizontal symmetry axis), while the line chart represents the T_L at measuring point RM1 under different loading conditions. It can be observed that for loading positions closer to the horizontal axis symmetry, the f_L in measuring point RM1 is above 12,000 Hz in M1–M3 cases. However, for H1–H3 and L1–L3 cases, the f_L is between 2000–4000 Hz, and for the loading cases closer to the right boundary, the frequency is relatively smaller. Let us define the longest duration frequency as f_L. The T_L changes of different cases have no obvious regularity, but they are all within 100 ms.

In Figure 16, a bar chart is presented that displays the f_L of vibration at measuring points RM2- RM6 for various loading cases. Compared to measuring point RM1, which is close to the boundary, these measuring points show a significant reduction in f_L, making them unsuitable for inclusion in the same bar chart. Notably, the results reveal that in the M3 case, except for measuring point RM3, which is symmetric to the loading position, the f_L detected by all other measuring points are significantly higher. Additionally, in both the H3 and M3 cases, measurement point RM3 exhibits a significantly lower vibration frequency compared to other measurement points at the same horizontal position, given that RM3 is located at an equal distance from the vertical axis of symmetry as the loading position. It is worth noting that for most shock conditions investigated, the frequencies with the longest vibration duration (fL) at measuring points RM3 and RM5 are significantly higher compared to other measurement points.

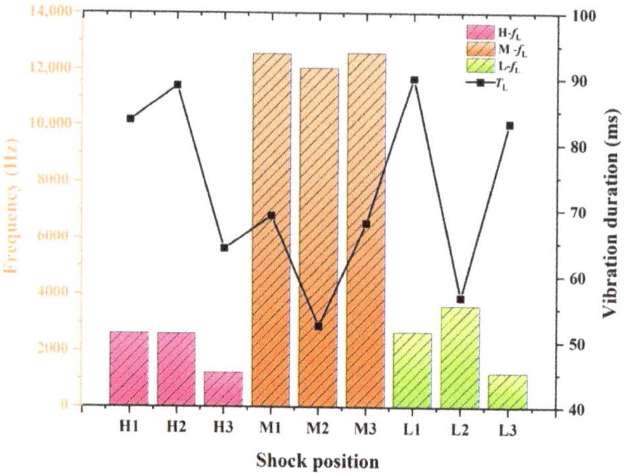

Figure 15. Vibration characteristic of row measuring point RM1.

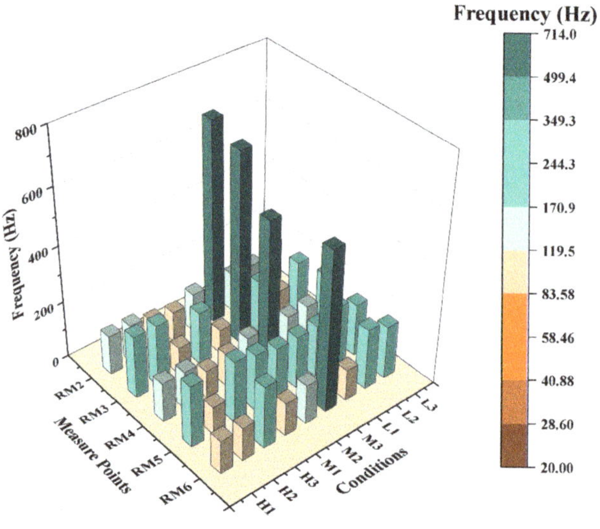

Figure 16. Frequency with the longest vibration duration at each measuring point.

5. Conclusions

This paper presents an experimental device consisting of a shock tube system and an adjustable steel plate, designed to investigate the transient shock vibration response of thin-walled structures. By applying shock loads at various positions and using accelerometers to collect data, a comprehensive understanding of the vibration response characteristics has been achieved. To analyze the vibration signals in the time-frequency domain, the research introduced two methods, the wavelet transform and the EEMD with pchip interpolation. The accuracy of these methods is validated, and the results provide information on the distribution and attenuation of vibration energy across various frequency bands. The study reveals several findings, which can be summarized as follows:

1. Improved Signal Analysis: We have refined time-frequency domain analysis by using the Ensemble Empirical Mode Decomposition (EEMD) method with pchip interpolation. This significantly enhances the accuracy, tackling mode mixing issues in traditional EMD and addressing redundancy and frequency-discontinuity problems in continuous wavelet transform (CWT). These enhancements will aid in accurately characterizing structure vibrations, thereby facilitating the development of more shock-resistant designs.
2. Boundary Vibration Attenuation: Our findings show that the boundary regions exhibit the fastest rate of vibration attenuation. This could guide the design and placement of sensitive components within a structure to minimize sustained vibrations.
3. Frequency Band Implications: The research reveals that the 100–300 Hz frequency band sustains the longest vibration duration for most shock conditions, providing pivotal knowledge for predicting potential failure areas or developing frequency-specific vibration dampening methods.
4. Influence of Shock Position: The study underscores that shock positioning significantly impacts vibration characteristics, and this insight can be incorporated into design strategies for specific shock or load conditions.
5. Asymmetric Shock Impact: We found that as the shock position moved towards the right side, the vibration amplitude and duration decreased. This implies that asymmetric shock loads can influence the overall vibrational characteristics, which is a critical consideration for improving equipment resilience and performance.

Author Contributions: Conceptualization, C.W.; Methodology, Y.S.; Data curation, X.L.; Writing—original draft, Z.L.; Writing—review & editing, W.X. All authors have read and agreed to the published version of the manuscript.

Funding: The research is supported by the National Natural Science Foundation of China (U20A2071, 12102233, 12221002) and the project of State Key Laboratory of Explosion Science and Technology (Beijing Institute of Technology). The opening project number is KFJJ23-17M, YBKT22-05, and LabASP-2019-03.

Institutional Review Board Statement: Not applicable.

Informed Consent Statement: Not applicable.

Data Availability Statement: The data generated or analyzed during this study are included in this published article. For further inquiries, please contact the corresponding author.

Conflicts of Interest: The authors declare no conflict of interest.

References

1. Ma, N.; Liu, L.; Meng, F.; Meng, J. Structural design and modal behaviors analysis of a new swept baffled inflatable wing. *Def. Technol.* **2023**, *24*, 382–398. [CrossRef]
2. Lu, W.; Li, B.; Li, J.; Hou, J.; Xu, X.; Zhang, L. Numerical simulation of dynamic response and failure of large-diameter thin-walled cylinder under vibratory penetration. *Ocean Eng.* **2022**, *249*, 110936. [CrossRef]
3. Li, H.; Shen, C.; Lu, G.; Wang, Z. Response of cylindrical tubes subjected to internal blast loading. *Eng. Struct.* **2022**, *272*, 115004. [CrossRef]
4. Carney, K.S.; DuBois, P.; Cudziło, S.; Jorgensen, G.A.; Binienda, W.K. The Effect of TNT Mass and Standoff Distance on the Response of Fully Clamped Circular Aluminum Plates to Confined Air-Blast Loading. *Int. J. Impact Eng.* **2022**, *170*, 104357. [CrossRef]
5. Patel, S.; Patel, M. The efficient design of hybrid and metallic sandwich structures under air blast loading. *J. Sandw. Struct. Mater.* **2022**, *24*, 1706–1725. [CrossRef]
6. Zhang, X.; Fu, X.; Yu, Q. Energy absorption of arched thin-walled structures under transverse loading. *Int. J. Impact Eng.* **2021**, *157*, 103992. [CrossRef]
7. Zhao, N.; Yao, S.; Zhang, D.; Lu, F.; Sun, C. Experimental and numerical studies on the dynamic response of stiffened plates under confined blast loads. *Thin-Walled Struct.* **2020**, *154*, 106839. [CrossRef]
8. Xu, M.C.; Song, Z.J.; Pan, J. Study on the similarity methods for the assessment of ultimate strength of stiffened panels under axial load based on tests and numerical simulations. *Ocean Eng.* **2020**, *219*, 108294. [CrossRef]

9. Liu, Y.; Zhou, Q.; Wei, X.; Xia, Y. Testing and modeling tearing and air effect of aluminum honeycomb under out-of-plane impact loading. *Int. J. Impact Eng.* **2019**, *135*, 103402. [CrossRef]
10. Chung Kim Yuen, S.; Nurick, G.; Langdon, G.; Iyer, Y. Deformation of thin plates subjected to impulsive load: Part III—An update 25 years on. *Int. J. Impact Eng.* **2017**, *107*, 108–117. [CrossRef]
11. Wang, Q.; Ren, H.; Li, J. Dynamic response of contact-blast-loaded free metal plate: Theoretical model, experiments and numerical simulation. *Thin-Walled Struct.* **2022**, *175*, 109228. [CrossRef]
12. Xu, Z.; Liu, Y.; Shi, Z.; Huang, F. Typical characteristics of counterintuitive behavior in thin aluminum plates under blast loading. *Int. J. Impact Eng.* **2021**, *159*, 104037. [CrossRef]
13. Kaufmann, R.; Olufsen, S.N.; Fagerholt, E.; Aune, V. Reconstruction of surface pressures on flat plates impacted by blast waves using the Virtual Fields Method. *Int. J. Impact Eng.* **2023**, *171*, 104369. [CrossRef]
14. Curry, R.; Langdon, G. The effect of explosive charge backing in close-proximity air-blast loading. *Int. J. Impact Eng.* **2021**, *151*, 103822. [CrossRef]
15. Chung Kim Yuen, S.; Butler, A.; Bornstein, H.; Cholet, A. The influence of orientation of blast loading on quadrangular plates. *Thin-Walled Struct.* **2018**, *131*, 827–837. [CrossRef]
16. Yao, S.; Chen, F.; Wang, Y.; Ma, Y.; Zhao, N.; Zhang, D.; Lu, F. Experimental and numerical investigation on the dynamic response and damage of large-scale multi-box structure under internal blast loading. *Thin-Walled Struct.* **2023**, *183*, 110430. [CrossRef]
17. McDonald, B.; Bornstein, H.; Langdon, G.; Curry, R.; Daliri, A.; Orifici, A. Experimental response of high strength steels to localised blast loading. *Int. J. Impact Eng.* **2018**, *115*, 106–119. [CrossRef]
18. Geertsma, R.D.; Negenborn, R.R.; Visser, K.; Hopman, J.J. Design and control of hybrid power and propulsion systems for smart ships: A review of developments. *Appl. Energy* **2017**, *194*, 30–54. [CrossRef]
19. Ruan, Y.; Liang, X.; Hua, X.; Zhang, C.; Xia, H.; Li, C. Isolating low-frequency vibration from power systems on a ship using spiral phononic crystals. *Ocean Eng.* **2021**, *225*, 108804. [CrossRef]
20. Zare Estakhraji, S.I.; Wall, M.; Capito, J.; Allen, M.S. A thorough comparison between measurements and predictions of the amplitude dependent natural frequencies and damping of a bolted structure. *J. Sound Vib.* **2023**, *544*, 117397. [CrossRef]
21. Yan, Y.; Li, Q. A pyroshock signal characterization method based on shock-waveform dictionary. *Int. J. Mech. Sci.* **2023**, *249*, 108251. [CrossRef]
22. Yan, Y.; Li, Q. A general shock waveform and characterisation method. *Mech. Syst. Signal Process.* **2019**, *136*, 106508. [CrossRef]
23. Nguyen, V.D.; Vu, Q.V. Dynamic vibration analysis of double-layer auxetic FGP sandwich plates under blast loads using improved first-order shear plate theory. *Alex. Eng. J.* **2023**, *69*, 135–165. [CrossRef]
24. Panda, S.; Barik, M. Transient Vibration Analysis of Arbitrary Thin Plates Subjected to Air-Blast Load. *J. Vib. Eng. Technol.* **2019**, *7*, 189–204. [CrossRef]
25. Yan, Y.; Li, Q. Low-pass-filter-based shock response spectrum and the evaluation method of transmissibility between equipment and sensitive components interfaces. *Mech. Syst. Signal Process.* **2018**, *117*, 97–115. [CrossRef]
26. Park, S.; Beak, J.; Kim, K.; Park, Y.-J. Study on Reduction Effect of Vibration Propagation due to Internal Explosion Using Composite Materials. *Int. J. Concr. Struct. Mater.* **2021**, *15*, 30. [CrossRef]
27. Wu, T.; Jiang, N.; Zhou, C.; Luo, X.; Li, H.; Zhang, Y. Experimental and numerical investigations on damage assessment of high-density polyethylene pipe subjected to blast loads. *Eng. Fail. Anal.* **2021**, *131*, 105856. [CrossRef]
28. Wu, S.; Li, H. A data-driven design method of distributed dynamic vibration absorber for broadband vibration suppression of thin-walled structures. *Thin-Walled Struct.* **2023**, *182*, 110264. [CrossRef]
29. Pandey, A.M.; Gopal, K.N. Transient vibration and sound radiation analysis of simply supported functionally graded sandwich plates. *Compos. Struct.* **2022**, *290*, 115520. [CrossRef]
30. Vieira, R.; Virtuoso, F.; Pereira, E. Buckling of thin-walled structures through a higher order beam model. *Comput. Struct.* **2017**, *180*, 104–116. [CrossRef]
31. Xu, X.; Carrera, E.; Augello, R.; Daneshkhah, E.; Yang, H. Benchmarks for higher-order modes evaluation in the free vibration response of open thin-walled beams due to the cross-sectional deformations. *Thin-Walled Struct.* **2021**, *166*, 107965. [CrossRef]
32. Pan, X.; Wang, G.; Lu, W.; Yan, P.; Chen, M.; Gao, Z. The effects of initial stresses on nonlinear dynamic response of high arch dams subjected to far-field underwater explosion. *Eng. Struct.* **2022**, *256*, 114040. [CrossRef]
33. Shen, Z.; Xia, J.; Cheng, P. Geometrically nonlinear dynamic analysis of FG-CNTRC plates subjected to blast loads using the weak form quadrature element method. *Compos. Struct.* **2018**, *209*, 775–788. [CrossRef]
34. Saidin, S.S.; Kudus, S.A.; Jamadin, A.; Anuar, M.A.; Amin, N.M.; Ibrahim, Z.; Zakaria, A.B.; Sugiura, K. Operational modal analysis and finite element model updating of ultra-high-performance concrete bridge based on ambient vibration test. *Case Stud. Constr. Mater.* **2022**, *16*, e01117. [CrossRef]
35. Grządziela, A.; Kluczyk, M. Shock Absorbers Damping Characteristics by Lightweight Drop Hammer Test for Naval Machines. *Materials* **2021**, *14*, 772. [CrossRef]

36. Guo, T.; Zhang, T.; Lim, E.; Lopez-Benitez, M.; Ma, F.; Yu, L. A Review of Wavelet Analysis and Its Applications: Challenges and Opportunities. *IEEE Access* **2022**, *10*, 58869–58903. [CrossRef]
37. Wu, Z.; Huang, N.E. Ensemble empirical mode decomposition: A noise-assisted data analysis method. *Adv. Adapt. Data Anal.* **2009**, *1*, 1–41. [CrossRef]

MDPI AG
Grosspeteranlage 5
4052 Basel
Switzerland
Tel.: +41 61 683 77 34

Materials Editorial Office
E-mail: materials@mdpi.com
www.mdpi.com/journal/materials

www.ingramcontent.com/pod-product-compliance
Lightning Source LLC
Chambersburg PA
CBHW061440210926
51850CB00008B/1568
9783725845842